CORPORATE STRATEGY, PUBLIC POLICY AND NEW TECHNOLOGIES

TECHNOLOGY, INNOVATION, ENTREPRENEURSHIP AND COMPETITIVE STRATEGY SERIES

Series Editors: John McGee and Howard Thomas

Published

STEFFENS
Newgames: Strategic Competition in the PC Revolution

BULL, THOMAS & WILLARD
Entrepreneurship

SANCHEZ, HEENE & THOMAS
Dynamics of Competence-Based Competition

BOGNER
Drugs to Market: Creating Value and Advantage in the Pharmaceutical Industry

Forthcoming titles

PORAC
The Social Construction of Markets and Industries

MOROSINI
Cross-Cultural Acquisitions and Alliances

Other titles of interest

DAEMS & THOMAS
Strategic Groups, Strategic Moves, and Performance

DENNING
Making Strategic Planning Work in Practice

DOZ
Strategic Management in Multinational Companies

FREEDMAN
Strategic Management in Major Multinational Companies

MCNAMEE
Developing Strategies for Competitive Advantage

Related journals—sample copies available on request

European Management Journal
International Business Review
Journal of Retailing and Consumer Services
Long Range Planning
Scandinavian Journal of Management

CORPORATE STRATEGY, PUBLIC POLICY AND NEW TECHNOLOGIES

PHILIPS AND THE EUROPEAN CONSUMER ELECTRONICS INDUSTRY

Xiudian Dai

Pergamon

UK Elsevier Science Ltd, The Boulevard, Langford Lane, Kidlington, Oxford OX5 1GB, UK

USA Elsevier Science Inc., 660 White Plains Road, Tarrytown, New York 10591-5153, USA

JAPAN Elsevier Science (Japan), Tsunashima Building Annex, 3-20-12 Yushim, Bunkyo-ku, Tokyo 113, Japan

---

First edition 1996

Library of Congress Cataloging in Publication Data
A catalog record for this title is available from the Library of Congress.

British Library Cataloguing in Publication Data
A catalogue record for this title is available from the British Library.

ISBN 0-08-042581-X

Printed and bound in Great Britain by BPC Wheatons Ltd, Exeter

ACKNOWLEDGEMENTS

This book is based upon my doctoral thesis which was completed at the University of Sussex. During the process of completing the thesis and preparing for the book many people generously offered their help one way or another. Although it is impossible to acknowledge everybody in person here, thanks are due to all of them.

Among others, I wish to express my most heartfelt gratitude to Professor Alan Cawson for his supervision and kind help with every aspect of my research and writing up the doctoral thesis and subsequently turning it into this book. Recalling the years at Sussex, I feel deeply grateful to Alan for his constant encouragement, his confidence in my ability and the many hours he spent in reading and discussing my work. My thanks are due to Professor Kevin Morgan and Dr Mike Hobday who read the manuscript and made comments in detail, when they were examining my doctoral thesis.

My thanks must go to the Sino–British Friendship Scholarship Scheme, which provided important financial contribution to the research period of my doctoral programme. The Great Britain–China Educational Trust also offered their generous help during the late stage of the research.

I'm grateful to Dr Margaret Sharp at SPRU for her timely advice in the early stage of my doctoral research, and Dr Peter Holmes at the Sussex European Institute, who spent his time reading and commenting on some of the draft chapters. Dr Amala de Silva and Dr Mike van Duuren carefully read some parts of the thesis and offered invaluable suggestions. Mike also helped translate some important documents from Dutch into English, which in part forms the reference to this book. My discussions with Dr Shiji Gao and Yaoxing Hu were stimulating and encouraging in the process of completing my doctoral programme.

I'm also very grateful to Dr Thanassis Chrisafiss of the European Commission, who helped me set up useful contacts with European Union officials during my field research in Brussels. My thanks also go to Peter Forster and Stephen Atkinson for their kindness in providing access to many Philips documents in London and Eindhoven.

This research benefited to a great extent from a number of industrial conferences. I sincerely appreciate the generosity of a great number of industrial conference organisers and managers, who offered me the opportunity to attend many international events relevant to the consumer electronics industry. One of the organisers once joked to me, "As a poor student you have taken a very expensive research topic!" In retrospect, I feel deeply indebted to a large number of senior managers from Philips and, more broadly, the information and communications technology sector as well as many European Union officials, who agreed to be interviewed. All of my interviewees generously offered their valuable time and provided invaluable contributions towards my doctoral thesis and, certainly, the current book.

In this Acknowledgement, many thanks are due to Andy Bottomley. Andy sacrificed many of his leisure hours to proof-read and comment on some parts of this book. Indeed, my discussions with Andy were in part a rich intellectual resource as well as a joy in refining many of the arguments presented in this book.

In the meantime, I'm grateful to both Series Editors, Professor John McGee and Professor Howard Thomas at Warwick Business School, for choosing this book as a part of the Technology, Innovation, Entrepreneurship and Competitive Strategy Series. My discussions with Professor McGee proved very helpful in clarifying some of the main arguments in this book during the last stage of writing. My thanks must also go to the Publishing Editor Tony Seward and Katherine Woolway as well as the Marketing Department at Elsevier for their hard work in bringing this book to the reader.

Also, I'm deeply grateful to all members of the Department of European Studies and many colleagues from the School of European Languages and Cultures at the University of Hull for their moral support and encouragement in the process of finalising this book. Among others, my warmest thanks are due to the Departmental Secretary Jacky Cogman and School Secretary Pat Escreet for their various assistance in producing the manuscript to this book before publication.

Finally, but not least, this book would not have been possible if it had not been for the love and constant encouragement from my wife, Liqun, and my parents as well as my brother and sister. I should like to conclude this Acknowledgement by dedicating this book to them all.

Hull, England
November 1995

Xiudian Dai

CONTENTS

LIST OF FIGURES

LIST OF TABLES

LIST OF ABBREVIATIONS

ACATS	Advisory Committee on Advanced Television Services
ACTV	Advanced Compatible Television
AEA	American Electronics Association
AHD	Audio High Density
ATRC	Advanced Television Research Consortium
ATSC	Advanced Television System Committee
AT&T	American Telephone and Telegraph
ATTC	Advanced Television Testing Centre
BBC	British Broadcasting Corporation
BOM	Board of Management
BREMA	British Radio and Electronic Equipment Manufacturers' Association
BRITE	Basic Research in Industrial Technologies for Europe
BSB	British Satellite Broadcasting
BSkyB	British Sky Broadcasting
BT	British Telecom
BTS	Broadcasting Television Systems (Philips/Bosch joint venture)
BU	Business Unit
CCIR	Consultative Committee for International Radio
CD	Compact Disc
CD-DA	Compact Disc-Digital Audio
CD-G	Compact Disc-Graphics
CD-i	Compact Disc-interactive
CDIA	CD-I Association (of North America)
CDICJ	CD-I Consortium Japan
CD-ROM	Compact Disc-Read Only Memory
CDTV	Commodore Dynamic Total Vision

CD-V	Compact Disc-Video
CECO	Change Evaluation Committee (of the Eureka 95 project)
CED	Capacitance Electronic Disc
CEO	Chief Executive Officer
CLV	Constant Linear Velocity
CRT	Cathode Ray Tube
CTV	Colour Television
DARPA	Defense Advanced Research Projects Agency (of Pentagon)
DAT	Digital Audio Tape
DBS	Direct Broadcast by Satellite
DCC	Digital Compact Cassette
DG	Directorate General (of the European Commission)
DRAM	Dynamic Random Access Memory
DTI	Department of Trade and Industry
DTP	Desk Top Publishing
DTT	Digital Terrestrial Television
dTTb	digital Terrestrial Television broadcasting
DVB	Digital Video Broadcasting
DVD	Digital Video Disc
DVI	Digital Video Interactive
EACEM	European Association of Consumer Electronics Manufacturers
EBU	European Broadcasting Union
ECDIA	European CD-I Association
EDTV	Extended Definition Television
EEC	European Economic Community
EEIG	European Economic Interest Grouping
EIAJ	Electronic Industries Association of Japan
EIIA	European Information Industry Association
ELG	European Launching Group
ELGDVB	European Launching Group for Digital Video Broadcasting
ESPRIT	European Strategic Programme for Research on Information Technology
ETSI	European Telecommunications Standard Institute
EU	European Union
Eureka	European Research Cooperation Agency
FCC	Federal Communications Commission
FDI	Foreign Direct Investment
FMV	Full Motion Video
FSFM	Full Screen Full Motion
GATT	General Agreement on Tariffs and Trade
GE	General Electric (US)
GEC	General Electric Company (UK)
GI	General Instrument
GII	Global Information Infrastructure
GMC	Group Management Committee

HD-Divine	High Definition Digital Video Narrowband Emission
HD-MAC	High Definition Multiplexed Analogue Component
HDTV	High Definition Television
HDVS	High Definition Video System
IBA	Independent Broadcasting Authority
IBC	International Broadcasting Convention
IBC	Integrated Broadband Communications
IBM	International Business Machines
ICT	Information and Communications Technology
IDTV	Improved Definition Television
IMA	Interactive Media Association
IMPACT	Information Market Policy Actions
IMS	(Philips) Interactive Media Systems
Ion	Image Online Network
ISO	International Standards Organisation
IT	Information Technology
ITC	Independent Television Committee
ITU	International Telecommunications Union
JESSI	Joint European Submicron Silicon Initiative
JPEG	Joint Photographic Experts Group
JVC	Japan Victor Company
LCD	Liquid Crystal Display
LEP	Light-Emitting Plastic
LVAP	LaserVision Association Pacific
MAC	Multiplexed Analogue Component
MD	Mini Disc
MEDIA	Measures to Encourage the Development of the Audio-visual Industry
MITI	Ministry of International Trade and Industry (Japan)
MMCD	Multimedia Compact Disc
MMPM	MultiMedia Presentation Manager
MNE	Multinational Enterprise
MoF	Ministry of Finance
MPC	Multimedia Personal Computer
MPDC	Motorola Philips Design Centre
MPEG	Motion Picture Experts Group
MPT	Ministry of Posts and Telecommunications
MUSE	Multiple Sub-nyquist Sampling Encoding
NAP	North America Philips (or Philips North America)
NBC	National Broadcasting Corporation
NHK	Nippon Hoso Kyokai (Japan Broadcasting Corporation)
NII	National Information Infrastructure
NO	National Organisation
NTL	National Transcommunications Limited
NTSC	National Television Systems Committee

OEM	Original Equipment Manufacturer
PAL	Phase Alteration by Line
PC	Personal Computer
PD	Product Division
PDA	Personal Digital Assistant
PIE	(Apple) Personal Interactive Electronics
PIMI	Philips Interactive Media International
POI	Point of Information
POS	Point of Sale
RACE	Research on Advanced Communications for Europe
R&D	Research and Development
RAI	Italian State Broadcasting Organization
RCA	Radio Corporation of America
RIIT	Research Institute of Industrial Technologies
SD	Super Density (Disc)
SECAM	*Système Électronique Couleur avec Mémoire*
SMPTE	Society of Motion Picture and Television Engineers
SRAM	Static Random Access Memory
SVR	Super Video Recorder
UKLDA	UK Laser Disc Association
VADIS	Video-Audio Digital Interactive System
VCR	Video Cassette Recorder
VER	Voluntary Export Restraint
VHD	Video High Density (Disc)
VIS	Video Information System
VLSI	Very Large Scale Integrated (Circuits)
VoD	Video on Demand
VR	Virtual Reality
VTR	Video Tape Recorder
WARC	World Administration Radio Conference

FOREWORD

In the second half of the twentieth century the world has witnessed an astonishing acceleration in the pace of innovation in the electronics industry, giving rise to a whole new area of economic activity that we now know as information technology. From the simple radio sets and record players of the 1950s, the consumer is now faced with a bewildering array of products, from the literally hundreds of different models of personal stereo set, to 'home cinema' systems incorporating five channels of sound.

Accompanying this spurt in technology are two critical factors central to Xiudian Dai's study. The first is the increasing complexity of products and their embodiment into systems incorporating critical linkages between hardware and software. The successful introduction of such product systems onto the consumer market requires an unprecedented degree of coordination to overcome the 'chicken and egg' problem whereby the investment in software requires an installed base of equipment in the home, but consumers will not purchase new products unless there is an adequate supply of software and programmes. Dai's detailed case studies of high definition television (HDTV) and Compact Disc-Interactive (CD-i) are excellent examples of the different ways in which existence from this impasse were sought. Dai shows us with great skill how even a major multinational company like Philips is unable to solve the innovation problem for these new products without a network of collaboration with public and private actors.

The second critical factor is the spread of digital technology from the computer industry to drive convergence *at a technological level* between consumer electronics, computing and telecommunications. This factor might well be called 'the digitalization of everything', since there is no area of the electronics industry which is untouched by its ramifications. As Dai shows, it was a breakthrough in digital technology—through which digital television pictures could be compressed to fit high definition signals into the bandwidth of a conventional signal—which scuppered the European analogue HDTV system.

Xiudian Dai's study reveals the interrelationship of technological change, corporate strategy and public policy through a fascinating case study of one major firm's experience of bringing a range of these new products to market. Philips is a very important case in its own right, and this book provides a rich source of information on the lead player in the European consumer electronics industry. But by placing Philips firmly within the context of the industry, and the firm's dealings with national governments and the European Union, Dai also reveals a great deal about the cross-cutting alliances and strategic collaborations which are now characteristic of the world industry.

This book can be read with profit by readers with a number of different concerns. Those who find the technology impenetrable will find this an excellent introduction to how and why digital technologies are having such a profound effect on the electronics and communications industries. Those who are concerned with European industry's competitiveness in the electronics sector will find much food for thought in Dai's conclusion that it will be adjustments at the level of the firm, rather than changes in public policy, which will secure the future of the industry. He shows how, albeit sometimes painfully, Philips has learned this lesson and has emerged as a stronger competitor. Finally, those interested in management and corporate strategy will find that Dai eschews a jargon-laden and prescriptive approach in favour of a realistic and complex analysis which reveals the constraints on management, and the way in which corporate strategy is inextricably linked to a political role for the firm.

University of Sussex Alan Cawson

SERIES EDITORS' PREFACE

Technology and technical change have always posed problems in the analysis of markets and of competition. Technical change is concerned with changes in the state of scientific knowledge and with the way in which that knowledge can be brought to bear in markets by firms.

However, the analysis of markets and of industries has been marked by a predominance of static modes of analysis—what is the nature of the market structure, how does this affect the commercial policies of firms and what implications does the present structure have for the performance of firms and the efficiency of markets? This has proved a taxing enough agenda for industrial economists trying to understand the implications for efficiency as well as for strategic management researchers attempting to explain the relative success of firms. Technical change, however, generally requires investment and always requires some form of change within firms, with consequences for the way in which competition evolves. Industry analysis, which might be thought of as a sort of meeting ground for economists and strategy researchers, requires a frame of reference which is more truly dynamic in character. In other words, the framework of industry analysis should contain within it a way of explaining the paths by which an industry can change and the mechanisms by which the alternative paths might be triggered. By contrast, the existing frameworks enable us to interpret industry structures in terms of their implications for efficient resource allocation and for competition on the implicit assumption of no change in the attendant underlying conditions, specifically the state of technical knowledge.

Of course economists have not been unconscious of this challenge. For many economists the notion of long-run marginal cost is essentially a dynamic one where expected future technical change not only affects the future course of prices but is also a crucial factor in determining the current price. The simplest and most direct incorporation of dynamics has been through models where technical change is treated as exogenous to the system and its implications explored. A more complex view allows for technical change to be treated endogenously, that is that technical change is induced or retarded by economic variables which are themselves determined within the economic model. Thus the level of prices will itself have an effect in stimulating the rate of technical change in a way hypothesized and tested within the model under consideration.

For strategy researchers even these endogenous models of technical change are insufficient. For the strategist the questions of interest concern the major decisions taken within the firm that condition the resource structure and behaviour of the firm and further affect the competitive dynamics in the market. For a strategy researcher it is essential to know how technical change is incorporated into the strategic thinking of the firm, how it is translated into investment decisions and into the value chain of the firm (and of the industry), and thereafter how it affects competitive dynamics. Traditionally the strategy literature treats technology as an implementation issue. The firm determines its strategy and this, in turn, defines how technology will be used. This ignores two problems—how does technology enter the strategy formulation process, and how are technological capabilities fostered and managed so as to create the basis for competitive advantage? To make progress along these dimensions requires addressing several points: (1) what is the relationship between technology and the economic conditions within which strategy is formulated, (2) what characteristics of technology and innovative activity affect the choice of and implementation of strategies, (3) how might technological issues be incorporated into strategic thinking and planning, and (4) how does technology affect the underlying capability of firms and how is this translated into competitive advantage? These issues form the agenda of the present series.

Two other books in the series (Steffens, *Newgames: Strategic Competition in the PC Revolution*, and Bogner and Thomas, *Drugs to Market: Creating Value and Advantage in the Pharmaceutical Industry*) focus on the successive waves of change and transformation that occur in industries as a result of technological change. Industry boundaries shift, competitive strategies have to be recast and the industry moves progressively from pioneering behaviour to competitive interaction and to the management of maturity. This study focuses on how specific major changes in technology are dealt with by one company (Philips) where the strategic stakes not only involve the industry and corporate winners and losers but are so high as to involve governments and their positions in the international political arena. In this study we are shown in considerable detail the interaction between the complexity and significance of new technologies, the corporate strategy of Philips and the competitiveness of the European consumer electronics industry, and public policy towards competitiveness.

Xiudian Dai's three case studies take us from the disappointment (to Philips) of the failed VCR investment through to the implications for public policy and for Philips itself. Following this and to some extent because of it Philips went through a fundamental restructuring. The second case involves High Definition Television where the implications for Philips in a world of highly interconnected major players and its concern to construct a defensible global position are well described. Although Philips had clearly learned from its earlier experiences, policy within the EU was dominated by political considerations of trade protectionism and government–industry coalition. Contrary to their avowed intention of picking winners, the EU authorities seem merely to have fostered losers. An important conclusion here might be that Philips and Thomson seem to have been capable of becoming leading players in this new game without any government help. This has great significance for public policy and suggests that the role of picking winners is less valuable and much more risky than that of fostering economic and technological infrastructures. The third case study, on the development of CD-i, finds little government intervention in terms of attempts to impose any legislative or protective procedures on the industry. Thus there has been more of a *laissez-faire* environment—so far. This technology has not yet reached any form of standardization but Philips has created a huge network of international collaboration and partnerships in order to secure a wide range of industrial support for hardware and software development. This is in stark contrast to Philips' experience with its V2000 VCR technology, where it suffered from the lack of any such support.

The interaction of technology, corporate strategy and corporate organization is a principal theme in this study, showing how Philips has fundamentally changed its strategies towards highly complex technologies. The second theme, of the interaction of technology and public policy shows all the dangers of undertaking long-term commitments to public support when the complexity of technology makes informed judgements difficult, and when the economic power of standards and dominant design result in economic power resting in private, corporate hands rather than in the public domain. This theme remains for further exploration.

University of Warwick John McGee
University of Illinois at Urbana-Champaign Howard Thomas

1

INTRODUCTION

The importance of technology to industrial and economic growth and, ultimately, the improvement of human life has already been widely accepted by many leading authors from a variety of disciplinary areas. In the 20th century, the development of the consumer electronics industry, one of the most dynamic technology sectors, has changed the way we live. With every significant technological breakthrough and major innovation in the consumer electronics industry, such as the introduction of the transistor radio, the gramophone, black and white and then colour television, the video cassette recorder (VCR), the compact disc digital audio player and, most recently, multimedia and digital TV, the industry has diversified and market has changed. In the same process of technical change, some old companies have disappeared and new ones have arrived. Among a group of others, Philips, the Dutch electronics giant, has survived radical competition driving every stage of innovation since the early days of the consumer electronics industry and indeed the company has often been a pioneer in the sector. In addition, it is due to this that Philips has made some strategic mistakes at times when radical technical changes reshaped the industry. This book is not intended as a historiography of Philips company, rather, my main concern is to examine some technological areas in which Philips has pioneered or been heavily involved, so that the role of corporate strategies in managing new technologies can be established.

Technical change is by no means the exclusive concern of corporate strategic planners. With the introduction of a range of new concepts and theories such as 'strategic technology' (Tyson, 1992) and 'strategic trade' (Krugman, 1986), policy-making bodies at national or supranational government level have also attempted to assume increased roles in managing new technologies. In Europe, trade policies, such as antidumping duties and those concerned with 'local content' requirement for manufacturers, and large scale close-to-market research and development (R&D) projects, etc., have been undertaken with a primary concern of promoting indigenous technologies and protecting domestic firms. By identifying 'strategic technologies', it is argued that government finds it easier, in political terms, to intervene in the process of technical change and innovation. Whilst certain cases pointed to the negative impact of government technology policy on innovation (e.g. the European Union's (EU) HDTV policy since the mid-1980s), there are also technological areas in which government involvement is considered vital and bare market forces are not the best solution. It would be too simplistic to completely reject government interventionism or to rely totally on free market forces.

Facilitated by the rapid development of digital technologies, the consumer electronics industry is experiencing a transitional period from the analogue to the fully digital stage. Consequently, the consumer electronics industry is becoming an integral part of a much wider industrial domain, i.e. the information and communications technology (ICT) sector. In response to this large scale technological convergence, the U.S. government has launched the National Information Infrastructure (NII) initiative and the EU has proposed the Information Society Programme.[1] The realisation of these new policy initiatives, if effectively implemented, would, it is argued, create a new technological society in which computers and TV screens will merge into a single information display board. It is anticipated that, with the fast-growing broadband networks all over the world, in addition to conventional voice and data communications, high quality interactive entertainment content will be delivered to the home and displayed on a 'hung-on-the-wall' flat panel where required by consumers. However, this process is causing tremendous regulatory confusion as the telecoms and broadcasting industries are still regulated by separate laws mostly at national level. It is hoped that a study, as presented in this book, about the interaction between corporate strategies at the firm and industry levels and public policy at the government level, with regard to managing new technologies in the consumer electronics industry since the 1970s, would contribute to an understanding of the same issue arising from the ICT sector, which is undergoing a radical global restructuring process.

KEY ISSUES

Since the late 1970s the world consumer electronics industry has witnessed fierce competition in either maintaining or gaining the leadership in trade and production involving new technologies among the most industrialised countries, particularly the Triad power of the U.S., Japan and Western Europe. The competition for new technologies has been joined by not only the firms, who actually carry out most R&D work to bring about major technological breakthroughs, but also the national governments whose policies and regulations interact with the industry and the market.

The central focus of this book is the role of corporate strategies, interfirm collaboration and public policy in determining the success and failure of new technologies and the competitiveness of firms, particularly the largest ones, against a background of increasingly intensified global competition. Although key factors are examined at different levels—the firm level, the industry level and the government level, it is argued that the success or failure of new technologies and the competitiveness of the firm is, in many cases, determined not by a *single* factor but a combination of different forces.

Given the above focus, I will examine a range of interrelated problems on the basis of a literature review and a series of empirical studies. The former will draw upon a wide range of academic literature, and the latter will be focused on the European consumer electronics industry, in which Philips' corporate strategies for new technologies will be analysed through detailed case studies. Despite previous efforts undertaken by a few authors, the central issue cited above has not been tackled on the basis of detailed case studies of a single leading firm and its key technologies. This study was initiated partly to fill that gap.

More specifically, the first problem to be examined in this book is the relationship between corporate strategies and the success or failure of new technologies. What is corporate strategy? To what extent can a variety of corporate strategies and related factors such as corporate structure, corporate culture, etc., affect the commercialisation of new technologies? Why and how do firms' corporate strategies change from time to time in accordance with the development of new technologies? How does a firm's organisational structure and corporate culture affect the success of its own technologies? Does leadership in precompetitive R&D necessarily lead to a firm's competitiveness in the marketplace? These issues will be tackled through detailed case studies about Philips' own history and its own technologies including VCR (Video Cassette Recorder) and CD-i (Compact Disc-interactive).

Influential authors, such as Ansoff and Chandler, have attempted to answer these questions in their writings about corporate strategies and managerial studies since the 1960s. Subsequent studies of corporate strategies accumulated in the 1980s and early 1990s. However, most of these studies were confined to the managerial level, and none of them was dedicated to detailed analysis of a single firm against a rapidly changing technological and competitive environment, in the same way that this study is organised.

Secondly, the importance of interfirm collaboration for the success of new technologies will be addressed. It is argued that, although strategic alliance and industrial collaboration is not a new phenomenon,[2] it was not until the 1980s that regional and global cooperation for precompetitive R&D projects or close-to-market product development became a common way for firms to do business. Why have many big firms opted to do business, particularly in the area of new technologies and new products, in partnership with their competitors over the last decade? Has the nature of market competition between firms been changed by the spreading practice of interfirm collaboration? Why do firms sometimes choose to collaborate with partners from the same country or region, and with foreign partners on other occasions? Is interfirm collaboration or strategic alliance a more effective way for the success of new technologies compared to a single firm working alone?

As a buzzword of the early 1980s, interfirm collaboration has drawn the attention of not only business and industrial consultants but also a large number of academics. Part of Chapter 2 will be devoted to showing differentiated views of a number of academics over the nature of and the motives for industrial collaboration and its relevance to competition and government regulations. As far as the national differences in industrial collaboration are concerned, Chapter 2 will also discuss the uniqueness of the Japanese system, as embodied in the *Keiretsu*. A discussion of the works by some authors, including Porter and Freeman, will help us understand the *Keiretsu* system and how it is different from the way industrial collaboration is organised in the West. The belief that the Japanese system is fundamentally different from the European environment actually forms part of the foundation on which some European industrialists lobby government officials to implement protectionist policies.

Whilst there is a lack of consensus on the effectiveness and legality of strategic alliances and industrial collaboration within academic discussions, Chapter 5 will demonstrate that, in practice, Philips' corporate structure and strategic changes from the early 1980s have been directed towards collaborating with its competitors, particularly the Japanese firms within the consumer electronics industry. Such a strategic move has been effectively extended to each of the company's new and core business areas: to collaborate with JVC, whose VHS defeated the Philips/Grundig V2000 format, in VCR production and developing a CD-i karaoke system; to enter alliance with Thomson and some American firms in order to develop a rival HDTV (High Definition TV)[3] system competing against its own European system; to campaign for getting wide industry support from both the hardware manufacturers and software[4] developers for the CD-i technology, etc.[5] This point will, in addition to Chapter 5, be further explored through detailed case studies on VCR, HDTV and consumer multimedia centred on CD-i in Chapters 4, 6 and 8, respectively.

The third essential area to be investigated is the impact of public policy (e.g. industrial policy and competition policy) on the development of new technologies and the competitiveness of the firms. As an important phenomenon of the 1980s, trade protectionism was gaining ground in each pillar of the Triad power. This has partly provoked new academic debate about industrial policy theories as represented by Krugman (1986), Tyson (1992), etc. Arguments against strategic technology and strategic trade theories have also been developed over the last few years (e.g. Carliner, 1986).

What characterises the discussion in many parts of this book is an analysis of the EU's policy changes towards the industry over the last two decades. When the European consumer electronics industry began to see the challenges from increased Japanese exports of finished durable consumer goods in the 1970s, the Community's role was less developed than it is now, and the much fragmented European market reflected the greater influence of national governments within the EU. The failure of European VCR technology represented by the Philips/Grundig V2000 format heralded further Japanisation of the European consumer electronics industry in the 1980s. Moreover, the process of European economic and political integration towards the single market has necessitated a policy-making shift to a larger extent from the national government level to the EU level since the mid-1980s. Partly due to these changes the EU has implemented a series of new policies towards the consumer electronics industry over the last decade. These include, on the one hand, purely protectionist measures (e.g. tariffs, antidumping duties, import quotas, etc.), and, on the other hand, intervention arrangements including, for example, the launch and coordination of a number of EU programmes, and subsidies to domestic firms.

Drawing upon the experience of the European VCR industry, this book will argue that protectionist government policy (e.g. antidumping) alone may not be effective in helping domestic firms to maintain or regain competitiveness; quite the opposite, long-term competitive advantage and efficiency could be better achieved by forces generated from factors inside the firm (e.g. sound corporate strategy, flexible and effective corporate structure, etc.). Through the HDTV case this book will question the ability of government bureaucrats to pick winners and promote new technologies and examine the consequences of attempting to do so. It is hypothesized that government industrial policies towards new technologies may easily be influenced by the corporate strategies of big and politically influential domestic firms. The book also argues that, while government policies are in most cases intended to protect domestic firms through limiting imports of foreign goods, or denying foreign firms access to local cooperative programmes, domestic firms are often quick to enter partnership arrangements with foreign competitors at home or abroad.

Overall, this book is intended to provide an empirical foundation for the understanding of how the success of new technologies and the competitiveness of the firm is affected by a number of important factors including corporate strategies, interfirm collaboration and public policy.

SCOPE OF THE STUDY

The turbulent changes occurring in the European consumer electronics industry since the 1970s have proved to be an exciting area for academic research. As an attempt to shed some light on the understanding of such an industry facing constant restructuring, this research is not designed to cover every aspect of the industry. On the contrary, in order to explore the major issues discussed above, I have chosen three different but interrelated angles to analyse: corporate strategies of the European firms, interfirm collaboration between European and foreign firms, particularly Japanese ones, and public policy of the EU towards the consumer electronics industry. This is done by giving particular emphasis to the Philips company.

From early days to the present time, Philips has been the leading domestic firm in the European consumer electronics industry. To some extent, each of Philips' strategic moves has, at any given time, inevitably affected the *status quo* of the European consumer electronics industry as a whole. On the technology side, Philips has been involved, and has sometimes taken the lead, in most of the important breakthroughs in the world consumer electronics industry. As a technology leader, Philips has experienced a substantial change in its corporate strategy from 'do-it-yourself' to collaborating with its European and Japanese competitors. Politically speaking, Philips plays an important role in influencing the process of European policy-making towards the consumer electronics industry. Altogether, these elements justify the choice of the Philips company as the focus of the empirical studies upon which this book is based. Of course, any study of Philips would not be comprehensive without referring to its relationships with other European and foreign firms.

However, Philips, originally a European manufacturer in the lighting industry, is a very diversified multibusiness company; consumer electronics, although a core business since the 1980s, is only a part of it. Among many other consumer electronics technologies, Philips' V2000 system, HDTV activities and the CD-i project have been chosen as the areas for detailed case studies. It is worth mentioning that such a choice of technologies for case studies is not a random one. On the contrary, this choice was made on the basis of my understanding of the major characteristics of the consumer electronics industry.

First, technical standardisation[6] has been one of the predominant features of the consumer electronics industry over the last few decades. In other words, any significant technological breakthrough in the history of this sector could hardly achieve commercial success without fierce competition among the companies involved. Moreover, to achieve world standard status, technological superiority may not be the only determining factor; nontechnical elements play their role in the competition as well. The Philips V2000 home video format is a widely cited classic case where the supposedly superior European technology lost the format battle to the Japanese.

More recently, enormous R&D work has been undertaken by companies all over the world to bring about HDTV and multimedia products in an effort to revitalise the saturated and declining consumer electronics industry. This has prompted two new format battles. The controversial

European HDTV system, HD-MAC, and the ambitious Philips multimedia system, CD-i, exemplify the participation of European technologies in the on-going processes of standardisation.[7] Studying the processes of technical standardisation in these three product areas offers the opportunity to address the role played by nontechnical factors, namely, corporate strategies, interfirm collaboration, and public policy.

The second major characteristic of the consumer electronics industry is its increased tendency towards technological convergence since the beginning of the 1980s. On the one hand, as Chapter 8 will show, this convergence suggests a combination or integration of traditionally independent products such as audio systems, video systems, computer games, etc. On the other hand, technological convergence through digitalisation is being extended beyond the consumer electronics industry itself; recent development reveals that the boundaries of different branches of the information technology (IT) sector, i.e. consumer electronics, telecommunications and the computer industries are becoming increasingly blurred. The study of emerging multimedia products, such as CD-i, will provide us with an opportunity to understand the nature of a new era in the history of the consumer electronics industry. It will also enable us to see how the major firms adapt themselves to the new competitive environment characterised mainly by technical change.

The consumer electronics industry over the last decade has developed an increased number of new products and new systems embodying digital technologies. The most dynamic force driving the development of technological convergence, as mentioned above, is digitalisation. Both multimedia products and the future HDTV sets and transmission technology will largely rely on the progress of digital techniques. Many of the technical controversies around HDTV have reflected differences between digital and analogue systems. Producers anticipate that digital VCR will be the next generation of home video technology.

Finally, the principle of synergy between hardware and software has largely characterised the consumer electronics industry since the 1970s. By studying the VCR industry, particularly the V2000 experience, the development of HDTV and the rapidly blooming multimedia industry, we will be able to see the important role of software or the synergy of software and hardware in determining the success of new technologies. Most recently, some of the leading Japanese manufacturers, such as Sony and Matsushita, have invested heavily in the American entertainment industry. This move is widely regarded as a measure of these companies' concern to ensure more support from the software industry for their new technologies. To some extent, as the empirical findings in this book suggest, consumer electronics is a software-led industry.

To return to the main issues, intensified international competition has accompanied the process of standardisation, technological convergence, digitalisation and hardware/software synergy: corporate strategies are being reformed or redesigned at the firm level; more regional or international interfirm collaboration has been initiated at the industry level; a range of public policies has been implemented at the governmental level and EU level in order to enhance the competitiveness of domestic firms.

Whilst Europe is the focus of the study as defined above, some discussion of technical changes, market development and public policies elsewhere in the world (e.g. the U.S. and Japan) will provide a context wherever necessary and appropriate in order to understand the European consumer electronics industry.

It should also be noted that, while the choice of a single firm, Philips, and three technological areas (VCR, HDTV, and CD-i) for intensive study provides some valuable insights for our understanding of the nature of the European consumer electronics industry and the major issues in question, any generalisation of the findings to be derived from the present study must be undertaken with care.

In order to provide a comprehensive and detailed understanding of the issues outlined above, the research for this book follows two lines of methodology: theoretical discussion and empirical studies. On the theoretical side, a wide range of literature across several academic disciplines, such as modern political economy, political science, managerial and innovation studies is reviewed. This multidisciplinary social science perspective allows us to look at the complicated issues of technical change and industrial competitiveness within a much wider context than is offered by any single theory.

In terms of theory, there is no attempt to devise a novel conceptual framework or to focus criticism on any particular traditional school. The purpose of the review, as will be shown in Chapter 2, is to provide a wide and synthetic academic perspective to look at the major issues to be touched on during the course of this study. This approach is deliberately interpretative, rather than aimed at 'testing' any of the theories founded on traditionally classified subject areas, such as managerial studies, political economy, and political science.

As far as the empirical work is concerned, I have chosen three different but intrinsically related areas to analyse: (1) the history of the Philips company (Chapter 3); (2) public policy changes within the EU towards the consumer electronics industry and the restructuring of Philips since the beginning of the 1980s (Chapter 5); and (3) new technology/product cases including the European VCR industry and Philips' V2000 format (Chapter 4), the European HDTV system (Chapter 6) and Philips' CD-i multimedia project (Chapter 8).

The major findings of this book derive from the author's face-to-face interviews with a wide range of individuals during the period 1990–1992. Interviewees include Philips managers based at either the company's headquarters in Eindhoven or its U.K. branches, and managerial staff from other consumer electronics or computer firms. Furthermore, fieldwork in

Brussels involved interviewing a number of EU officials in charge of public policy affairs at various levels. Some academics specialising in the consumer electronics industry or related areas were also approached during the course of the research.

The research for this book has also benefited from a number of international industrial conferences. At industrial conferences different, sometimes conflicting views, over the nature, the development and the future of a new technology or a new product were often expressed by representatives from corporate organisations and government authorities as well as by academics. This approach proved fruitful in building up an understanding of a single company, Philips, and the development of individual technologies or products against a wider international and industry background. Although changes in other areas such as Japan and the U.S. are not the primary concern of this book, they undoubtedly form an invaluable background to the understanding of the European consumer electronics industry.

STRUCTURE OF THE BOOK

The main part of the book comprises the following seven chapters.

The literature review in Chapter 2 introduces a range of interrelated issues and themes which have been widely discussed by different authors. The issues, such as corporate strategies and the management of new technology, strategic alliances between firms, and public policy and industrial competitiveness, are also the major concerns of the empirical studies from Chapter 3 to Chapter 8. Therefore, the discussion developed in Chapter 2 is intended to provide the analytical framework and theoretical background against which the series of case studies can be understood.

The empirical studies in the book begin with Chapter 3, which deals with the history of the Philips company, not only one of the top three leaders in the world consumer electronics industry but also the flagship of the European consumer electronics industry. The discussion in Chapter 3 reveals the historical roots of a number of the important and interesting issues, such as technical change and its effect on corporate structure, diversification *versus* concentration, interfirm collaboration *versus* cartelisation, which faced both Philips and the European industry from the late 1970s and early 1980s.

Chapter 4 covers the development of the European VCR industry with a focus on the role played by Philips and the *débâcle* of its V2000 system. It is argued that the bitter V2000 experience signalled the start of increased foreign competition, led by the Japanese, with the European industry and heralded an outcry by the leading European firms for government intervention. Chapter 4 also sheds light on the specific reasons why the technologically superior V2000 system lost the format battle to VHS. Moreover, some implications are drawn from the VCR format battle.

Chapter 5, first of all, broadens the scenario of Japanese challenges to the European consumer electronics industry. Secondly, I discuss from two different perspectives the European response to the process of intensifying competition from Japanese firms symbolised by the VCR case. On the one hand, public policy-making towards industry in Europe began to shift from the national government level to the EU level from the mid-1980s. On the other hand, the eclipse of European VCR technology by the Japanese led to a series of strategic changes and an organisational restructuring within Philips.

A combined European effort undertaken by influential European firms (e.g. Philips and Thomson) and public authorities (e.g. the European Commission, the French government and Dutch government) to revitalise the declining European consumer electronics industry and fight back another wave of Japanese competition is well demonstrated by the launch of the Eureka 95 HDTV project in 1986. Therefore, Chapter 6 shifts the focus of analysis of technological manoeuvring from corporate strategy level to government and EU level to investigate various issues concerning European technology policy towards the development of HDTV in Europe. This discussion provides an important case study of how public policy affects corporate strategy and *vice versa.*

The interplay between corporate strategies and public policy in the case of HD-MAC in Europe since the mid-1980s was not a success for EU policy-makers. Chapter 7 discusses European policy changes after the collapse of HD-MAC in favour of alternative technologies, mainly digital TV development. Compared with its heavy-handed interventionist technology policy on HD-MAC, the EU has, by and large, left the industry alone to develop fully digital TV technologies. This new policy approach, as Chapter 7 shows, has allowed room for the European Digital Video Broadcasting (DVB) group to develop a set of new standards associated with the launch of digital television broadcasting in Europe. It is widely recognised that the operation of the DVB group is bringing Europe into the mainstream of a global contest for fully digital TV development.

In contrast to the HDTV case, the emergent multimedia industry has been subject to little government intervention. Chapter 8 concentrates on another format battle between various interactive multimedia systems, in which Philips' CD-i is a major contender. In the process of developing and promoting its CD-i standard, Philips from the late 1980s adopted an internationally-oriented strategic alliance and partnership strategy in order to re-establish its technological leadership and international competitiveness. At the same time, Chapter 8 explores how the Philips management has come to acknowledge the importance of software for the success of a new consumer technology, a lesson which the company has learned from its own failures.

Chapter 9 concludes the book with a brief summary of the major findings of this study and some suggestions for future research.

NOTES

1. Note that parallel generic national initiatives/programmes of the same or similar nature have also been launched or proposed in most industrialised and many developing countries since the early 1990s.
2. As empirical evidence will suggest in Chapter 3, Philips was for a long time engaged in a number of cooperative arrangements with other firms from the world electric lightbulb industry.
3. 'HDTV' in this book is used to define the next generation of TV technology with a high picture resolution (over 1,000 horizontal scanning lines) and a widescreen aspect ratio (16×9). Technologically speaking, two substantially different approaches have been adopted to build HDTV systems (transmission): hybrid but mainly analogue solutions (e.g. MUSE in Japan and HD-MAC in Europe) and fully digital systems (e.g. the American standard being developed by the 'grand alliance'). As HD-MAC has now been dropped and MUSE is subject to increased technological uncertainty, fully digital TV has become the mainstream technology currently led by the U.S.
4. Note that there is a wide variety of 'software', such as operating systems and application programmes used in the computer industry; entertainment software including computer games, films, TV and radio broadcasting programmes, and prerecorded software on audio/video cassettes, audio/video discs, CD-ROM or multimedia discs, etc. As far as the consumer electronics industry is concerned, the term 'software' is used throughout this book to refer to a large variety of audio-visual programmes or titles created to interplay with the hardware in many different ways.
5. Philips also has a licence to manufacture Sony's Mini Disc system, which is in direct competition with its own DCC (Digital Compact Cassette) system.
6. Standardisation in many cases was achieved through fierce format battle among competitors with different products or systems to offer.
7. Note that the standardisation process for HDTV has now been shifted to digital TV including digital HDTV. Chapters 6 and 7 have detailed discussion about this.

2

TECHNICAL CHANGE AND THE COMPETITIVENESS OF FIRMS: CORPORATE STRATEGIES AND PUBLIC POLICY

The purpose of this chapter is to introduce, from an interdisciplinary perspective, a range of themes and issues which have been stressed in the literature of contemporary political economy, political science and management studies. The central point of this chapter is that the development of new technologies, new products and, eventually, the competitiveness of firms is affected by a number of factors. Among many others, three major aspects are considered here to be most relevant: corporate strategies, strategic alliances and interfirm collaboration, and pubic policy towards the industry. I will discuss these three aspects under three sections respectively: corporate strategy and the management of new technology; strategic alliances and industrial collaboration amongst firms; public policy and industrial competitiveness. Finally, in the fourth section, I will introduce two new models to show how new technologies are influenced simultaneously by forces from three different levels: the corporate/firm level, the industry/sectoral level and the national and supranational government (e.g. the EU) level, so that two analytical frameworks are provided for the book.

With an interdisciplinary approach, this chapter is not confined to discussing theories advocated by one particular school. On the contrary, I will review how a range of academic writers from different backgrounds understand and interpret the same issues. Of course, these issues will be further investigated in a great detail through a series of case studies included in the empirical chapters which follow.

CORPORATE STRATEGY AND THE MANAGEMENT OF NEW TECHNOLOGY

How is industrial innovation best managed at the corporate level? Corporate strategy is identified by some business analysts as one of the most important factors in determining whether a new technology or new product can be successfully commercialised by a single company or, sometimes, a group of companies. In the history of commerce and industry, many highly innovative firms, such as Philips, have experienced extreme difficulties in translating their technical innovations into competitive advantages. In this section, I discuss various issues concerning corporate strategy and its implications to the management of new technologies from a variety of academic perspectives.

The Nature of Corporate Strategy

It is argued by many authors in strategic studies that the concept of corporate strategy is an analogy of or originated from the classical as well as modern military art (e.g. Mintzberg and Quinn, 1991; Ansoff, 1965).[1] In its military sense, strategy is a vaguely defined 'grand' concept of a military campaign for application of large-scale forces against an enemy, and it is sharply contrasted to 'tactics', which is a specific scheme for employment of allocated resources (Ansoff, 1965, p. 118). Sometimes, 'strategy is when you are out of ammunition, but keep right on firing so that the enemy won't know' (Author unknown, cited in Ansoff, 1965). Military strategists conceived their strategies with an ultimate objective of winning the battle or strengthening yourself by concealing your own weakness and confusing the enemy. One can use the same term 'strategy' as an analogy to describe corporate policy at the highest level of a business organisation. However, the ultimate target of a well-defined corporate strategy should not be an overall policy to confuse the enemy but one of setting the right direction for the firm to develop in a given external environment. Sometimes, a firm needs to illustrate and publicise its general strategy and make it easy to understand for competitors so that competitors can become partners.[2]

Although strategies have been used throughout the history of commerce and industry, the study of corporate strategy as an academic subject did not receive serious attention until the 1960s (Kay, 1993).

The validity and usefulness of any particular approach to corporate strategy depends very much on the situation or competitive environment in which it is to be used. In other words, one cannot simply adopt a strategy devised in the 1960s as the most appropriate for the management in the 1990s because the world has changed (Pearson, 1990). In the 1990s, mass production of a standard product is hardly an adequate response; growth is no longer guaranteed positive; market dominance is no longer an end in itself and may even be a source of vulnerability (Pearson, 1990, p. 2). The world is changing and corporate strategy needs to change with it.

In *Strategy and Structure,* Afred Chandler formulated his well-known theory which suggests a company's structure follows its strategy (1962, p. 14). He identified corporate strategy and corporate structure as two major interrelated aspects as far as the growth of an industrial enterprise is concerned. Chandler's theory consists of three basic elements:

- Structure follows strategy and the most complex type of structure is the result of the concatenation of several basic strategies;
- Strategy is the determination of the basic long-term goals and objectives of an enterprise, the adoption of courses of action and the allocation of resources necessary for carrying out these goals;
- Structure is the design of organisation through which the enterprise is administered.

For the period following the second world war through to about 1970, every firm was diversifying and what Chandler suggested was that a strategy of broad diversification dictates a structure marked by decentralisation (Peters and Waterman, 1982). While Chandler's theory caused a revolution in management practice and initiated a period when diversification and expansion was in fashion, some authors in management studies found that strategy rarely seemed to dedicate unique structural solutions, and that the crucial problems in strategy were most often those of execution and continuous adaptation: getting it done, and staying flexible (Peters and Waterman, 1982, p. 4). In the mid-1960s, Ansoff also published his well-known study on corporate strategy, in which he defines corporate strategy as a rule for making decisions relating to a firm's match with its environment (1965, pp. 119–120). A strategy as such specifies four basic aspects: product–market scope; growth vector (or direction); competitive advantage (the leading strengths); and synergy (common thread of individual parts or aspects of the firm).

The publication of *Competitive Strategy* by Michael Porter in 1980 was widely regarded as the beginning of a new era for studies on corporate strategy. Porter (1980) argues that a firm's profitability was determined by the characteristics of its industry and the firm's position in it, so these should also determine its strategy. He describes the context in which a firm's competitive strategy is formulated as shaped by two categories of factors: (1) factors external to the firm—industry opportunity and threats (economic and technical), broader societal expectations; (2) factors internal to the company—company strengths and weaknesses, personal values of the key implementers. In its essence, 'formulating competitive strategy is relating a company to its environment' (Porter, 1980, p. 3).

Following this line of thinking, Porter (1980) believes there are three generic strategies:

- overall cost leadership—lower cost relative to competitors;
- differentiation—being unique within an industry;
- focus—particular strategic target.

Despite the fact that much attention of corporate strategy research in the 1980s was drawn to Porter's works, his understanding about the nature of corporate strategy does not differ very much from that of Ansoff's in the 1960s; on the contrary, both emphasise the point that the real function of a firm's overall strategy is to relate or match its internal resources to its external environment. As a matter of fact, many more other studies followed the same way to define corporate strategy in the 1980s (e.g. Johnson and Scholes, 1988).

In *The Strategy Process,* Mintzberg and Quinn (1991) define corporate strategy from another perspective. They believe that a strategy is "the pattern or plan that integrates an organisation's major goals, and action sequences into a cohesive whole"; and a well-formulated strategy "helps to marshal and allocate an organisation's resources into a unique and viable posture based on its relative internal competencies and shortcomings, anticipated changes in the environment, and contingent moves by intelligent opponents" (p. 5).

Most recently, Hamel and Prahalad have claimed that the essential element of corporate strategy is an aspiration that creates competitive advantages by designing a chasm between the ambition of top management and the resources of the company (1993, p. 84). In other words, a company's competitiveness is born in the gap between a company's resources and its managers' goals. Hamel and Prahalad (1993) believe that such a strategy has made it possible that companies such as Toyota, CNN and Sony of much smaller scales and less internal resources win the competition against General Motors, CBS and RCA, respectively. A classical example of JVC having won the format battle for home video cassette recorders against the industrial leaders including Philips and Sony (as well as its parent company Matsushita) in the late 1970s and early 1980s may be added to this list. This brand-new notion of corporate strategy is a complete deviation from the aforementioned traditional understanding about the nature of corporate strategy, which was conceived to primarily match a company's resources with its opportunities and external environment.

While many proponents of strategic studies believe that the primary target of corporate strategy is to beat the competition, Kenichi Ohmae thinks from another perspective. He suggests that the first principle of corporate strategy is to create and deliver value to customers (1990). Ohmae argues that managers too often and too willingly launch themselves into old-fashioned competitive battles because it is their familiar ground and they know how to fight; but what the managers have failed to understand is the fact that an effective customer-oriented strategy could avoid the battle altogether (1990, p. 34). To strengthen his argument, Ohmae (1990) has cited Sun Zi,[3] whose famous classical theories of military strategy suggest that the best strategy of war is to win the battle without fighting with the enemies.

Freeman (1982) has also emphasised the pivotal role of the need of customers to the success of an innovative firm. He argues that a firm which is closely in touch with the requirements of its customers may recognise potential markets for novel ideas (e.g. new discoveries and technical possibilities derived from the advance of scientific research) or identify sources of customer dissatisfaction, which lead to the design of new or improved products or processes (1982, p. 111).

The lack of a consensus over the nature of corporate strategy, as indicated in discussions above, does not suggest that the study of corporate strategy is worthless but is a reflection of the reality in which there is no single way for various firms to approach the future.

As far as terminology is concerned, corporate strategy has many other equivalents. For instance, in *Exploring Corporate Strategy*, Johnson and Scholes (1988) treat corporate strategy, business policy, management policy, corporate policy, and strategic management as equivalent terms to denote the most general levels of a company's strategy.

Corporate Structure and Corporate Strategy

Chandler's famous theory, as mentioned above, that structure follows strategy may prove valid in a sense that corporate strategy and corporate structure are two interrelated major concerns of a company's top management. However, one cannot conclude from this theory that structure is an insignificant issue for managers to consider compared to strategy. In fact, a sound strategy can only be effectively implemented through a well-designed corporate structure. Since the second world war, theories of corporate strategy have survived several generations of painful transformation and have grown appropriately agile and athletic; but attention to organisational development has not kept pace and managerial attitudes lag even further behind (Bartlett and Ghoshal, 1990). According to Bartlett and Ghoshal, there does exist a kind of incompatibility between the design of corporate strategy and the renovation of corporate structure:

> As a result, corporations now commonly design strategies that seem impossible to implement, for the simple reason that no one can effectively implement third-generation strategies through second-generation organizations run by first-generation managers (1990, pp. 144–145).

Even worse, many large multinational companies remain caught in the structural-complexity trap that paralyses their ability to respond quickly or flexibly to the new strategic imperatives (Bartlett and Ghoshal, 1990, p. 140).

To become a 'global' company, a firm needs to globalise itself from an organisational point of view. In various management studies, 'globalisation' has recently become a fashionable word. However, the meaning of 'globalisation' itself has changed over time. Companies used to globalise themselves by diversifying and expanding throughout the Triad (or more broadly) and setting up subsidiary branches or manufacturing sites in as many countries as possible. Raymond Vernon of the Harvard Business School is regarded as a leading proponent of such a 'United Nations' model of globalisation (Ohmae, 1990). Due to the fact that successful companies enter fewer countries but penetrate each of them more deeply, Vernon's model gave way by the early 1980s to a competitor-focused approach to globalisation (Ohmae, 1990, pp. 30–31). In other words, the pressure for globalisation today is driven not primarily by diversification of technologies and organisational operations but by the needs and performance of customers whose needs have been globalised.

In the history of corporate development, diversification resulted in not only industrial growth but also severe structural problems which were becoming increasingly beyond the control of the top management within many big firms. During the high industrial growth period since the second world war, according to Chandler (1991), many business executives in the West had come to believe that, if they were successful in their own industry, they could be just as successful in others; and many of them simply enjoyed empire building. By the 1970s, most large business enterprises started restructuring their organisations by creating integrated business units within the divisions that coordinated and controlled a single product or very closely related product lines. As a result, most big companies had three levels of autonomous planning and administrative offices—business unit, the division and the corporate HQ (Headquarters) or office (Chandler, 1991, p. 34). For these companies, the functions of the corporate HQ in designing and implementing strategies could hardly be compared with that of their past, when their business was almost homogeneous. To effectively monitor and coordinate the subheadquarters of the product divisions and business units has become a challenge to the corporate HQs.

It seems that the R&D efforts and marketing efforts within many multinational firms do not always lead to harmony, let alone synergy. Even worse, conflicts between these two activities sometimes push a globally stretched multinational firm towards more bureaucracy, slowness to respond, more inefficiency and, overall, less competitiveness.

Ohmae (1990) observes, for some firms, the flow of good ideas (namely R&D) substantially outpaces the understanding of customer need and market potential; while for others, the need of customers and the potential of market is clear but the idea flow is inadequate (p. 72). To illustrate his argument, Ohmae has analysed the case of Nissan, an automotive company that used to be severely troubled with product development, which languished for many years because the internal process of the company gave too strong a voice to marketing managers; despite that, the marketing team in Nissan used to complain that the products they were being asked to sell were not what the market wanted (1990, p. 72). As a result, Nissan's market share within the Triad (the U.S., Europe, and Japan) had been dropping. The empirical studies in this book also demonstrate a comparable and long-lasting managerial problem faced by Philips in launching new products, although in an opposite way, in its history as well as recent years' corporate restructuring.

Apart from the conflicts between different segments (sales or marketing departments and R&D and production activities), another structural problem facing the multinational companies is generated from the administrative relationship between the headquarters and the overseas local operations, in particular those national organisations of a big company. A classical example of corporate structure having severely undermined a company's global technology strategy was the relationship between the Philips headquarters in Eindhoven and its relatively independent national organisations, in particular North America Philips. More detailed discussion is presented later in this book.

As far as those globally stretched multinational companies are concerned, there are two major types of corporate structure design: highly centralised corporate structure with decision-making powers largely possessed by world headquarters, and decentralised corporate structure, which is normally embodied in the form of matrix structure (with highly autonomous national organisations as the rows and various product divisions as the columns).

Although not obviously in favour of the matrix structure, some scholars strongly link a decentralised organisational structure with flexibility. By rejecting the traditional 'headquarters mentality'—the headquarters of a firm inserts strict control and intervention over the management of its overseas operations, Ohmae (1990) advocates a decentralised and highly flexible management structure for multinational companies. He argues that a highly centralised managerial hierarchy structure does not fit in the running of an 'amoebalike' network of a global organisation, which needs a considerable level of autonomy and flexibility to reflect the quickly changing situation in the overseas local markets. In other words,

> No company can operate effectively on a global scale by centralizing all key decisions and then farming them out for implementation. It doesn't work. The conditions in each market are too varied, the nuances of competition too complex, and the changes in climate too subtle and too rapid for long-distance management. No matter how good they are, no matter how well supported analytically, the decision makers at the centre are too far removed from individual markets and the needs of local customers (Ohmae, 1990, p. 87).

Among others, Sony, Matsushita, Yamaha, Honda, Nissan, and Omron of

Japan have already decentralised responsibility for strategy and operations to each of the Triad markets and keep only corporate service and resource allocation functions at world headquarters, according to Ohmae (1990).

It is worth noting that decentralised strategic decision making and managerial autonomy does not mean that a global company faces the danger of losing its corporate identity and overall integrity. On the contrary, according to Ohmae (1990), branches or overseas operations of a global company are interlinked or integrated by shared corporate values rather than the headquarters. This argument seems to thoroughly reflect the idea of 'global localisation', a widely cited fashionable term coined by Akio Morita, the Chairman of Sony.

'Global localisation', the Sony slogan, is equally advocated by Bartlett and Ghoshal (1990) but in a shortened form '-ization'. This is derived from their analysis of the 'misleading' matrix management.

The matrix structure became fashionable in the 1970s and the early 1980s as the organisational solution to strategies that required multiple, simultaneous management capabilities (Bartlett and Ghoshal, 1990, p. 139). The matrix structure is characterised by its dual reporting system and overlapping responsibilities as a result of the coexisting national organisations and product divisions. Bartlett and Ghoshal believe that the matrix structure proved all but unmanageable—especially in an international context: the dual reporting system led to conflict and confusion, whilst overlapping responsibilities produced turf battles and a loss of accountability (1990, p. 139).

To get out of this organisational trap, the key organisational task facing top management of big companies, as Bartlett and Ghoshal (1990) suggest, is not to design the most elegant structure but to capture individual capabilities and motivate the entire organisation to respond cooperatively to a complicated and dynamic environment (p. 140). In other words, the top management need to develop a healthy 'organisational psychology'—the shared norms, values, and beliefs that shape the way individual managers think and act; and instead of searching for the ideal structure and simply imposing it on the company from top down, they should focus on building a shared corporate vision (Bartlett and Ghoshal, 1990). This is a process of generating local flexibility while maintaining corporate identity. In citing the successful case of Unilever, Bartlett and Ghoshal (1990) have noted that Unilever committed itself to the Indianisation of its Indian company, Australianisation of its Australian company, and so on; but the company also added programmes dedicated to the 'Unileverisation' of its local managers in each national company.

To further the discussion about corporate governance and corporate structure, Charles Handy published his 'new federalist paper' at the end of 1992, claiming that businesses in every country are moving in the direction of federalism which is the way to govern their increasingly complex organisations. In the same way that the term corporate strategy has a military origin, the theory of 'federalism' is directly borrowed from political discussions. Having adopted a political analytical approach, Handy (1992) believes that two of the major principles of the federalist theory are 'subsidiarity' and 'interdependence' or 'pluralism'. By subsidiarity he means that power belongs to the lowest possible point in the organisation; whilst interdependence clearly distinguish an organisation of federation from that of confederation in which individuals or branches yield no 'sovereignty' and autonomy for decision-making. In that sense, "[f]ederalism encourages combination when and where appropriate but not centralisation" (Handy, 1992, p. 65). Handy claims that pluralism is a key element of federalism because it distributes power, avoiding the risks of autocracy and the overcontrol of a central bureaucracy (1992, p. 66). The result of distributing power is the new 'dispersed centre' of federalism, a centre that is more a network than a place. To make his point clear, Handy (1992) says that, in business, federalism is neither simple decentralisation nor a simple divisionalisation—the former loses the advantage of scale and the benefits of central coordination over separate businesses, while the latter leaves too much power in the hands of those division heads and pays too little attention to local needs or to the knowledge and contacts of those out in the marketplace (1992, p. 61). On the contrary, in a federalist organisation, as Handy asserts, governance is the legislative function responsible for overseeing management and monitoring and, most important, for the corporation's future, for strategy, policy, and direction (1992, p. 67).

Federalism might be suitable for certain business organisations; but Handy's claim that "businesses in every country are moving in the same direction" (1992, p. 60) may have taken the concept too far. As Handy (1992) himself admits, federalism is not simple, and it "matches complexity with complexity" (p. 72). This universal panacea is not convincing to many other commentators. Just as one company can suffer from too simple an organisation if it is in a complex and turbulent environment, another can pay an efficiency penalty for adopting an organisation too complex for its environmental demands.[4] Moreover, within transnational companies, almost no two divisions are organised in the same way (e.g. Unilever's detergents business is now much more centralised than its food business) because of pressures for maximum responsiveness to differences in national markets.[5]

No matter what kind of organisational structure a firm has inherited from the past, the 1990s seem to be a new era of challenge faced by big businesses. Traditional economic and managerial theories about economies of scale and economies of scope (e.g. Chandler, 1990), which attribute a substantial role to the big firms in the history of economic growth, may not find it easy to explain the new turbulent business world in the 1990s. Having taken into account the spectacular failures of a number of giant companies (e.g. General Motors, IBM, Philips, Matsushita, Daimler-Benz, etc.), *The Economist* commented, in one of its recent surveys, that "[t]he era of corporate empire-building is over", because "in 1993 'big' no longer means, as it once did, 'successful'; before long it is likely to mean 'failing' " (17 April 1993, pp. 13–14).

After all, the size of the firm does matter but 'bigness' does not necessarily lead to efficiency or competitiveness.

Core Competence and Corporate Strategy

In the long run, the competitiveness of firms derives from an ability to build, at lower cost and more speedily than competitors, the core competence that spawns unanticipated products (Prahalad and Hamel, 1990, p. 81). The concept of 'core competence', strongly advocated by Prahalad and Hamel, refers to the collective learning in the organisation, especially how to coordinate diverse production skills and integrate multiple streams of technologies (1990, p. 82).

The core competence is embodied in a firm's 'core product', sometimes core skills. For instance, Philips' optical-media expertise and in-house R&D ability has enabled the company to lead the world industry in a number of product areas including CD-audio, CD-ROM, CD-i, etc. One aspect of Sony's core competence is miniaturisation.

Although the core competence is rooted in a firm's innovation capability, it is important to note that "cultivating core competence does not mean outspending rivals on R&D" (Prahalad and Hamel, 1990, p. 83). Sony may not have spent more on researching into the compact cassette technology than Philips did, but the success achieved by the former in miniaturising the product into the Walkman style was innovative and proved a huge success.

Dodgson (1989) and Peters and Waterman (1982) have used similar terms, such as 'core business' and 'core technology', to depict the concept of core competence. In a shift from the expansion- and diversification-based corporate strategy (e.g. pioneered by Chandler in the 1960s) to one heightened by rationalisation and concentration in the U.S. and Europe in the late 1970s and early 1980s, a large number of big multinational companies have explored the benefits of concentrating on their core businesses (Dodgson, 1989, p. 220). It is contended by Peters and Waterman (1982) that concentrating on core businesses and core technologies in part reflects the belief that a firm's comparative advantage lies in its ability to use its internal resources to compete only in a limited number of areas; and that a great number of problems have arisen from a firm's inability to manage its diversified businesses.

Considerable evidence presented in the following empirical chapters shows that a big company such as Philips has been rationalising and restructuring itself throughout the 1980s in an effort to overcome the problems generated partly from previous diversification.

The Concept of Synergy and Corporate Strategy

To concentrate on its core business or to build core competence does not mean that a firm will divest the rest of its noncore businesses. On the contrary, the competitiveness of a firm may not be achieved without a well-designed corporate strategy which effectively combines or integrates different parts, different activities of the company and makes them support each other. Here I briefly introduce the concept of synergy.

The importance of synergy to the running of a multidivisional or diversified firm has been and is being proved by the performance results of various corporations throughout the world. In particular, the development of the consumer electronics industry is a typical case, in which the principle of synergy between hardware and software to a great extent determines a firm's success or failure. This will be shown in great detail in the following empirical chapters, particularly Chapters 4, 6, 7 and 8.

In *Corporate Strategy*, Ansoff describes synergy as one of the major components of a firm's product–market strategy. In business literature, the concept of synergy is frequently described as the '2 + 2 = 5' effect to denote the fact that a firm seeks a product–market posture with a combined performance that is greater than the sum of its individual parts (Ansoff, 1965, p. 75). Ansoff regards synergy as a measure of a firm's ability to make good on a new market–product entry and, therefore, synergy is one of the five components of corporate strategy (1965, pp. 111–112).[6]

From an organisational point of view, Chandler (1990) asserts that one of the most critical tasks of top management in a multibusiness company has always been to maintain the organisational capabilities and to integrate the corporate facilities and skills into a unified organisation, "so that the whole becomes more than the sum of its parts" (p. 594).

Corporate Culture and Corporate Strategy

Corporate culture is another important concept closely related to corporate strategy.

Best (1990) argues that firms, apart from being economic organisations, are social institutions with unique cultures—a company's culture is not a commodity that can be bought and sold in the market like factors of production; it is more like an individual person's identity. Therefore, he believes that firms, as collectivities of people, have sophisticated means of resisting change, one of which is to deny the need for it (1990, p. 22).

Generally speaking, corporate culture refers to the corporate personality or ideology as perceived collectively by organisation members; in other words, it defines what the organisation is internally and explains how people in the organisation perceive the organisation and consequently determines how they behave (Pearson, 1990, pp. 30, 32). The importance of corporate culture to corporate strategy is illustrated by the fact that the latter can translate into the former, or the latter might take the form of a dominant culture (Pearson, 1990, p. 30; Porter, 1980, p. 41).

According to Pearson (1990), corporate culture is not simply homogeneous, but the sum of many subcultures. More specifically, there might be different subcultures on different levels of the organisation; differentiated subcultures may exist in different parts of the organisation. In particular, one branch operating in one country could develop a very strong subculture which is in sharp contrast with other subcultures developed in other country-specific environments. Therefore, with no effective managerial control, different subcultures may fall into conflict and competition with each other. It is up to the top management to ensure that the desired corporate culture norms are predominant over all subcultures.

Some companies, such as IBM and Marks & Spencer, have a powerful and identifiable corporate culture; one can admire their products and their achievements in the marketplace, but their corporate culture is not to everyone's taste (Kay, 1993, pp. 66–67). It is worth noting that acquisitions or mergers, as Milgrom and Roberts (1992) suggest, could generate problems due to the conflicts of corporate cultures from the companies involved. In this case, special costs would occur. The case study on Philips' restructuring process presented in Chapter 5 will demonstrate the complexity and importance of corporate culture.

Corporate Strategy and Industrial Innovation

Generally speaking, technology is a crucial resource of industrial firms, and its generation, acquisition and diffusion is a strategic concern for firms of all sizes in a wide range of industries (Dodgson, 1989, p. 1). In particular, for those global companies with a large R&D team and heavily involved in bringing about technical breakthroughs, to successfully manage the process of industrial innovation or commercialising their new technologies is one of the most important aspects of their corporate strategy. Under this theme, I will review some issues in connection with industrial innovation and the factors contributory to the success and failure of it.

First of all, let's look at the concept of innovation. A clear distinction between invention and innovation, first made by Schumpeter and since adopted in most economic analyses of technical change, served to highlight the role of the entrepreneur in the entire process and gave invention a somewhat inferior status (Freeman, 1992, p. 76). In commenting on Schumpeter's contribution and defining innovation, Freeman (1982) writes:

> We owe to Schumpeter the extremely important distinction between inventions and innovations, which has since been generally incorporated into economics theory. An *invention* is an idea, a sketch or model for a new or improved device, product, process or system. Such inventions may often (not always) be patented but they do not necessarily lead to technical *innovations*. In fact the majority do not. An *innovation* in the economic

> sense is accomplished only with first *commercial* transaction involving the new product, process, system or device, ... (p. 7).

Accepting the above distinction, the borderline is that the new device or process is not only potentially *marketable* but actually *marketed* (Dosi, 1984, p. 73, Note 2). In his discussion about invention and innovation, Parker (1978) believes innovation is rather more than the development of invention—marketing and general presentation to the customers are part of the innovation activity (p. 7). Also, in line with this distinction, the role of inventor, although of course acknowledged, was not comparable to that of innovator (even though sometimes these two roles might be combined in the person of inventor–entrepreneur) (Freeman, 1992, p. 76).

Having been aware of the difference between innovation and invention, Parker (1978) has reiterated the point that there is no tidy distinction between the two activities of technological change; on the contrary, they are inextricably linked. More specifically, invention is the first stage in the process of technological innovation; whilst innovation is the stage which transforms a promising technical possibility into a marketable reality.

There are different methods to categorise innovation. Among many others, according to West (1992), one way to do so is to separate innovation into those that are revolutionary or evolutionary (p. 45). Similarly, many other scholars (e.g. Jorde and Teece, 1992, p. 5) put innovation into two basic regimes, i.e. radical innovation and incremental innovation.

Second and more importantly, why have some innovations been successfully accepted, even become world standards, whilst others have ended up with bitter failure in the marketplace?

It is suggested that a 'failure' is an attempted innovation which failed to establish a worthwhile market and/or make any profit, even if it 'worked' in a technical sense; whilst a 'success' is an innovation which attained significant market penetration and/or made a profit (Freeman, 1982, p. 115).

According to Porter (1980), the development or success of an emerging industry is faced with a number of problems or uncertainties, of which the following aspects are worth mentioning:

Inability to obtain raw material and components. For example, acute shortages of colour picture tubes in the mid-1960s was a major strategic factor affecting participants.

Absence of product or technological standardisation. Failure to agree on product or technical standards can impede cost improvements. The lack of standard is usually caused by the high level of product and technological uncertainty that still remains in an emerging industry.

Customer confusion. Closely related to the second point, emerging industries are often beset by customer confusion, which resulted from the presence of a multiplicity of product approaches, technological varieties, and conflicting claims and counterclaims by competitors.

Erratic product quality. In an emerging industry, product quality is often erratic due to the lack of standards and technological uncertainties. This erratic quality, even if caused by only a few firms, negatively affects the images and credibility of the entire industry.

Regulatory approval. Emerging industries often face delays and red tape in gaining recognition and approval by regulatory agencies if they offer new approaches to needs currently served by other means and subject to regulation. Namely, government policy and state regulation can slow down the progress of an emerging industry. On the other hand, they can put an emerging industry on the map almost overnight (and, therefore, can possibly distort competition among competing product approaches by making the wrong technology choice).

High cost. The emerging industry is often faced with unit costs much higher than firms know they will eventually be. Therefore, firms sometimes price below cost initially or the industry development is severely limited.

The resistance to technical breakthroughs and product innovations from within the firm undertaking the innovative activities should not be underestimated. When innovations are first introduced, as Freeman (1992) argues, just because they mark a break with past production practice and experience, both the management and workforce are unfamiliar with radical new products and processes and sometimes resist their introduction; even with the best organised R&D, it is sometimes impossible to eliminate all 'bugs' in the R&D stage (pp. 73–74).[7]

In the process of technological innovation, according to Parker (1978), there are two essential risks, technical and market, which are closely related to the success or failure of commercialisation. Technical risk is largely determined by the competence of a company's R&D team; market risk, on the other hand, is more likely to be influenced by external factors. Parker (1978) argues that, under normal circumstances, the commercial success of a company's innovation will depend on "generating an adequate market for the new product", and a high proportion of "non-technical failures are found to be due to lack of interaction between the marketing and the R&D functions [of the company]" (p. 58). In other words, without adequate market analysis and strong marketing effort, "money is likely to be spent on projects which are successful in a technical sense, but which fail because of lack of market outlets" (Parker, 1978, p. 59).

Despite all their uncertainties about innovation, most firms still commit huge effort to innovation. The reason is simple: not to innovate is to die. A firm which fails to introduce new products or processes in an industry, whether the chemical or instruments or electronics industry, cannot survive, because its competitors will pre-empt the market with product innovations, or manufacture standard products more cheaply with new processes (Freeman, 1982, pp. 169–170).[8]

STRATEGIC ALLIANCES AND INDUSTRIAL COLLABORATION BETWEEN FIRMS

Compared to the situation several decades ago, new technologies become generally available more quickly since the 1980s; time has become even more of a critical element in corporate strategy and nothing stays proprietary for long and no one player can master everything (Ohmae, 1990). Thus, "operating globally means operating with partners—and that in turn means a further spread of technology" (Ohmae, 1990, pp. 5–6).

In many cases it is beyond the capability of a single firm to successfully commercialise a new technology or product. Instead, strategic alliances and industrial collaboration have become the most adopted means to promote new technologies and new products on a global scale, and the most important component of corporate strategy in the 1980s (Mytelka, 1991, p. 15). This section discusses issues related to interfirm collaboration such as why and how new technologies have been managed by firms through cooperative activities at the industry level as well as government involvement in strategic alliances between firms. It is also the concern of this section to touch on the controversies about competition and collaboration.

New Era of Competition: the Phenomenon of Strategic Alliance

According to Hagedoorn (1993), strategic (technology) alliances[9] and partnerships are cooperative agreements which are aimed at improving the long-term perspective of the product–market combinations of the companies involved (p. 375). In a slightly different way, Dodgson (1993) defines collaboration as any activity where two or more partners contribute differential resources and know-how to agreed complementary aims (p. 13). Major forms of strategic alliances include: joint ventures and research corporations—combinations of the economic interests of at least two firms in a 'distinct' firm, in which profits and losses are usually shared in accordance with equity investment; minority investment—e.g. those equity investment made by a large firm in a smaller high-tech firm; joint R&D agreements—refer to joint research packs and joint development agreements which establish joint undertaking of R&D projects with shared resources;[10] technology exchange agreements—e.g. technology sharing and cross-licensing. Dodgson (1993) categorises the variety of these forms of alliances into two basic groups: vertical collaboration and horizontal collaboration. Vertical collaboration occurs throughout the chain of production for particular products, from the provision of raw materials, through all manufacture and assembly of parts, components and systems, to their distribution and servicing; whilst horizontal collaboration occurs between partners at the same level in the production process. No matter which form they take, strategic alliances always preserve a relatively large degree of formal decision-making independence for the initiating organisations (Gerlach, 1992, p. 3). This is the major reason why they are difficult to manage (Hagedoorn, 1993, p. 375).

Strategic alliance itself is not a new phenomenon; on the contrary, it has been a long-standing feature of a number of industries (e.g. the oil exploration, mining and lighting industry) However, it is one of the most widespread trends of recent years that the popularity of strategic alliances has reached a high point.[11] Before the early 1980s, big multinational firms such as IBM and Philips would prefer doing business themselves. It is interesting to note that these companies have substantially changed their strategies and chosen strategic alliances as the way of handling new technologies and new products. In describing the change of European industrial policies, Sharp (1989) states that, while the 1960s and 1970s could well be called the Age of National Champion, the 1980s may earn the title of the Age of Collaboration (p. 202).

Motives for Strategic Alliance and Industrial Collaboration

There are a number of reasons why firms, particular those big ones, have joined forces through strategic alliances and industrial collaboration. Among many others, the following aspects are widely discussed in the literature.

First of all, technical change is tending to undermine sectoral boundaries, and this has led to new combinations of products and markets[12] which can make the existing strategies of firms ineffective. Consequently, "firms are increasingly looking towards cooperative solutions in the form of mergers, joint ventures, and alliances" (Cawson *et al.,* 1990, p. 377). This is particularly the case of the European consumer electronics and semiconductor industries, and increasingly so of the European telecommunications industry. To put it in another way, it is technological or industrial convergence that has made cross-sectoral interfirm collaboration necessary; to bring about a new product which involves several key aspects of technologies needs the joint action of different firms with different specialities. Some authors (e.g. Teece, 1986; Jorde and Teece, 1992a; Dodgson, 1993) have suggested that strategic alliances, including consortia and joint ventures, are often an effective and efficient way for firms to obtain access to those specialised assets needed to organise for technical innovation, particularly when an industry is fragmented. It is also argued that new knowledge is the product and the goal of the alliance; and the alliance brings into the corporation new expertise concerning product marketing strategies, organisational know-how, new tacit and explicit knowledge (Ciborra, 1991, p. 59). Through partnerships, systematic combination of distinct specialised skills is achieved, and this enables participating firms to remain competitive by creating the flexibility needed to react to a changing environment and by enhancing the capacity to manage the process of change itself (Delapierre and Zimmerman, 1991, p. 108). Technical change extends not only sectoral boundaries but also the boundaries of the firm and, therefore, without participating in interfirm cooperative arrangements, even the most advanced companies may lose their leadership position (Dodgson, 1989, pp. 6–7).

Secondly, due to the rapid increase of cross-border commercial activities, a kind of 'internationally linked economy', a term used by Kenichi Ohmae, is emerging, various kinds of industrial alliances are being made increasingly attractive and viable for big firms. The other side of the coin is that the increased number of international industrial alliances is becoming one of the driving forces to the development of an 'internationally linked economy'. Therefore, "the pattern is obvious: a product, non-equity-dependent set of arrangements through which globally active companies can maximize the contribution to their fixed costs. These alliances are an important part of the way companies get back to strategy" (Ohmae, 1990, pp. 127–128).

Thirdly, as technology is increasingly becoming stateless, internationalisation through alliance rather than acquisition is advocated by some influencing writers. In their joint article discussing corporate strategies, Hamel and Prahalad suggest that in making acquisition, the acquirer must pay both for the critical skills it wants and for skills it may already have hence the costs and problems of integrating cultures and harmonising policy prove much larger in an acquisition than that in an alliance (1993, p. 80). Against such a background characterised by internationalisation, the growth of world competition in technology-related areas makes the strategy of sole reliance on internally financed and internally conducted R&D insufficient and perhaps suicidal (Link and Tassey, 1987, p. 10).

Fourthly, as Hamel and Prahalad (1993) have observed, industrial collaboration (or 'co-option') has proved an effective route through which a company seeks to absorb its partners' skills and make them its own or to enrol others in the pursuit of a common objective:

> Enticing a potential competitor into a fight against a common enemy, working collectively to establish a new standard or develop a new technology, building a coalition around a particular legislative issue—in these and other cases, the goal is to co-opt the resources of other companies and thereby extend one's own influence (p. 82).

To illustrate this point, the authors say that "Philips has a knack for playing Sony and Matsushita against each other, enrolling one as a partner to block the other" (p. 83). At least, Philips co-opted its major competitor Sony to establish a world standard for compact disc by the end of 1970s. Philips also absorbed some of Sony's technical resources to finalise its compact disc system.

Another reason for firms to globalise themselves through alliances is, as suggested by some writers, related to political concerns. Although the postwar growth of multinational firms and foreign direct investment raised the prospect of 'global firms' to whom national boundaries would mean little or nothing, much of the current wave of international joint-venture activities reflect the opposite phenomenon (Mowery, 1991, p. 96). In some countries, local representation is still a matter of law and nationalism fights back against the increasing globalisation of everything, hence it is "cheaper and safer to expand one's scope by a series of alliances and ventures" (Handy, 1992, p. 63). By comparison with foreign direct investment, licensing or, in particular, export, collaborative ventures lower both the financial and political risks of innovation and foreign marketing (Mowery, 1988, p. 10). In the last two decades, some Japanese consumer electronics firms successfully penetrated the European VCR market through alliances with local companies (see Chapter 4). Since the early 1980s, discussions about 'strategic' technologies or industries have become a politically sensitive topic, and to set up a wholly-owned business by a foreign firm in these industries may not prove an easy task. Whether they are called 'Trojan Horse' or not, industrial alliances or strategic partnerships or joint ventures have been the effective ways for foreign firms to get into the local market with less chance to be caught by protectionist measures.

Moreover, the reduction and sharing of increased costs of R&D in a number of technological fields is another key argument for collaboration (Hagedoorn, 1993, p. 373). This point is elaborated by many authors including Dodgson (1989):

> The development of new technological systems is extremely costly. ... It is very difficult for one firm independently to cover the risks of such a financial burden. This provides another reason for the increased emphasis on collaboration within industry. Firms increasingly are looking for complementary funds from within industry and from government (p. 4).

Finally, it seems that strategic alliances and partnerships are one of the most effective ways for companies to get into a highly competitive foreign market. While many Western companies and their governments claim that the Japanese market is too self-closed to foreign competitors, examples of successful penetration into the Japanese market by some Western firms are soundly praised by some observers. IBM, amongst others, may not have many allies in the U.S., but it has teamed up with many partners in Japan. To be sure, IBM (Japan) has actually become a mini-Mitsubishi group, and entered formal corporate relations with nearly one thousand local companies as extensions of their sales, services, engineering, and manufacturing operations and hence become an all-around player in the eyes of Japanese consumers.[13]

In summarising the various motives for strategic alliances, Hagedoorn (1993) puts them into three groups. The first group of motives are those related to the sharing and further advancement of research and the diffusion of some basic scientific and/or technological knowledge amongst participating companies.[14] The second group of motives are more closely related to concrete innovation processes involving two or more companies, who explore the possibility of secretly capturing some of the capabilities, knowledge or technologies of partners. The third group of motives are associated with market access and technology development through a combined effort of companies.

Although there are a large number of motives for firms to engage in partnership, Hagedoorn, drawing on his empirical research, suggests that two basic categories, i.e. market and technology-related motives are predominant (1993, p. 381).[15]

Apart from all sorts of reasons or motives for firms to group together as mentioned above, some commentators have also stressed the importance of collaboration to technological or product standardisation badly needed by those innovating firms. In his recent study, Dodgson has discussed the importance of interfirm collaboration to the adoption of particular technical standards:

> With many new technical standards being created, both domestic and international, firms may feel their cases for the adoption of particular technical standards may be improved by their promotion by a number of firms, rather than singly. ... Collaboration provides an effective mechanism for the joint creation and promotion of standards (1993, p. 30).

Mytelka (1991b) also argues that collaboration in precompetitive R&D plays an important role to ensure harmonisation of technical solutions and enable participating firms to develop compatible products; and that collaboration is a prerequisite to the creation of new markets in those dynamic sectors (p. 197).

Since the 1980s we have witnessed fierce competition over technological standards at a global scale (at least within the Triad) as new technologies and products themselves become increasingly globalised. Time and again, competition in the consumer electronics industry has been dominated by standards battles. The competition between technological standards within the consumer electronics industry is so fierce that no single firm, either big or small, seems to be able to set a world standard alone without some form of collaboration with its competitors, as shown by most of the empirical studies in this book.[16] In other words, to make its new product accepted by the consumers, a company would have to get as much industrial support as possible through alliances. It is also true that, because of this particular reason, the best technology is not necessarily the one to be accepted by the market.

In his comment about strategic partnerships and alliances between global companies, Ohmae argues that "[p]reparing a company effectively for partnerships necessary in a borderless economy is a decade-long process. Real, sustained, and committed effort is needed. But alliances are worth the effort. Properly managed alliances are among the best mechanisms that companies have found to bring strategy to bear in global markets" (1990, p. 136).

Keiretsu: The Japanese Interfirm Alliance System

Of the many forms of strategic partnerships and industrial alliances, the *Keiretsu* in Japan are widely regarded as a unique way of interfirm cooperation. According to Gerlach (1992), there are two types of *Keiretsu*: 'vertical *Keiretsu*' and 'intermarket *Keiretsu*'. Vertical *Keiretsu* organise suppliers and distribution outlets hierarchically beneath a large, industry-specific manufacturing concern (e.g. Toyota's chain of upstream component suppliers are known as this type of vertical interfirm organisation); while intermarket *Keiretsu* refer to those large groupings involving trading companies and large banks and insurance companies, which provide for their members reliable sources of loan capital and a stable core of long-term shareholders (Gerlach, 1992, p. 45). The principal function of the *keiretsu* structure is to facilitate interchange among related companies—companies loosely linked in Japanese groups look to each other for guidance and input on new products, new processes and new businesses (Porter, 1990, p. 408). *Keiretsu* are believed, particularly by the Americans and Europeans, to be one of the major forces to make the Japanese market closed to foreign companies hence lead to domestic collusion and distort the principles of free market competition.

Are *keiretsu* fundamentally different from the cooperative arrangements in the U.S. and Europe? Cutts (1992) regards *keiretsu* as evidence of Japan's basic nature and, therefore, Japanese capitalism differs greatly from typical business practice in the West (p. 48).

More specifically, some scholars believe that the legal and legislation systems in the West (U.S. and Europe) are fundamentally different from those in Japan. Although American antitrust legislation, to some extent, has become more relaxed over the last decade,[17] cartels have long been treated as informal and usually illegal arrangements among companies to control prices and curb competition among themselves in the West. This is in sharp contrast with the Japanese environment, where hundreds of cartels and *keiretsu* networks are established to set prices, rationalise industries and respond to depressed markets, and their establishment and practice have been permitted by law and even supervised by the government (Cutts, 1992, p. 49). To broaden this view, Cutts (1992) suggests that cartels are a way of life in Japan—the Japanese society is largely linked together by a web of informal cartels and formal *keiretsu,* which account for about 90%

of all domestic business transactions and *keiretsu* a structural vehicle that ensures their continued success; cartel principles are likely to dominate business in Japan for a long time to come. By saying this, however, Cutts does not believe the Japanese *keiretsu* system is organized at the expense of competition. On the contrary, competition among cartels and *keiretsu* in Japan is ruthless; these cartelised industrial groups or *keiretsu* compete with each other in a way just as competing companies elsewhere in the West (Cutts, 1992, p. 49).[18]

It is suggested in some literature that the Japanese *keiretsu* system differs essentially from the strategic alliance practice in the U.S. and Europe. Strategic alliances, according to Gerlach (1992), create a framework within which companies are able to cooperate in a set of specific business activities (e.g. the development of new technologies without greatly altering the relationships those companies have directly with each other, their own shareholding structure, or the basic strategic constraints under which they operate); in contrast, Japan's *keiretsu* comprise direct and indirect linkages among banks, industrial firms, and commercial enterprises that shape a complex web of interests affecting each participating company as a whole (pp. 7–8). Gerlach (1992) believes that the *keiretsu* system engages in a wide variety of activities such as opening up sources of capital flows between banks, corporate borrowing, exchanging raw materials and information, intermediate product trade, and defining the underlying ownership structures of the participants though the pattern of share crossholding.

Some scholars do not share the above view and claim that the *keiretsu* network system is not uniquely Japanese. According to Freeman (1992), subcontracting arrangements (although not necessarily cross-shareholdings) were characteristic of both American and British steam-powered industry in the 1880s (p. 90). Moreover, as the empirical studies in the following chapter show, the early days of the American and European lighting industries were also characterised for decades by cartelisation involving quotas of production, fixing prices, market fragmentation, etc.

While the *keiretsu* system remains important, the role of other forms of industrial collaboration should not be underestimated in promoting science and technology as well as industrial competitiveness. Apart from the interfirm relations established within each *keiretsu* family, most interfirm research is undertaken by unrelated firms, especially in the most innovative sectors. Of nearly 200,000 cases of joint patent applications by 29 different firms, most were among otherwise unrelated partners (Levy and Samuels, 1991, pp. 122–123). As far as the development of science and technology is concerned, according to Levy and Samuels (1991), research consortia are an important feature of Japan's 'national innovation system', although they are not the only one (p. 143).

In analysing the shift of multinational companies from traditional international trade, multinational production and globalisation to networking, Michalet (1991) argues that the old internalised structure of these firms was based on hierarchy, whilst the 'network firms' or 'hollow corporations' favour horizontal relationships like those developed in Japanese companies (p. 46). He also suggests that the 'Fordist' model of production is being replaced with a 'Toyotist' one (1991, p. 49, Note 8).[19] Michalet (1991) concludes that, first, the complex structure of network firms and alliances will determine the emerging 'contractual economy'; second, competition will shift from firms struggling directly against one another for market shares to a new type of cartel structure based on technology; and, finally, market access for a firm will be determined by its position as a partner in a network or alliance.

The State and Strategic Alliances

In many cases industrial collaboration is either launched or sponsored by the state rather than the firms themselves with the excuse of maintaining or enhancing the country's competitiveness in the industry concerned.

It is believed that the 1980s witnessed a 'veritable explosion' of government-supported research (Levy and Samuels, 1991); national and supranational governments have used a variety of methods to actively promote technological collaboration, including the formation of research associations and consortia, the relaxation of legislative restrictions, the creation of a variety of technology transfer organisations, and taxation policies (Dodgson, 1993, p. 87). At the forefront of government-promoted collaboration have been cooperative research programmes, as indicated by the experience of the Triad—the U.S., Japan and Europe.

Government involvement in organising industrial cooperation or strategic alliances is a typical form of protectionism.[20] From the beginning of the last decade, the EU has made substantial efforts to develop a common industrial policy by means of launching a series of interfirm cooperative programmes. In 1980, according to Alic (1990), the European Commission proposed a European strategy for the electronics industry that would have included government-funded programmes to develop semiconductor processing equipment, as well as an advanced EU-wide communications network (p. 324). Although this proposal failed to be implemented due to the French government's objection to it as insufficient, it served to mark the beginning of a new wave of inter-European collaboration, starting with ESPRIT, RACE, BRITE, and Eureka (a pan-European programme with member countries from outside the EU as well), etc. (Alic, 1990, p. 324).[21] In the U.S., the Reagan administration launched the Strategic Defense Initiative (Star Wars) in the early 1980s; Sematech was formed with U.S. government backing as a cooperative venture to produce semiconductor chips; and most recently, the new Clinton administration has proposed a $17bn budget to support a large scale 'information superhighway' programme aimed to enhance the competitiveness of American electronics and information industries. Japan is no exception: the International Superconductivity Technology Centre is a jointly funded government–industry R&D establishment; although it failed to achieve its original goal—to create a thinking computer capable of inferring new knowledge from its database and writing its own software (Levy and Samuels, 1991, p. 139), the Fifth-Generation Computer Project was also initiated by the Japanese government. Whereas vertical technological collaboration is very common and occurs spontaneously in Japan, horizontal collaboration frequently requires government stimulations (Dodgson, 1993, p. 136).

In the 1980s, it is the development of technology policies—through defense-related institutions in the U.S., through ministries of finance (MOF), industry (MITI) and telecommunications (MPT) in Japan,[22] through ministries of industry and of R&D in France and DG XIII in the European Union—that best exemplify the role states are coming to play as promoters of strategic partnership (Mytelka, 1991a, p. 25). Generally speaking, every legislative initiative regarding technology policy in the 1980s was directed at least in part to stimulating interfirm research collaboration (Levy and Samuels, 1991, p. 123).

Delapierre and Zimmerman (1991) believe that a systematic combination of the different specialisation of European industrial groups, backed by the European industrial policy, will serve as a European springboard to take on the competition from the U.S. and Japan. They have also observed that, while the old Europeanism remains deeply entrenched as an industrial strategy, rooted in nationalistic trends and policies favouring domestic manufacturers, European firms, particularly French ones, are entering into alliances with leading firms from the U.S. and Japan (1991, p. 108).

In the future, Michalet (1991) argues, on the one hand, the world economy will likely witness an even greater number of state entities participating in alliances alongside private firms; on the other hand, it is also likely that states will be needed to guarantee or supervise the rules of the game developed by the members of the alliances and insure a minimal compatibility between the various standards set up by competing business coalitions (p. 48).

Competition and Collaboration

As we have seen, companies engage simultaneously in processes of competition and collaboration. In one respect, cooperating while competing poses difficulties for both day-to-day management and long-term strategy: the sheer difficulty of organising and managing a collaborative R&D venture generates a built-in set of limits because of disagreements on priorities and financing from the partners (Alic, 1990, p. 320; Dodgson, 1993, p. 25). On the other hand, alliance or collaboration also has limitations in terms of antitrust legislation or competition policy, although cooperative R&D programmes/projects in many cases are government mediated or sponsored. Neoclassical theory and American antitrust law hold that imperfect competition and interfirm cooperation, including cartels, increase prices and reduce innovation (Best, 1990, p. 17).[23] Therefore, it is widely recognized that the ability of innovating firms to cooperage by striking necessary vertical and horizontal agreements, or entering into alliances, often raises issues in antitrust as do other elements of business strategy (Jorde and Teece, 1992, p. 6).

To a large extent, business cooperation in recent years is encouraged by policy-makers in each country to enhance economies of scales and by creating linkages between their domestic firms rather than between their own firms and foreign firms (Alic, 1990, p. 320).[24] But some scholars argue that cooperation among competitors itself will not enable firms to rebuild their competitiveness or, even worse, will undermine competitive advantages (Alic, 1990, p. 330; Porter, 1990, p. 122).

In the EU, according to Sharp (1989), EU-sponsored collaboration for precompetitive research under its own scheme such as ESPRIT are given block exemption under EU competition rules; whilst collaboration for R&D, i.e. collaborative R&D undertaken by firms necessary for the launch of a new product, require case by case exemption (p. 218).

As a matter of fact, international interfirm collaboration extends beyond sharing the heavy financial (and intellectual) burden of R&D, and includes manufacturing and marketing (Dodgson, 1989, p. 5). In other words, collaboration can promote cartelisation and oligopoly and raise entry barriers to new entrants (Dodgson, 1993, p. 25). Even at the precompetitive stage, interfirm collaboration may have an anticompetitive element. This is well indicated by the political nature of technical standards creation through collaboration. It is true that parallel standards can exist side-by-side and there is obviously the case of *de facto* standards controlled by single firms. However, as Dodgson (1993) argues, there are also examples of collusion on the part of firms and governments to provide exclusionary standards; this exclusion of nonparticipants (competitors) in the process of standards formation may explain why firms feel the pressure to collaborate to avoid the high costs of them not doing so (p. 78).

Compared to Europe and Japan, U.S. antitrust legislation used to be, perhaps, the most restrictive and effective restraint on interfirm collaboration in the private sectors. In the 1960s, the U.S. Department of Justice charged the American automakers with conspiring to slow the pace of R&D progress towards controlling exhaust emissions; this resulted in a consent decree banning joint R&D among automobile firms directed at emissions control (Alic, 1990, p. 322). However, after constant lobbying and pressure from the private firms and heightened concern over American industry's competitiveness, U.S. antitrust enforcement has been gradually relaxed since the 1980s. According to Alic (1990), the Department of Justice has consistently signalled that it would view cooperative activities by taking into account not only national but also worldwide context; and it would act to bar cooperative R&D activities only when the proposed arrangements posed clear dangers of restraint of trade (p. 323).

It has also been recently recommended by the former American President's Commission on Industrial Competitiveness that the U.S. change its antitrust law to reflect the new global markets within which American firms operate, and that the uncertainty as to what constitutes antitrust violations has deterred actions that could have desirable competitive effects (Jorde and Teece, 1992, p. 3).

Some scholars argue that alliances or partnerships are not necessarily established at the expense of competition. To take the information technology sector as an example, Delapierre and Zimmerman (1991) suggest that partnerships are desired to offer some level of complementarity, hence compatibility for manufacturers so that the combination of their products can be virtually turned into customised systems (p. 115). They believe that the principal partners continue to remain competitors with respect to end market applications, although their agreements on common standards ensure larger potential markets for all of them (1991, p. 115). Levy and Samuels (1991) also argue that research consortia actually increase competition (p. 143).

PUBLIC POLICY AND INDUSTRIAL COMPETITIVENESS

The development of high technologies and the impact on international competition has been increasingly drawing the attention of government policy-makers all over the world. Market forces have in many cases been superseded by public authorities intervening to try to bolster economic success. This section discusses why and how new technologies are affected at the government or supranational level in the Triad—Europe, the U.S. and Japan—by protectionist trade and industrial policies which favour domestic firms.

The Revival of Protectionism Since the 1980s

It is commonplace, on the one hand, to observe that national borders mean much less than they used to due to the internationalisation of trade, business, and technology; but, on the other hand, the last decade has witnessed a sharp increase in what has been called 'techno-nationalism', policies launched by governments with the objective of giving their national firms a particular edge in an area of technology (Nelson and Wright, 1992). Nelson and Wright have strongly argued that these policies no longer work very well; and it is increasingly difficult to create new technology that will stay contained within national borders for very long in a world where technological sophistication is widespread and firms of many nationalities are ready to invest in exploring new generic technologies[25] (1992, p. 1961). As an extension of state power, "protectionism, in one form or another, either active or passive, tariff or nontariff, and so on, is essentially part of the process of closing ranks against the threat from outsiders" (Cawson *et al.*, 1990, p. 22).

The belief in protectionism is obviously a violation of the traditional free trade principles. The absolute advantage theory of Adam Smith explains international trade based on differentiated endowment of natural factors. David Ricardo extended the absolute advantage notion to that of comparative advantage which asserts that a nation's resources can be allocated by market forces to those industries where it is relatively most productive. The Ricardian theory rests on technology differences in a broad sense, and this type of thinking has been developed by the more recent 'technology gap' theories of trade that suggests nations export in industries in which their firms have a lead in technology (Porter, 1990, p. 17). Both Adam Smith and David Ricardo believed the market mechanism was an 'invisible hand' sufficient to govern international economic activities. Contrary to this classical belief, managed trade of various kinds (e.g. tariffs and quotas on imports, antidumping policy, voluntary export restraint agreement, 'buy-domestic' requirements, and getting access to foreign markets through home government diplomacy, etc.) has been extensively practised by national governments in close collaboration with their domestic industries (particularly those big firms) over the past few decades.

Many American and European political and industrial leaders blame foreigners (e.g. the Japanese) too often for their own failure. This seems to be an attempt to escape from being responsible for their declining industries and trade deficits. By maintaining that the Japanese do not trade fairly, American and European industrial policies in many cases protect domestic firms, which failed to produce and market competitive products over the last decade.[26]

The 1980s witnessed a rapid growth of FDI (Foreign Direct Investment). Between 1983 and 1988 global FDI increased by over 20% per annum in real terms compared to the annual growth rate of 5% for world trade volumes during the same period (Julius, 1990, p. 99). This growth of FDI might be in part the result of various protectionist policies mentioned above. Moreover, new forms of trade protectionism in terms of government intervention (e.g. R&D or production subsidies to domestic firms) have come into fashion since the early 1980s. These measures were intended to improve the competitiveness of domestic firms (particularly those 'national champions' picked by politicians and bureaucrats) through a strong government hand. However, the effectiveness and justification of government action towards the industry remains controversial.

The following are some major forms of protectionist trade policies, which have been boldly practised in the industrialised countries, particularly the EU, since the beginning of the last decade

Tariffs and quotas on imports. Tariffs and quotas on imports are two of the most traditional measures used by countries to protect their home market against foreign competition. Quotas, as a rising category of trade restraint, are restrictions on the quantity of foreign imports (Weidenbaum,

1986, p. 287). In Europe, the EU has applied tariffs on imported durable consumer goods from Japan and other East Asian countries since the early 1980s. According to Schuknecht (1992), 8% of imports in industrialised countries are covered by quantitative restrictions, and the EU maintains over 500 quota restrictions against imports from nonmember countries (p. 62).

VER (Voluntary Export Restraint) agreements. VER agreements are another form of quantitative restriction on imports. The terms of a VER agreement are normally negotiated between two governments, and 'voluntarily' accepted by the exporting companies. Failing to confine their export to the quantity defined by the VER agreement would inevitably provoke the receiving country to apply a high tariff on them. In May 1981, MITI (Ministry of Trade and Industry) of Japan, under pressure from the American government, entered an agreement to restrain Japanese automobile exports to the U.S., and this agreement remained in force until 1988 in order to allow the American auto industry time to become competitive.[27] In the mid-1980s the EU entered VER agreements with Japan to restrain VCR (Video Cassette Recorder) imports from the latter.

VERs proved to be an innovative means of evading international policy constraints under the GATT. As the legal basis of VERs is extremely obscure (they are sometimes called 'memorandum of joint action' or 'verbal notes'), their political nature can hardly be disguised (Schuknecht, 1992, pp. 63–64). The International Monetary Fund suggests 261 VERs in the world, of which 138 are in the EU, and most of them are negotiated in the textile and clothing, footware, steel, vehicles and consumer electronics industry.[28]

Antidumping policy. 'Dumping' is a very loosely defined term, and the lack of universally accepted technical measurement has led to nations implementing antidumping policies at their discretion. It is argued that, when dealing with integrated global industries, "the impossibility of defining and then measuring the 'dumping' of products or the 'potential for predatory behaviour' must simply be accepted" (Julius, 1990, p. 99). Antidumping policies have been repeatedly applied by the EU to those foreign firms exporting to the Community at prices claimed to be lower than that of the home market throughout the 1980s. As far as the consumer electronics industry is concerned, European antidumping investigations have been undertaken in most product areas following complaints from European firms.

Since the last decade antidumping has proved to be the EU's most powerful trade policy instrument for some sectors, such as the consumer electronics industry. During the period 1980–90, there were over 400 cases and over 900 decisions on the record with the EU; and this compares to 389 cases initiated by the U.S. for 1980–87, one of the other major users of antidumping policies (Schuknecht, 1992, pp. 59–60).

'Screwdriver plants'. Antidumping policy is also applied to the so-called 'screwdriver plants', whose major operation is to assemble parts or components designed and produced outside their host country. To thwart 'screwdriver plants', the host government usually set a quantified minimum 'local content' requirement.

Buy-domestic requirement. Many countries nowadays maintain their 'buy-domestic' practices (Weidenbaum, 1986). In the case of the European telecommunications industry, there had grown up in each European country a 'club' of nationally based private-sector equipment suppliers who were guaranteed a steady flow of equipment orders as the state-owned public network service expanded, and this cosy, interlocking relationship served to make life pleasant and easy for both sides (Morgan, 1989, p. 20). One of the recent bilateral trade arguments between the U.S. and the EU was centred on each side's 'buy-domestic' regulations. The Community maintains its utility procurement regulation, Article 29, which largely excludes foreign firms from supplying products for public procurement, while the American 301 Congress bill mainly favours domestic firms for Federal contracts.

Access to foreign markets. Government action to get access to foreign markets on behalf of domestic manufacturers is not rare at all in the history of international commerce.[29] However, the direct involvement of top government officials in bilateral trade negotiations since the beginning of the 1980s has marked a new era of trade protectionism. A typical and widely quoted example of this new wave of protectionism is the American–Japanese semiconductor agreement first signed in 1986. According to this agreement, Japan would have to buy from American suppliers up to 20% of the amount of semiconductors needed for domestic industry. A failure to meet this target might lead to retaliation from the U.S. Moreover, at the beginning of 1992, the then President George Bush, accompanied by a group of 21 American industrial leaders (such as Lee Iaccoca of Chrysler) rather than ordinary diplomats, visited Tokyo and urged the Japanese government to open its market for American firms and their products.

Government intervention (e.g. subsidy) is partly based on the appeal for 'level playing field'. The so-called 'level playing field' argument means the domestic government should match foreign subsidies so that domestic firms or industry is not put at a disadvantage by foreign governments (Spencer, 1986, p. 81).

The concept of 'business environment'. Protectionism or managed trade procedures may not prove sufficient for governments to protect their domestic markets. One important characteristic of these protectionist policies is that domestic firms too often play an important role to pressurise their home governments to erect tariff or nontariff barriers to reduce foreign competition. It is interesting to note that domestic firms frequently use unfair competition as an argument when they lobby the government. Since the 1980s, terms such as 'business environment', 'fairness' and 'level playing field', etc., are the buzzwords frequently cited by proponents of industrial policy and corporate lobbyists. The meaning of these terms themselves appears neutral, but they are sometimes used in a sense persuading policymakers to implement trade policies or regulations favourable only to domestic firms or erect trade barriers to keep foreign competitors outside.

Government Intervention

In his discussion about the variety of government–industry relationships, Cawson (1986) summarises them into three forms. First, the *market* mode of state intervention. This form of intervention is intended to preserve and reinforce market mechanisms but gives no priority to any particular firm(s) or industry. Second, the *bureaucratic* mode of intervention which occurs at the sectoral level. This form of intervention allows the state bureaucracy to be directly involved in many aspects of major decision-making in the industry in isolation from market pressures. The third form of intervention is *corporatism*[30] (including *mesocorporatism* and *microcorporatism*). Corporatist intervention is selective and discretionary on the basis of the existence of a developed interest association in a particular industry (mesocorporatism). In many cases corporatist intervention takes the form of bipartite negotiations between the state agencies and leading firms when a monopoly association is absent in a particular sector (microcorporatism). This book is mainly concerned with the discussion of various issues related to the substance of the last form of government intervention in the so-called high-tech or new industries. These issues include subsidies to firms through government initiated precompetitive R&D and close-to-market activities, managed trade, public procurement , etc., to either help those beleaguered big firms or pick winners for new industries.

Throughout the 1980s government intervention has been widely seen as the 'visible hand' in many high-tech industries across many parts of the world. As an example, the development of the IT (Information Technology) sector in Europe has been strongly influenced by the EU and its member states. Through various IT programmes (e.g. RACE, ESPRIT, JESSI, etc.), European firms have received substantial financial help from the EU and national governments for precompetitive and close-to-market activities; the scale of this European support in forms of direct loans, R&D grants, procurement policies is very large (Hobday, 1991, p. 91). With the belief that government support is legitimate and necessary for their survival, the American and Japanese manufacturers in the semiconductor industry have also benefited from government subsidies on different scales (*ibid.*)[31].

The role of national government can hardly be denied but many agree that it should be limited to improving or creating a competitive environment favourable to domestic firms, as argued by Porter (1990). Excessive involvement or shadowing corporate decision-making by the government through its policies may end in failure. There are two reasons to explain this. Firstly, the national government could never be as close to the market as the firms. Secondly, government policies are normally influenced by political forces that distort market forces.

> Governments have been notably unsuccessful in managing firms and in responding to the fluid market changes that characterize international competition. Even when staffed with the most elite civil servants, governments make erratic decisions about the industries to develop, the technologies to invest in, and the competitive advantages that will be the most appropriate and achievable. Examples of tendencies toward shadowing corporate decision making by governments can be drawn from Japan, Korea, Singapore, the U.K., France, and many other nations. Government simply cannot be as in tune with market forces as industry participants, nor can it practically isolate its decisions from political forces that distort them (Porter, 1990, pp. 619–620).

Porter sees government subsidy, extensive collaboration under the auspices of government, and any other temporary protection as counter-productive (1990, p. 30). Rather than providing help, the proper role of the government is to push and challenge its industry to advance. More broadly speaking, government (through its policy and regulation) can influence the major determinants of national competitiveness either positively or negatively, but government itself is not a direct participant in the market place nor a major determinant of national competitiveness (Porter, 1990). To put it in another way, “government intervention can provide opportunities for firms, and can constrain their behaviour, but it cannot substitute for their role” (Cawson *et al.*, 1990, p. 376).

Based on their empirical studies about the European electronics industry including telecommunications and consumer electronics, Cawson *et al.* draw their conclusion suggesting that even where governments were acting strategically to promote industries and products, outcomes were ultimately decided by the strategies of firms themselves (1990, p. 316).

Overall, government policy of any kind should seek to create a fertile environment for innovation and entrepreneurship and to offset evident market failures (Grossman, 1986, p. 67).

The practice of government intervention and various arguments either in favour of or against it are closely related to different perspectives of the so-called ‘new thinking’ on strategic trade and industrial policy as discussed in the following section.

Strategic Trade and Strategic Technologies

‘Strategic’ is originally a military term which has been recently used by proponents (including academics and politicians) of industrial policy theories to justify their arguments. In many cases, 'strategic' simply means 'important'. Generally speaking, a strategic sector is either a sector where there is substantial 'rent',[32] or where the return to labour or capital is exceptionally higher than other sectors (introduction to Krugman, 1986, p. 16; Tyson, 1992, p. 39); or a sector which yields important external economies (or spillovers) because of the increased role of technological competition and producers are not in fact paid the full social value of their production (or investment) (introduction to Krugman, 1986, p. 15; Tyson, 1992, p. 39). Most high-tech industries are important or strategic in both ways (Tyson, 1992, p. 39). A strategic view of the necessity and nature of government intervention in technology development improves industry's case for support (Dodgson, 1989, p. 10).

In connection with these two assumptions, strategic trade theory suggests that there are oligopoly profits to be made by oligopoly producers in certain industries, and if government intervention can shift a greater share of these global oligopoly profits to its own producers, the nation may benefit by trade intervention for this purpose (Cline, 1986, p. 230). To put it in another way, the competitive struggle between nations to secure a bigger share of the global rents from high-technology production is inherently zero-sum in nature—what one nation gains, others lose and, therefore, one nation's measures to promote its high-wage, high-tech producers can easily raise the ire of its trading partners (Tyson, 1992, pp. 39–40).

Because of the higher returns or spillovers or bigger share of global oligopoly rents generated by high-tech sectors, idealised models or traditional free trade theories cannot be applied to these strategic sectors hence a 'new thinking'—activist trade and industrial policy—is suggested as necessary by those proponents of the above new theories.

To be sure, the popularisation of new theories on strategic trade and industrial policies to some extent reflects a reality that protectionism and interventionism has been broadening in international trade, in particular, the high-tech based or 'sunrise' sectors. However, there is also a strong voice in favour of the principles of free trade and against any theoretical justification of appeal for industrial targeting in terms of intensive state subsidy and export promotion. In arguing against export promotion by government, Grossman (1986) claims that,

> ... subsidies distort the allocation of resources from market-determined to less productive uses, and that they lower the price that our country's exports command on competitive world markets (p. 47).

The importance and their economic externalities of industries and technologies have been widely recognised as a crucial factor affecting the potential productivity and growth of many other industries. Therefore, these industries and technologies do deserve special attention but the number of these is very limited. Policies to upgrade such industries must be strictly conditional and not seek to guarantee firm profits or ensure foreign share through quotas (Porter, 1990, p. 624). Porter believes that national economic prosperity need not come at the expense of other nations, and many nations can enjoy it in a world of innovation and open competition (1990, p. 30).

Based on different views of the above strategic trade and industrial policy theories, Krugman (1986) has suggested four 'generic' policy positions:[33]

(1) *Immediate activism.* Despite uncertainties about the appropriate trade policies, proponents of this view (e.g. Brander, 1986) are in favour of strong and immediate action by the government simply because other countries are already playing the game.
(2) *Cautious activism.* This policy position equally calls for immediate government action but only on a selective base—to intervene in areas where the cases are clear-cut or necessary (e.g. the high-tech sectors). This view is favourably discussed by Tyson (1992) and Borrus, Tyson and Zysman (1986), and Rugman and Verbeke (1990).[34]
(3) *Cautious nonactivism.* This position argues that any immediate government action or substantial policy change may bring about uncertainties and political risks; but careful monitoring of possible harm caused by trading partners while maintaining the basic principles of free trade may prove more feasible. Proponents of this position include authors such as Grossman (1986).
(4) *Strong nonactivism.* Proponents of this view (include most authors in favour of traditional free trade principles) argue that free trade will eventually make a country better off than any programmes of government intervention. They believe that the potential gains from activist trade policy might be insignificant, and the increased flexibility offered by the interventionist trade policy will be abused for domestic political reasons and makes international economic relations deteriorate.

Overall, supporters of industrial policies, policy positions (1) and (2), claim that some industries (e.g. the high-tech ones) are special and they develop too slowly if government help (e.g. subsidies, protection of domestic market) is not provided, hence domestic economic growth will suffer. It is important that some proponents of strategic trade theories have also argued that industrial policy should be conditional, i.e. only if the government agency involved in policy-making has the capacity to function independently from pressure exerted by firms seeking shelter and politicised decisions are avoided can bureaucratic discretion lead to efficiency (Rugman and Verbeke, 1990, p. 120). Contrary to these views, critics of industrial policies argue that targeting high-tech industries is a bad policy for several reasons (Carliner, 1986). Firstly, market forces are more qualified than government policy-makers to pick winners. Secondly, the spillovers or external economies to be generated by those targeted emerging industries are exaggerated. Finally, and related to the second point, industrial targeting has little to contribute to a nation's overall economic strength and growth.

A Political Impasse: Industrial Policy *Versus* Competition Policy

Industrial policy is a confusing term which everybody uses differently (Bangemann, 1992), i.e. there is no standard definition for the term although its use goes back to the 1960s (Weidenbaum, 1986). As a relatively recent concern of governments, industrial policy arises when macroeconomic policy instruments (e.g. fiscal and monetary policies) fail to discriminate sufficiently between the different needs of various branches of industry (Cawson, 1986, p. 113). Therefore, industrial policy, by nature, is sector-oriented. In his study on industrial policy in Britain, Grant (1982) defines industrial policy as the following:

> ... industrial policy is taken to refer to a set of measures used by governments to influence the investment decisions of individual enterprises—public or private—so as to promote such objectives as lower unemployment, a healthier balance of payments and a generally more efficient industrial economy. Investment decisions are taken to cover not only the creation and expansion of production capacity and re-equipment, but also decisions about R&D and product development (p. 2).

Substantially different from the above understanding, another school of academics in the U.S. defines industrial policy by including competition policy as an integral part of the definition:

> We define a nation's industrial policy as the aggregate of policies that directly and indirectly affect industrial performance through its impact on microeconomic variables. Key elements of industrial policy in the U.S. are trade policy (including adjustment assistance), certain aspects of tax policy, government procurement policy, regulation, science and technology policy, and competition (antitrust) policy. Antitrust policy ... ought to be assessed not just on a stand-alone basis, but also on how it integrates with other aspects of U.S. industrial policy (Jorde and Teece, 1992, p. 12).

In this chapter, and the rest of the book, I use the term 'industrial policy' in a sense closely related but essentially opposite to 'competition policy'.

Generally speaking, industrial policy has two objectives: "to assist in the restructuring of declining industries and to facilitate the positive adjustment of the economy to new technologies, new products, and new markets" (Weidenbaum, 1986, p. 263). Some advocates assert that industrial policy should focus on older and declining sectors in order to keep them competitive; while some claim that the government ought to intervene, particularly invest, in newer, high-tech fields. In 'an appeal for a European industrial policy', Bangemann (1992) has argued that the future European industrial policy should no longer support individual industrial sectors but the competitiveness on a large scale (p. 20).

The conflict between industrial policy and competition policy, similar to that between protectionism and free market principles, has never been resolved at either a national level or international level. In the U.S., for instance, antitrust policy and trade policy are often at odds, and it has been urged by some academics that this tension needs to be addressed and resolved (Jorde and Teece, 1992, p. 13).

It is true that competition policies (e.g. antimonopoly and antitrust laws) have been in existence in all major industrial countries; and, among many other arrangements, interfirm collaboration has commonly been seen, particularly in the U.S., as an anticompetitive tool used to frustrate free international trade (Dodgson, 1993, p. 169). However, competition policies have come under pressure as protectionism, particularly through industrial policy, has gained ground in the major industrial regions. Of various factors tilting the balance towards industrial policy, forging 'national champions' is worth mentioning. National champions are those government-favoured companies that are given a secured home base to compete in the world market. Although there are sound economies of scale to be brought about by national champion companies, many authors (e.g. Porter, 1990; Oppenheim, 1992) distinguish sharply between forging world-class companies through the selective and toughening process of international competition and artificially creating internationally competitive companies via the choices of bureaucrats and politicians; and the latter rarely works in practice.

Compared to the U.S. and Japan, the conflict between industrial policy (e.g. technology policy) and competition policy is more difficult to resolve in the EU. The Treaty of Rome failed to establish a general policy framework for either industrial policy or technology policy, and this left the Commission to operate only through unanimous decisions of the Council of Ministers often seized by conflicting national interests (Sharp, 1991, p. 61).

When conflicts are generated at the national government level (or within the European Commission), the political interests assumed by different policy domains seem to be evident. As a widely recognised difficulty, the responsibility for policies which affect technical innovation is normally divided between different government departments. These government departments involved in policy-making are often only imperfectly aware of the implications of their departmentalised policies for innovation elsewhere in the economy or, even when they are aware of these indirect connections, they will not usually attach any great weight to them simply because they are usually preoccupied with their primary mission (Rothwell and Zegveld, 1981, p. 47).

It may be that various protectionist policies are actually illegal under the U.S. antitrust and European competition laws, but the Community and the U.S. government manage to get round this illegality by making politically acceptable, although in many cases ambivalent, arguments. For instance, it is argued that the term 'precompetitive' has been used by EU authorities to be politically convenient, as in the case of various EU R&D programmes which have been designed to some extent to provide a block exemption to Community competition rules (Dodgson, 1993, p. 172). This confusion is sometimes intensified by the conflicting political interests and policy orientation of different departments within the Commission.

Lobbying: The Role of Firms in Government Policy-Making

The term 'lobbying' derives from American politics, and originally described attempts to influence elected representatives during the passage of legislation through Congress.[35] The term's contemporary meaning is broader, referring to the practices of *interest groups,* directed not only at seeking support from elected members, but also from political parties, and from the general public through the mass media.[36]

A company's lobbying function, as Weidenbaum (1986) analyses, includes two aspects: 'offensive' lobbying and 'defensive' lobbying. Offensive lobbying is designed to get the company's views on pending legislation of special interest across to politicians and government bureaucrats to opposing or at least amending the flow of government legislation that results in great government control over business decision making; while defensive lobbying is geared to avoid embarrassing investigation of and attacks on the company (Weidenbaum, 1986, p. 385). The empirical studies included in this book are concerned primarily with firms' offensive lobbying.

Corporate strategists have found ways to turn the political process to their own advantage through their active lobbying and interaction with government bureaucrats and politicians and, in this manner, public policy moves from the pursuit of national goals towards a more commercial and self-serving set of competitive entry barriers (Rugman and Verbeke, 1990, p. 2). Many big firms have a department or bureau devoted to various kinds of regular contact with government officials. In the meantime, most government policies or regulations towards the industry and business world have been made with close consultation with firms. Therefore, the firms, particular those big ones, play a significant role in government policy-making, and can try to enlist government support to bolster their position in the market:

> Firms can defend and advance their interests in a variety of ways, some of which involve enlisting the state's capacity to coerce other actors, which implies revision of the existing patterns of exclusion and privilege (Cawson *et al.*, 1990, p. 12).

Cawson *et al.* also argue that firms are essentially political as well as economic actors, "not only in a sense of seeking to influence public policy, but also in the sense of exercising their market power and resisting or accommodating to the market power of other firms" (1990, pp. 12–13). This book provides ample evidence of the political role of firms, and the extent to which the European consumer electronics industry is affected by political and economic factors.

Instead of using their economic power exclusively to compete against their rivals in the marketplace, the dominant firms "may try to achieve the same result at lower cost by in effect 'borrowing' the authority of the state" (Cawson *et al.*, 1990, p. 23). On the other hand, as a result of lobbying, the firm(s) (often those domestic ones) may be granted privileged status by the government in order to achieve the latter's policy goals or to enhance its political power (Cawson *et al.*, 1990; Porter, 1990). In other words, to grant privileged status to the lobbyists is not only beneficial to the firms but can also be in the political interest of the government.[37]

Many believe that the success of industrial lobbying depends on the economic power of the firms which lobby. "Money will go where the political power is. Anyone who thinks government funds will be allocated to firms according to merit has not lived or served in Washington very long".[38] But against this, Weidenbaum (1986) asserts that industrial policies heavily influenced by corporate lobbying are implemented only in favour of old and politically established firms and at the expense of new but economically strong firms:

> The future is underrepresented politically. New and growing firms, although economically strong, are politically weak. The successful companies are too busy to politic. In contrast, the past is overrepresented politically. Many old, established firms are politically strong but economically weak. The 'sickies' have incentive to lobby for government help (p. 249).

The bargaining power of the old and well-established firms, as Cawson (1985) contends, is embodied in the fact that state agencies often pursue specific sectoral policy aims by negotiating planning agreements with individual firms from sectors where a considerable degree of monopoly is obtained, and where priority intervention has been identified by state actors as necessary to prevent collapse or to induce growth (p. 16). He also argues that, inevitably, the sectoral aim of industrial policy as such might be pursued through a 'national champions' or 'backing winners' or 'nursing lame ducks' policy (1985, p. 18).

The role played by firms in framing public policies, according to Dodgson (1989), is demonstrated by the fact that both the Alvey programme (of the U.K.) and the ESPRIT programme (of the EU) were formulated on the basis of representations by firms, who presented arguments on the need for public support (p. 10). Considerable evidence has also been showed by Sharp (1989) and Cawson *et al.* (1990) about the consumer electronics industry, in which major European firms, led by Philips, have played a significant role through industrial lobbying during the process of EU policy-making against foreign competition since the beginning of the 1980s (e.g. the EU quotas on video cassette recorders and the 19% 'infant industry' tariffs on compact disc players). As a result, the corporate strategy of companies in the Triad economies of the U.S., Japan and Europe, to some extent, are regarded as the trade policies of these blocs (Rugman and Verbeke, 1990, p. 5). This point will be further investigated through detailed empirical studies in the following chapters (e.g. Chapters 4–7).

Overall, in an age when business success depends on staying close to customers, successful companies spend time in close touch with what's going on in the marketplace, while unsuccessful companies run to Washington, Brussels or Kasumigaseki (the district in Tokyo where all the Japanese governmental agencies are located) (Ohmae, 1990, p. 103).

The Price of Protectionism

Protectionist trade and technology policies, in most cases, are implemented at a high price. Various negative effects of these policies have been suggested.

First of all, it is argued that the desire for protection from foreign rivals is usually justified by citing the 'unfair' advantages possessed by them; but this could doom a national industry in the long run (Porter, 1990). In other words, firms sometimes propose and support government policies that are not necessarily in their own long-term interest. More specifically, the negative effect of protectionist government policies, as suggested by Porter (1990), are twofold. First, government 'assistance' will allow firms or an industry to be slow or refuse to change in accordance with the development of competitive circumstances. Second, selecting only policies that please local firms may well be counterproductive:

> Choosing policies based on unanimity of corporate or union support, as politicians under short-term pressure for reelection or reappointment are prone to do, may do as much harm as good (p. 625).

In one of his empirical studies on the European semiconductor industry, Hobday (1991) has made a similar argument to that of Porter. Hobday argued that one of the potential problems facing the European firms in the semiconductor industry is their dependence upon government and EU subsidies:

> ... subsidy cannot be a substitute for genuine improvements in management, productivity and competitive behaviour. If subsidies serve to camouflage poor R&D practices, production inefficiencies and antiquated organisational structures, then the large-scale finance being channelled to firms would only postpone the decline of Europe's SC [semiconductor] industry. An unhealthy continuous dependence on state support for finance, combined with the bureaucratic influence of government upon firms, could be debilitating in terms of performance, dynamism and the ability to respond to market requirements (1991, pp. 91–92).

In short, it is highly possible that the short-term benefits gained by domestic firms from government actions are often realised at the expense of the firms' long-term competitiveness. As far as Europe is concerned, "there are substantial gains from integration through the introduction of competitive forces into sectors of European industry where inefficiency has survived because of regulation, public procurement rules, or other protectionism" (Kay, 1993, p. 365).

Secondly, the costs of maintaining the privileged position of a firm (or industry) is usually covered by public money collected from tax-payers (Cawson *et al.*, 1990, p. 23). Proponents of strategic trade and industrial policy theories argue that when profits to the domestic firms rise by more than the amount of the subsidy or the benefit to domestic firms exceeds the cost to tax-payers, some of the cost will be recovered from income taxes on shareholders (Brander, 1986, p. 29). The latter view appears far from convincing because new ventures under government subsidy do not always become successful (particularly when big risks inherited from uncertain generic advanced technologies are taken into account). In other words, compensation for government subsidies by income taxes from shareholders is not guaranteed.

Thirdly, it is accepted, not only by those in favour of free trade, but also by some proponents of new theories on strategic trade and industrial policy, that measures or policies designed to promote or protect some particular sector within one country means promoting or protecting that sector at the expense of other sectors (Krugman, 1986, p. 14). The benefits received by the protected industries are obvious, but the costs are partly shifted to other companies that buy from the protected industries (Weidenbaum, 1986, p. 285). An interesting case to illustrate this point is the long-lasting American–Japanese semiconductor trade conflict. Shortly after the signing of the 1986 U.S.–Japan semiconductor accord, prices of semiconductor products (e.g. D-RAMs) increased by about 40% due to shortages caused by price floors for import in the U.S. (Oppenheim, 1992, p. 110). This result might be in the short-term interest of those leading American manufacturers of semiconductors but definitely detrimental to the computer manufacturers who use D-RAMs in large quantities. As a matter of fact, leading computer firms such as IBM and Apple made their protests against government measures hostile to D-RAM imports from abroad, particularly Japan.

Finally, the largest part of the cost of protectionism is paid by individual consumers of various final products. Various forms of trade barriers erected against foreign goods have been widely regarded as one of the major factors raising prices to consumers. It was estimated that the cost of car import restrains in the U.K. alone amounted to £2 bn in 1988.[39] A study by the National Consumer Council (NCC) in 1990 revealed an extra cost of £74 on average to every computer printer, and £20 to the price of a VCR bought by U.K. consumers; and the total annual cost of antidumping measures to EU consumers was at £1,170 million (equivalent to more than a 5% price increase). The internal barriers between the EU countries cost the consumers more than £83 bn a year, as estimated by the European Commission. The cost of the American trade barriers accounts to a total of $80 bn a year. In another report published on 6 May 1993, the NCC argues that it is the EU governments that have imposed high antidumping duties who have made the Japanese form export cartels (in terms of VER agreements) against the EU consumers without public debate.[40]

CONCEPTUAL FRAMEWORKS FOR MANAGING TECHNICAL CHANGE

To summarise the discussions presented in the last three sections, new technologies or new products and, ultimately, the competitiveness of the firm are mainly influenced by a combined force, rather than any single element. In recognising the role of other factors affecting the process of technical change, I would particularly stress the importance of three forces: a firm's corporate strategy; strategic alliances and industrial collaboration between firms; and public policy, especially technology policy of national governments and supranational bodies such as the EU and the interplay between them.

Figure 2.1 shows that, while there is an interplay, corporate strategy, industrial collaboration and public policy each have different roles to play in promoting new technology or the process of technical change. A malfunction with or between any of these forces could possibly lead to the failure of a particular new technology at either a developmental stage or marketing stage. This will be investigated in later chapters in the book. The European VCR technology, i.e. Philips' V2000 system, failed to become the dominant standard for home video machines in the early 1970s, given its purported technical superiority. It is argued in Chapter 4 that the main

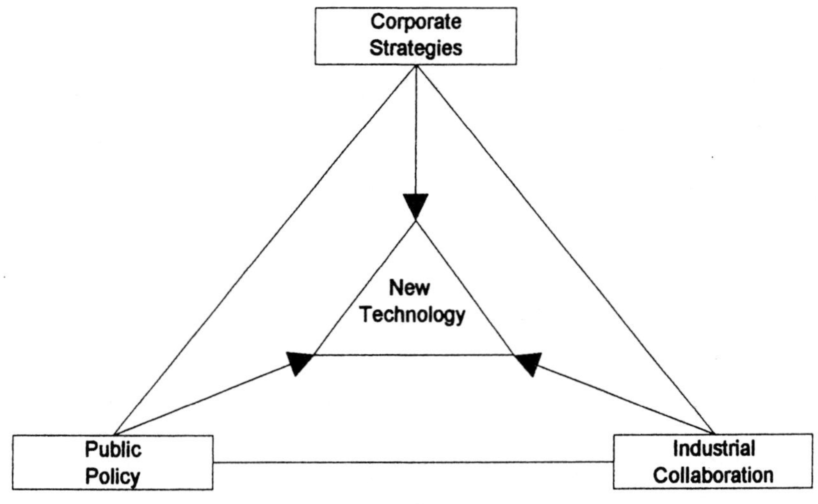

Fig. 2.1. Managing Technical Change (1)

reason for the failure of Philips' V2000 system was the company's corporate strategy which did not match the new competitive environment in the early 1980s. The study of the multimedia sector and Philips' CD-i project (Chapter 8) shows that, to a certain extent, the strategic alliance between Philips and Sony for launching the interactive multimedia system in the early 1990s did not appear to be as effective as the two companies' collaboration for the compact disc audio system in the early 1980s. Partly because of this, the CD-i system has not become a global standard for consumer multimedia. Effective use of government intervention may be important in terms of promoting the so-called 'strategic technologies' and new technological areas where 'market failure' exists (such as telecommunications infrastructure, new broadcasting systems, etc.). However, misuse or excessive use of government policy and regulatory power may lead to failure of important technologies. The development of high definition television (HDTV) technologies since the second half of the last decade, as Chapters 6 and 7 will show, has seen government intervention at its highest level in the history of the consumer electronics industry. While the U.S. government, *via* the Federal Communications Commission, opened the bidding process for HDTV technologies by inviting domestic as well as foreign firms to join the competition, the EU chose to pick winners from the early stages of the development of technology. The effect of divergence between American technology policy and EU technology policy was embodied in the subsequent breakthrough in digital HDTV technology in the U.S., and the collapse of the European HD-MAC system, an analogue format.

The interplay of the three factors in the process of managing technical change, as shown in Fig. 2.1, will also be seen in most of the case study chapters included in this book. For instance, in the case of European HDTV development, the EU's MAC legislation and other HD-MAC strategies proposed by the European Commission were established to back the large scale European industrial collaboration, i.e. the Eureka 95 Project. In the meantime, leading European consumer electronics firms such as Philips and Thomson adopted a dual-commitment strategy—they used public financial support to develop the analogue HD-MAC system in Europe but, in the meantime, were also heavily involved in digital HDTV projects in the U.S. The nature of European corporate strategies in the cases of Philips and Thomson is that the two companies were anticipating their short-term interests to be realised *via* HD-MAC while getting a strong foothold in the more futuristic digital HDTV technology *via* participating American projects. In criticising the failure of the EU's HDTV policy, one important factor cannot be denied: corporate strategies adopted by some leading European firms have, to a certain extent, benefited from Brussels' public policy as far as HDTV is concerned. In other words, the design of the EU's technology policy is, sometimes, biased towards corporate interests, either deliberately or unintentionally. It is also worthwhile noting that, when government technology policy is in favour of protecting indigenous infant industries, the benefit of regulated market competition should not be ignored. Philips' successful involvement in interactive digital TV, including digital HDTV, development in the U.S. and Thomson's multi-channel digital satellite television business in some segments of the North American market since the early 1990s were not helped by government protection.

Figure 2.2 shows at which levels new technologies are manoeuvred and the interrelationships at different levels. It is obvious that the tripartite relationship shown in Fig. 2.2 is another inversion of the tripartite relationship indicated in Fig. 2.1.

It is important to bear in mind that the choice of corporate strategies, industrial collaboration and public policy and the three levels (firm/corporate level, industry/sectoral level and government or supra-government level) corresponding to the three factors should not be treated as exclusive. On the contrary, other factors, such as market forces or consumer choice, national and international economic environment, etc., are all relevant to the success or failure of new technologies. However, the interrelationship between the three forces included in Fig. 2.1 and the three corresponding levels shown in Fig. 2.2 are established as the major conceptual frameworks to help analyse the process of technical change in the European consumer electronics industry. To a certain extent, each of the case study chapters presented in this book reflects these two broad analytical frameworks one way or another.

Finally, the first three sections of this chapter have reviewed the general issues about the way in which new technologies, new products and, most fundamentally, the competitiveness of the firm are affected by forces from

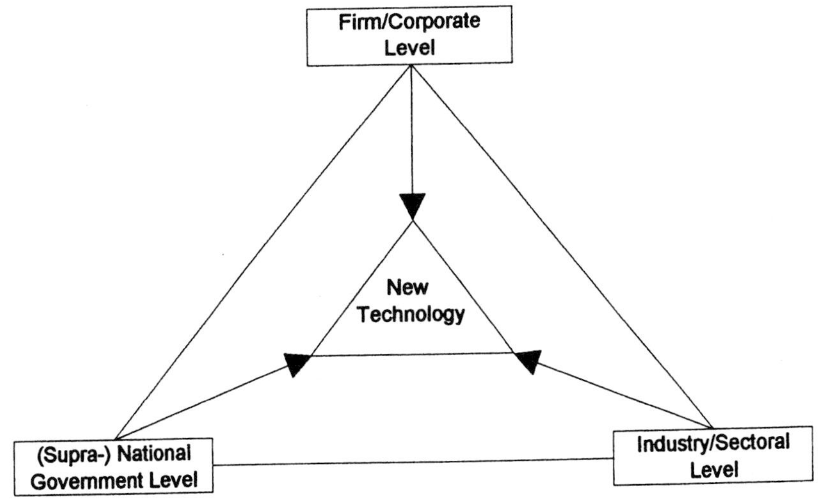

Fig. 2.2. Managing Technical Change (2)

the corporate, sectoral and governmental levels. Although these factors have been discussed separately under three different themes, they are intrinsically interrelated. This interrelationship is well illustrated in the conceptual frameworks represented by Figs 2.1 and 2.2. To summarise the discussions developed in this chapter, I cite a remark by Silverstone and Hirsch (1992) on technology:

> Technology is produced in environments and contexts, as a result of the actions and decisions, interests and visions, of men and women in organizations and institutions of complex and shifting politics and economics. These organizations and institutions often provide the framework for alliances and networks of actors to engage in the potent work of the R&D, production and distribution of technologies. Technologies emerge, we now increasingly clearly see, as a result of these complexes of actions and objects, politics and culture (p. 3).

To better understand the general issues touched on during this literature review, I turn now to the empirical materials largely drawn upon detailed case studies beginning with the history of the Philips company, which has grown from a small Dutch family business in the early days of the electric lightbulb industry to a large and diversified global company over a period of about 100 years.

NOTES

1. The Greek origin of the word 'strategy' is 'strategos', which means the art of the army general (Mintzberg and Quinn, 1991, p. 13). The equivalent of 'strategy' in Chinese is 'zhanlue', which means 'the planning (or guidelines) of war'.
2. Such as Japanese companies do with corporate slogans, e.g. NEC's 'Computers and Communication' *versus* Fuji's 'Imaging and Information'.
3. Sun Zi, the ancient Chinese philosopher and military strategist, has long been known in the West as Sun Tzu. 'Sun Zi' is used here in accordance with the modern Chinese Pinyin pronunciation system.
4. Ghoshal, S. and Nohria, N. (1993), 'Horses for Courses', *Sloan Management Review,* Winter, Reprint 3422, quoted in *Financial Times,* 19 March 1993.
5. *Financial Times,* 19 March 1993.
6. The other four components of corporate strategy, according to Ansoff (1965), are product–market scope, growth vector, competitive advantage, and 'make or buy' policy.
7. Parker (1978) has also discussed the dynamic resistance to change. He sees this resistance as an insidious force that can emasculate companies' vitality (p. 102). A typical example of resistance from inside the firm is the case of Philips' V2000 home video system. Detailed empirical study of this case will be presented in Chapter 4 of this book.
8. Based on the understanding that innovation immediately affects firms' survival, Freeman (1982) has suggested six alternative corporate innovation strategies which firms may choose to follow. The six are: *offensive strategy*—designed to achieve technical and market leadership by being ahead of competi-

tors in the introduction of new products; *defensive strategy*—to profit from the mistakes of early innovators and from their opening up of the market; *imitative strategy*—to take advantage of early mistakes and differentiate products by minor technical improvements; *dependent strategy*—to accept an essentially satellite or subordinate role in relation to other stronger innovative firms; *traditional strategy*—a 'traditional' firm sees no reason to change its product because the market does not demand a change, and the competition does not compel it to do so; *opportunist strategy*—to prosper by finding or identifying an important 'niche' market, and provide a product or service which consumers need but nobody else has thought to provide.

9. Dodgson (1993) treats 'alliances', 'collaboration, 'cooperative agreements' and 'network' as synonyms.
10. Government subsidies toward the so-called strategic technology areas are often channelled to domestic firms through joint R&D programmes or projects since the early 1980s.
11. *The Economist* reported that around 90% of these alliance agreements have been made between firms from the few industrialised countries, i.e. the U.S., Japan and Western Europe (27 March 1993, p. 20).
12. As Morgan (1989) has noted, at the forefront of technological change, the convergence of computer and telecommunications has made the previously single-product service—voice telephony—grow into a diversity of services—voice, data, video, facsimile (p. 19).
13. Miyamoto, N. (1988), *IBM's Alliance Strategy* [in Japanese], Kodan-sha, Tokyo, quoted in Ohmae (1990).
14. A number of more specific motives of the first group are the reduction, minimising and sharing of the uncertainty, rather than the risk, which is inherent to performing R&D, according to Hagedoorn (1993). It is probably the unknown likelihood of success in research that leads some companies to combine their efforts.
15. Other literature, such as Alic (1990, p. 327), suggest that the most important among the factors leading to cooperation in R&D are the risks of independent efforts and the costs of continued participation in some technologies.
16. It is also briefly suggested that many of the strategic partnerships in the information technology sector are also centred on standardisation (Delapierre and Zimmerman, 1991, p. 114). The information-processing products (e.g. complex systems, computing, data storage, data capture, data retrieval and data transfer units) are characterised to a large extent by the nature that even the largest integrated industrial groups are no longer capable of covering the entire range of applications. Therefore, strategic partnerships are believed to be a positive sum game.
17. For instance, the U.S. Congress passed a new legislation, called the National Cooperative Research Act (NCRA), in 1984, which has taken significant steps to remove legal disincentive to cooperative innovation. However, the NCRA still precludes joint manufacturing and production of innovative products and processes, which are often believed necessary to provide the cooperating ventures with significant feedback information to aid in further innovation and product development, and to make the joint activity profitable (Jorde and Teece, 1992a, pp. 56–57).

18. While presenting the view of Cutts (1992) on the Japanese alliance systems, the author would like to distinguish between 'cartels' and '*keiretsu*'. Cartels set price and divide market for the industry and, therefore, exclude competition; whilst, there are normally more than one *keiretsu* in a given industry, and it is hardly possible for one *keiretsu* to set the price for the entire industry. In other words, while emphasising the Japanese alliance systems in terms of cross shareholding and sharing of information, etc., one should not underestimate the rivalry among members of each *keiretsu* and that between different *keiretsu*.
19. The issue of whether the Japanese *keiretsu* system is anticompetitive has also attracted the attention of the media. In a recent survey, *The Economist* reported that Japan's networks (*keiretsu*) tolerate, indeed encourage intense domestic rivalry: they are undeniably 'competitive'. An alliance of small firms in the Japanese paradigm leaves much more scope for internal competition than a single firm of equivalent size. In the Japanese paradigm, on the one hand, firms buy similar parts from more than one supplier and this leaves scope for the suppliers to compete; on the other hand, the suppliers do not supply parts exclusively to only one firm, so they have far more independence than those affiliates of big firms in the West. The comparison between General Motors of the U.S. and Toyota of Japan may well indicate this difference. Whereas GM is an 'alliance' in which the principal partners are not merely wholly owned subsidiaries but elements within a single integrated management hierarchy, Toyota is a looser confederation which offers much greater scope for rivalry. See *The Economist,* 6 March 1993, p. 19.
20. As in most R&D or close-to-market projects in each part of the Triad, the U.S., Western Europe and Japan, foreign companies are often excluded from participating.
21. Actually more EU programmes, such as JESSI, were launched during the course of the 1980s.
22. According to Levy and Samuels (1991), cooperative research consortia (each of them is often participated by more than dozens of firms) and other industrial alliances in Japan are also coordinated or/and funded by other public authorities such as the STA (Science and Technology Agency), the Ministry of Education, Mombusho (Science and Culture) (p. 122).
23. In the meantime, Best suggests that the implementation of a sector strategy, i.e., industrial policy, is a partial means of realising the joint benefits of Shumpeterian nonprice forms of competition: strategically managed interfirm associations can promote the long-term development and competitiveness of a sector; non-strategically managed interfirm associations will likely have the opposite effect (1990, pp. 18–19).
24. Alic (1990) also mentioned that R&D 'cooperation' among private firms has sometimes shaded off into behaviour seemingly aimed at retarding rather than accelerating technological progress (p. 322).
25. 'Generic technology' has been used to express the ways in which some new technologies open up a wide range of possibilities for further innovations in many sectors of the economy (Freeman, 1992, p. 81).
26. Note that protectionist policies are sometimes extended to areas where domestic firms are absent. For instance, the ITC (International Trade Commission)

of the U.S. ruled that Japanese manufacturers of flat panel computer displays should be subject to antidumping duties of up to 63% in 1991, when there were simply no volume manufacturers of such products in the U.S. (Oppenheim, 1992, p. 110). As a result, major American manufacturers of notebook computers, such as IBM and Apple, decided to move some of their production abroad to avoid their home country's antidumping duties on flat panel screen imports. Typically, protectionism of any kind is not called protectionism or border-closing; it is cloaked behind a 'long-term view' of a country's competitiveness and how to improve it. See *The Economist*, 27 March 1993, p. 28.

27. As a protectionist measure, VER agreements were used by the American government to protect its domestic industry much earlier. For instance, in the postwar era President Eisenhower adopted 'voluntary' restraints on Japanese exports of cotton textiles to launch America into a new sphere of protectionism in the 1950s, and his policy was extended by President Kennedy with mandatory controls on all textile imports (Oppenheim, 1992, p. 54).
28. IMF (1988), 'Issues and Developments in International Trade Policy', *Occasional Paper* No.63.
29. In the old days, countries with strong military forces often opened up foreign markets through war fairs.
30. Corporatism, according to Cawson (1985), is a specific sociopolitical process in which organisations representing monopolistic functional interests engage in political exchange with state agencies over public policy outputs which involve those organisations in a role that combines interest representation and policy implementation through delegated self-enforcement (p. 8).
31. The amount of government subsidy towards the semiconductor industry in the U.S. is partly seen in the 1986 Sematech Programme towards which the American government has provided the private firms with \$1.5 bn funding (Hobday, 1991, p. 91). In total the U.S. state funding for R&D in 1987 was \$60 bn, of which \$41 bn went to defence-related industries, and the remaining \$19 bn was received by purely commercial research programmes participated by private firms (Oppenheim, 1992, p. 90).
32. 'Rent' means profit above the minimum amount necessary to keep the owners of the firm in business (Brander, 1986, p. 24).
33. In slightly different terms, Grant (1982) has also suggested four strategic approaches to industrial policy: pure market approach, social market approach, selective approach, and socialist approach. The major difference between the social approach and the selective approach is that advocates of the former believe in a policy for industry rather than the comprehensive industrial policy advocated by selective interventionists (Grant, 1982, p. 14). Moreover, advocates of the social market approach recognise the inevitability of some intervention by government in industrial affairs, but they see intervention as inherently undesirable and, therefore, should be infrequent, temporary and limited in scope.
34. In a similar way to the proponents of the so-called 'cautious activism', Rothwell and Zegveld (1982) have also suggested that government assistance towards industrial innovation is needed, and it must be directed towards certain strategic technologies such as biotechnology, energy technology, communications technology, and microelectronics, etc. (pp. 239–240).

35. *The Blackwell Encyclopaedia of Political Institutions,* V. Bogdanor, Ed., Basil Blackwell, Oxford, 1987, p. 337.
36. *Ibid.*
37. It is suggested that, on the one hand, state is sometimes not powerful enough or has insufficient specialised knowledge to formulate and implement policy without the agreement of its partners (including firms and their interest associations); on the other hand, the interest organisations not only represent the interest of their respective interest categories, but also play a part in the governance of those categories (Cawson, 1985, pp. 6–7).
38. Senator William Proxmire of Wisconsin, quoted in Weidenbaum (1986).
39. Figures in this paragraph are mainly from various sources such as *What to Buy for Business*, IMF, the National Consumer Council, the European Commission, and the U.S. Institute for International Economics, etc., quoted in Oppenheim (1992). For more detailed discussion on the costs of protectionism paid by the consumers in terms of 'hidden tax', see Weidenbaum (1986). He suggests that protectionism can be viewed as a hidden tax on the consumer (p. 285).
40. Quoted in *Financial Times,* 6 May 1993.

3

PHILIPS' HISTORY: FROM FAMILY BUSINESS TO THE OPERATION OF A MULTINATIONAL ENTERPRISE

INTRODUCTION

N.V. Philips' Gloeilampenfabrieken (Philips) celebrated its 100th anniversary in 1991. Philips' historical development can be seen as comprising the following stages: (1) family business; (2) limited liability company; (3) a federation of national organisations; (4) a product division structure. During the course of a century's development, Philips diversified its product areas, enlarged its production capacity, increased its international presence, and, so, improved its competitive position gradually. In order to analyse and understand Philips' present corporate structure and strategic manoeuvring for new technologies in the consumer electronics industry, it seems necessary to study the process of Philips' historical evolution. As will be shown later in the book, some aspects of the former have their origin in the latter. However, due to constraints on this book, a detailed chronological elaboration is not embarked on. On the contrary, only those aspects which are closely related to the later chapters fall within the range of this chapter.

This chapter covers the history of Philips from its origin as a family business in The Netherlands up to the 1970s when Philips' competitive position in the consumer electronics industry began to be challenged by newly emergent East Asian competitors, mainly the Japanese. More specifically, the main objectives of this chapter are the investigation of the following issues.

Firstly, the parallel hierarchies of engineering or production management and commercial or marketing management within both the top management team and each national organisation are widely regarded as a unique characteristic of Philips' managerial structure. When and how did this aspect start to characterise the company? What effect did it have on Philips' capacity to compete on the world market?

Secondly, another important characteristic of Philips before the 1970s was its development as a loosely organised federation of national organisations. Did this structure correspond to the consumer electronics industry comprising fragmented national markets in Europe before the 1970s? How could Philips survive the fierce competition from other multinational enterprises (MNE) in the European as well as the world market before the 1970s? What strategies and policies had Philips adopted? Did the creation and growth of the EEC[1] have any significant impact on Philips' strategic planning and organisational structure? What was the company's response to the course of European integration since the late 1950s?

Thirdly, it will be argued in this chapter that both Philips' geographical expansion and its product diversification were stimulated by a long held expansionist policy.

Fourthly, I will examine the question of whether Philips' central weakness has been its inability to successfully commercialise its innovations.

Finally, I will also examine Philips' attitudes towards and involvement in the 'cartel movement' which happened in the electric lighting industry before the second world war. How should we assess Philips' international collaboration with other firms before the Japanese consumer electronics manufacturers entered the European market?

THE ORIGIN OF PHILIPS AND ITS EARLY DAYS

Philips' origin and early history may not seem to be immediately relevant to the discussion about the company's contemporary performance in the consumer electronics industry. However, a brief review of Philips' historical development would provide a rich historical and cultural context in which the company's organisational structure and managerial tradition have been fostered. To a great extent, the major aspects of Philips' structural problems and managerial inefficiency shown in the last three decades are closely linked to the company's corporate history, in particular the early days dominated by the key figures of the Philips family.

Background: The European Lighting Industry in the Late 19th Century

On October 21, 1879, the first working incandescent lamp was successfully tested by the inventor Thomas Edison. This invention symbolised a major revolution in both technology and industry all over the world.

In 1881, Siemens and Halske founded a small experimental factory for the manufacture of incandescent lamps in Europe. Emil Rathenau had meanwhile acquired Edison's patent. In 1883, the Deutsche Edisongesellschaft fur angewandte Elektrizitat (the German Edison Company for Applied Electricity) was founded, with a capital of five million marks. Later, this company cooperated with Siemens and Halske to establish the Allgemeine Elektrizitat-Gesellschaft (General Electric Company). In addition, at that time, there were two small incandescent lamp factories in Italy, one in Austria, and one in Hungary (Bouman, 1970, p. 19).

The Establishment of Philips & Company

Frederik Philips, a Dutch businessman mainly running a gas works at Zaltbommel, was the founder of Philips. From 1889, Frederik began to finance his eldest son Gerard Philips, an engineer in electrotechnics, to conduct private experiments in making a carbon filament lamp better than those offered for sale by the German and British factories at that time.

Based on Gerard's experimental work, a private agreement was drawn up between Frederik Philips and Gerard Philips by which the firm, called Philips & Company, was brought into existence as a limited partnership on the 15 May, 1891. Under this agreement, Frederik Philips invested 75,000 guilders (approximately £6,200). At the same time, a production site of approximately 1,450 square metres was bought at Eindhoven in the southern Dutch province of Brabant. As it turned out, Eindhoven was an ideal location for the Philips family to start their business because the low labour costs in this region enabled Philips to survive competition from other bigger and more established concerns like AEG and Siemens in Europe. Anton Philips, Frederik Philips' youngest son, once described the situation of the first few years within Philips in part as the following:

> In May, 1891, Messers Philips & Company, in which my father and my brother were partners, began manufacturing filament lamps. The first year was spent in installation and experimentation, and only in 1892 did actual production begin. The firm had its full share of teething troubles. A German foreman had to be engaged and also a commercial traveller for the sales. In 1893, losses were so large that it was decided to give up the factory.[2]

As far as the newly established Philips family company was concerned, the following characteristics should be taken into account.

Firstly, the parallel hierarchies. From the very beginning of the business, a structure of parallel hierarchies was roughly designed even though all the initial investment was made by the father. Within this dual structure, Frederik Philips was in charge of the commercial side of the company and this duty was passed to Anton Philips on 1 April, 1899, a year before Frederik Philips died. Meanwhile, Gerard Philips was responsible for the technical and engineering side of the company. This dual structure was deeply rooted in the process of the company's development and was not fundamentally changed until the 1980s, when a radical restructuring process was started.

Secondly, experiment-oriented. As mentioned above, the Philips concern was divided into two basic hierarchies—technical management and commercial management. This kind of division, to some extent, indicates that scientific research and technical experiment were highly regarded within the company.

Even before the setting up of the company, Frederik Philips had already supported Gerard Philips to establish a laboratory in the parental home at Zaltbommel to conduct experiments. And in 1890, Gerard Philips reckoned that he had found a manufacturing process suitable for mass-production. As he declared himself in the same year, "I am able to produce extremely homogeneous and equal cellos filaments on a business scale" (Bouman, 1970, p. 21).

Before the actual production started, as mentioned by Anton Philips, the company spent one year in installation and experimentation. It is true that Philips would never have been able to survive the fierce competition without having increasingly enhanced its scientific and technological experimentation as its strong foundation for expansion and development. In the electric incandescent lamp manufacturing industry, Philips' R&D work gained a high reputation, and worldwide recognition after the 1920s. The following comment is taken from an early study by Bright:

> Of all the European companies, Philips seems to have been the most outstanding in R&D. It pioneered in the design of photoflash lamps, in the use of the heaviest inert gases for gas-filled lamps, in machinery development, and in other directions of incandescent-lamp design and production (1949, p. 311).[3]

Not only in the lighting industry but also in other areas, such as X-ray technology, audio-visual technologies, semiconductors, optical discs, etc., Philips has been one of the leading corporations as far as scientific research and technical change are concerned. This is a major theme of both this and subsequent chapters.

Thirdly, export-oriented. At the beginning of production, Philips' annual output was not as high as had been expected. In 1892, the output was 11,000 lamps; in 1893, 45,000. The first profits were not made until 1895 when the company's production rose to 500 lamps a day. As soon as Philips' daily output reached 500 lamps, a serious difficulty emerged in the area of business development. Due to the size of the country, the domestic market for Philips was extremely limited. Holland could not offer much scope for Philips' constantly increasing volume of production. Anton Philips once described the national disadvantages in the sense of infrastructure and the size of the country, under which the company had laboured:

> Electric power stations in Holland were few. The Amsterdam power station had 7,500 light points. When that figure is compared with our modern

> factories at Eindhoven which have 50,000 light points, it is obvious that Holland did not offer much scope for sales. Rotterdam had a power station with approximately 3,000 light points; The Hague and other large towns had none. Small private electric power stations were operated by the breweries, the big yeast and methylated spirits factory, the Gouda candle works and a few shipping firms; and that was about all.[4]

Due to the limitation of the domestic market, Philips was forced to sell its light bulbs abroad.

Apart from its advantage of low labour costs, Eindhoven was also an ideal site from which Philips salesmen could do business with the neighbouring countries. In particular, customers in the north of Belgium, Westphalia and the Ruhr industrial area were within easy calling distance of Eindhoven.

In the last few years of the 1890s, Philips was exporting 97% of its products.[5] Thus we can see that Philips was an export-oriented enterprise right from its early days.

Table 3.1 shows the rapid soaring of the Philips' light bulb business in the early years. Ten years after the establishment of the company, Philips became one of the largest electric light bulb manufacturers in Europe. Thanks to Anton Philips' international commercial vision and his great efforts, the company, by that time, had already appointed 48 sales agents in 28 large foreign cities and most of its light bulb products were sold through these agents.[6]

From Family Business to a Limited Company

After 20 years' development, especially after completion of the five-storey concrete building in 1911 and the acquisition of the costly American machines necessary for large-scale production of drawn-filament lamps, the Philips brothers, Gerard Philips and Anton Philips, realised that an appeal to the capital market was inevitable. Therefore, on 29 August, 1912, a limited company based on the previous family business was set up with the following resolution:

- Total capital: 6,000,000 guilders (approximately £500,000);
- Ordinary shares: 3,500,000 guilders (approximately £292,000);
- Preference share: 2,500,000 guilders (approximately £208,000).

Table 3.1. Philips' Light Bulbs Output ('000), 1895–1899.

Year	1895	1896	1897	1898	1899
Output	109	280	630	1,200	1,800

Source: Based on various documents.

In accordance with the establishment of the limited company, the name of the concern was changed from Philips & Company to N.V. Philips' Gloeilampenfabrieken (Philips' Incandescent Filament Lamp Factories).

The reason why Philips was converted from the original family business to a limited liability company may be understood from the following aspects: (a) the rapid increase of output could only be realised by sufficient supply of raw materials which cost more and more capital; (b) it was necessary to have adequate investment in new or more advanced equipment to manufacture the new generation of electric lamps for the sake of keeping the concern competitive in the European light bulb industry; (c) because Gerard Philips had no children he wanted to split the capital so that he would be able to withdraw from the business if he wished.[7]

As a matter of fact, the conversion to a form of limited liability company provided the concern with the chance to collect sufficient capital to purchase production materials and finance new technical development more rapidly. Meanwhile, the issue of shares made it possible for other firms and individuals to participate in Philips business and hence made the connection between the business and the Philips family much looser. Needless to say, this conversion laid down the foundation for the company's further expansion in the future.

THE EXPANSIONIST POLICY

Having become a fully-fledged electric lamp manufacturer in the European lighting industry, Philips entered a new era of development from the beginning of the 20th century to the second world war in 1939. On the one hand, this era is strongly characterised by the concern's aggressive strategy aimed at rapid expansion; on the other hand, the company assumed a defensive strategy concurrently, which will be discussed in the following section. Apart from the business plunge during the 'great recession', these two strategies worked satisfactorily in parallel. In this section the former aspect of Philips' strategic planning is explored.

Roughly speaking, Philips' aggressive strategy could be interpreted as an 'expansionist policy'. There is no doubt that the expansionist policy played an indispensable role for the company to develop into an industrial organisation with the status of a multinational enterprise. To understand the history of the company's corporate strategies, it is very important to bear in mind Philips' expansionist policy at every stage, from a family business to a limited company, then to a national enterprise and finally to a multinational group.

As far as Philips' expansionist policy was concerned, corporate activities like establishing national and international subsidiaries, acquisitions, industrial collaboration (e.g. cooperations and joint ventures), product and business diversifications, etc., were the major aspects. In short, during the whole process of the company's expansion, Philips' policy was concentrated on two extremely ambitious lines: one was that Philips endeavoured to strengthen its links with as many places as possible in the world; another was that the scope of Philips' product range had to be expanded as far as possible.[8]

The Race Between the Two Brothers

As mentioned above, the parallel hierarchies began to characterise Philips from the outset. In the early days, Anton Philips was responsible for the commercial side of Philips management, while his brother Gerard was in charge of the engineering aspects of the company. Anton was trying to sell more than the firm could produce and Gerard was trying to do his best to produce more than Anton could sell. The race between the brothers came to symbolise the interaction between the two sides of Philips at the managerial level. To some extent this race or form of interaction promoted the rapid expansion of the company during its earliest days.

Expansion in Holland

It was not until the 1920s that Philips began to set up foreign subsidiaries, apart from its foreign trade agents. Before that, Philips' expansion was mainly conducted within Holland. Table 3.2 indicates the rapid growth of Philips' employees before the 1920s.

Apart from the rapid increase of employees within Philips, the firm extended its interests in acquiring shares in other companies in the same sector. In 1920, for example, Philips reached an agreement with the N.V. Metaaldraadlampenfabriek 'Volt' (Metal Filament Factory 'Volt' Limited) at Tilburg, and with Vitrite in Middleburg. In the same year, the majority of shares in N.V. Pope's Metaaldraadlampenfabriek (Pope's Metal Filament Lamp Factory) at Venlo was also acquired by the Philips factories in Holland. These takeovers greatly enhanced Philips' competitive position, and made it possible for the company to invest abroad. In 1921, Philips management decided to buy factories together with Osram in Austria and Switzerland.

Table 3.2. Philips' Employees (Person), 1896–1917.

Year	Employees	Year	Employees	Year	Employees
1896	60	1900	400	1911	2,500
1897	170	1907	500	1914	2,100
1898	120	1908	1,050	1915	3,100
1899	280	1910	1,900	1917	3,750

Source: Adapted from Bouman (1970).

The Boom of Foreign Subsidiaries

The Philips brothers, especially Anton Philips, were not satisfied with the status Philips had gained as a national concern. Immediately after the establishment of the Societe Lumiere Economique, the first foreign branch in Brussels in 1919, more than 20 other foreign subsidiaries were set up by Philips in many parts of the world. Table 3.3 gives a list of the Philips subsidiaries established during the period 1920–1930.

As we have seen, all Philips' production activities were concentrated in Holland, particularly in Eindhoven before the 1920s. Mainly through the establishment of the foreign subsidiaries, as shown in Table 3.3, Philips established its first international production structure—with Eindhoven as the centre of the concern continuing to export products, foreign subsidiaries began producing locally for the local markets.

Criticisms of the Expansionist Policy

Philips' expansionist policy promoted by Anton Philips' unexhausted ambition and continuous optimism was not always untouched by criticism.

Table 3.3. Foreign Subsidiaries of Philips, 1920–1930.

Time		Subsidiaries	Country
Oct.	1920	The S.A. Philips Eclairage et Radio	France
	1923	Warsaw Factory	Poland
	1923	Norsk A/S Philips	Norway
	1923	Swenska A/B Philips	Sweden
May	1923	The Philips S.p.A	Italy
Apr.	1924	The Finska A/B Philips	Finland
Jun.	1924	The Philips Gliihlampen Vertriebgesellschaft m.b H.	Czech
Jul.	1924	The S.A. Philips do Brasil	Brazil
Jan.	1925	The Philips Electrical Ltd.	England
Jul.	1925	The Philips Industri og Handels A/S	Denmark
Nov.	1925	The Philips G. m. b. H. Vien	Austria
Mar.	1926	The Philips Iberiea S.A.E.	Spain
Apr.	1926	The Philips Radio und Elektrizitats A.G.	Hungary
Nov.	1926	The Deutsche Philips G.m.b.H.	Germany
Dec.	1926	The Philips Electrical Industries Ltd.	Australia
Apr.	1927	The Philips Electrical Industries of New Zealand Ltd.	New Zealand
Sep.	1927	The Philips S.A. Romana	Romania
Apr.	1928	The Philips Nord-Africaine Eclairage et Radio of Algiers	Algiers
Jun.	1929	The South African Philips Ltd.	S.Africa
Sep.	1929	The Philips Electrical (Ireland) Ltd.	Ireland
Feb.	1930	The Philips Electrical Company (India) Ltd.	India

Source: Adapted from Bouman (1970).

After the crisis of October 1929, a general recession of business and industrial practice began to spread all over the world during the first few years of the 1930s. Philips could not avoid being affected by this worldwide crisis. As the annual report for 1930 records,

> ... in September 1930 a marked retrogression set in, expressed in decreased sales and a marked reduction of the profit margin.[9]

Due to the worsening economic slump, all countries were experiencing difficulties and were trying to restrict their imports as much as possible. In order to avoid high tariffs, Philips was forced to transfer many production plants from Holland to abroad. The transfer of production made the reduction of employees in Holland inevitable.

In 1932, a great number of Dutch personnel in Philips were dismissed and the salaries of those who remained were reduced accordingly. The high share capital of the company was also reduced by 25%, namely, the capital of the enterprise was reduced from 250,000,000 guilders (approximately £33,000,000) to 187,500,000 guilders (approximately £24,670,000) at the general meeting of shareholders on 29 May, 1933.[10]

As a reaction to the above and other measures undertaken by Philips, the company and its leaders were seriously criticised by the press. The N.V. Gloeilampenfabriken, once extolled to the sky, was now termed 'a giant with feet of clay' by a professor at Delft. Dr Anton Philips was accused of having entered upon an irresponsible expansionist policy for which he was now being called to account.[11]

However, it was exactly in the 1930s that Philips established its second international production structure, which was characterised by the following aspects.

Firstly, production was transferred to foreign countries on a world scale.[12]

Secondly, the number of workers outside The Netherlands increased from 13,000 (32.5% of total) in 1929 to 18,000 (49.3% of total) in 1934 and 28,000 (58.9% of total) in 1937; while the Philips workforce in The Netherlands was correspondingly reduced in the 1930s. Detailed changes of Philips' employment are given in Table 3.4.

Thirdly, due to the above two aspects, the export function of Philips in Holland was reduced.

Philips' International Collaboration

In fulfilling the objective of preventing Philips from being left in a vulnerable position between the American and the German producers, the management of the company were not satisfied with the results achieved from its domestic expansion. On the contrary, Philips extended its aggressive strategy from a national level to an international vision. As far back as 1916, Philips had started to practise a new strategy—in which

Table 3.4. Philips Employment, 1929–1939.

Year	Holland (Number/%)	Abroad (Number/%)	Total
1929	27,000/67.5	13,000/32.5	40,000
1934	18,500/50.7	18,000/49.3	36,500
1935	16,500/47.1	18,500/52.9	35,000
1936	15,000/43.5	19,500/56.5	34,500
1937	19,500/41.1	28,000/58.9	47,500
1938	19,000/43.2	25,000/56.8	44,000
1939	19,000/42.2	26,000/57.8	45,000

Source: Adapted from Bouman (1970)

international collaboration was seen as a means of enhancing its international competitiveness. The 1916 collaborative agreement signed between Philips and General Electric of the U.S. provided the two sides with the opportunity to share each other's technological expertise and experience. Following this, as Table 3.5 indicates, a series of industrial collaboration agreements were signed between Philips and a number of other companies in the period 1916–1946. It seemed that seeking industrial collaboration either at a national level or at a European level or at an international level remained a crucial aspect of Philips' corporate activities prior to the mid-1940s. We will return to this theme in later chapters.

Philips' Aggressive Pricing Policy

Although later Philips took part in the creation of a European electric lamp cartel, it initially improved its market share by means of selling lamps at a lower price than its competitors.[13]

As indicated above, Philips was an export-oriented company in its early years due to the insufficiency of the Dutch market and 97% of its products were exported to neighbouring countries in those days. In order to develop its market share in Germany, Philips included a clause in all its sales contracts stating that Philips light bulbs would always be priced half a pfennig lower than those of AEG.[14]

Obviously, Philips' presence in German territory made its rivals[15] feel uneasy. For instance, Anton Philips visited the possible customers in the towns of the Ruhr and Westphalia and other German cities as well as the AEG agents in Russia frequently. This forced Berlin to consider ways and means to eliminate competition from Eindhoven.

As a matter of fact, Philips was not only taking on its German competitors but also became a challenging power to the American electric manufacturers.

Table 3.5. Philips International Cooperations.

Year	Name of Cooperator	Objective of Cooperation
1916	General Electric	Exchange technical know-how and experience; to grant each other production rights
1919	the International General Electric Company of Paris	Using each other's capital
1925	Julius Pintsch A.G [German]	Sharing manufacturing experience
1925	the Radiorohrenfabrik G.m.b.H.[German]	Offset the strong position held by the AEG and Siemens
1926	Deutsche AKkumulatorenfabrik A.G.[German Accumulator Factory Limited]	To advance the manufacture of rectifiers
1931	Telefunken [German]*	Exchange patents for radio values and radio sets
1946	Deutsche Grammophon Gesellschaft [German]	Sound recording
1946	Siemens and Philips Phonographic Industry [Philips Phonografiche Industrie]	Sound recording

Note: *According to Bouman (1970), this collaboration was extended in June, 1932 into a Weltinteres-sengemeinschaft, a world community of interests, which gave the Deutsche Philips G.m.b.H. the desired freedom to sell radio values in Germany upon the same conditions as Telefumken.
Source: Based on various documents.

Under the management of the Philips brothers, particularly Anton Philips, the business of Philips boomed rapidly. Firstly, Philips became one of the few large European producers of electric lamps; and secondly, when Philips grew into a fully-fledged producer it started to challenge General Electric, the world leader who had never been vulnerable to any other company anywhere in the world. Prior to the establishment of Phoebus—the global electric lamp cartel—in the 1920s, J.M. Woodward, the head of International General Electric's[16] European incandescent lamp business, once seriously warned A.W. Burchard, the president of International General Electric, of the danger of increasing competition from Philips in the American market. J.M. Woodward wrote to his company president in part as follows:

> I may again say by way of emphasis that I desire to effect an arrangement which will be to the absorbing interest of the Philips Company throughout as much of the remaining 'useful life' of Anton Philips as possible. He, and practically he alone, constitutes a danger to our American profits.

> Should our patent protection at any time become weak in all or part of America he will be our greatest menace, the least vulnerable and the most resourceful of our competitors (Stocking and Watkins, 1946, p. 326).

Woodward's warning was also accompanied by another forecast about Philips' competitive strength and its capacity to challenge the industry leaders in the study of Bright (1949):

> The most important potential foreign competitor [to GE] is Philips, which is already on the scene in the form of the North American Philips Company (p. 315).

Both of the above warning and forecast proved not to be exaggerations. Shortly after Philips became the biggest exporter of electric lamps, the company made another step forward and took over the world leadership from GE to become the largest company in the world lighting industry.

NATIONAL INDEPENDENCE AND INTERNATIONAL COMPETITIVENESS

In the process of expansion, it was always the main preoccupation of the central administration of the concern to maintain Philips' national independence and improve its international competitiveness. This is the second aspect of Philips' corporate strategies mentioned above—the company's defensive strategy.

Philips wished to retain its national independence, i.e. its national character as a Dutch company, by means of preventing any major portion of the share capital of the limited liability company falling into foreign hands. At the same time, to improve a company's international competitiveness Philips needed to participate in international technical and industrial collaborations with its rivals. These two apparently contradictory aspects set up tensions within the management of the company.

Security: The Establishment of the Holding Company

On 20 March, 1920, under the decision of Philips' directors, a holding company, namely, the N.V. Gemeenschappellijk Bezit van Aandelen Philips' Gloeilampenfabrieken (Joint Ownership of Shares in the Philips' Incandescent Filament Lamp Factories Limited) was founded. The principal aim of this holding company was stated as follows:

> The buying and in other ways acquiring of shares in N.V. Philips' Gloeilampenfabrieken and furthermore the taking of such measures as may assist the said company to maintain its national character (Bouman, 1970, p. 93).

Since then, Philips' national independence has been adequately guaranteed through the establishment of the Joint Ownership Company which was intended to be able to control Philips' shareholders efficiently.[17]

The role and function of the holding company in maintaining Philips' independence or national character was also discussed by Stocking and Watkin (1946). It seems that it would be difficult, if not impossible, for any other company to wrest ownership from Philips, because,

> The holding company owns 96% and 86%, respectively, of Philips' outstanding preferred and common stocks. It in turn is closely controlled by the holders of only ten preferred shares, representing six thousandths of one% of the paid-in capital stock of Bezit. These shares are believed to be owned exclusively by the Philips family (Stocking and Watkins, 1946, p. 318).

Philips and the 'Cartel Movement'

The term cartel, according to Stocking and Watkins (1946), is defined as an arrangement among, or on behalf of, producers engaged in the same line of business designed to limit or eliminate competition among them. Cartels "range from loosely defined gentlemen's agreements or informal understandings among business rivals—by which any one of the numerous elements affecting the flow of goods to market, and hence market price, are brought under joint producer control—to formal compacts providing administrative machinery for regulating output, sharing markets, and fixing prices" (Stocking and Watkins, 1946, p. 3). The so-called 'cartel movement' in the world electric lightbulb industry first started in the U.S. at the end of the 19th century and then followed in Europe at the beginning of this century. Finally, cartelisation was realised at a global level in the 1920s. As an international movement, cartelisation came to an end at the time of the second world war. As a fast growing producer of electric light bulbs, Philips participated in both the European cartel and the global cartel.

In explaining the historical development of Philips' expansionist policy, a Philips company historian identifies the company's industrial or technical cooperation with its competitors by means of participating in the cartel movement as its independence strategy:

> During the period 1919 to the late 1930s, Philips adopted a kind of independence strategy. This independence strategy refers to the company's business practice through cartelisation and know-how exchange agreements. The benefits of this strategy is that Philips did not have to pay royalties to other companies.[18]

Philips and the Euro-Cartel

Since the last decade of the 19th century, the world electric light bulb manufacturing industry was experiencing more and more competition. Due to the fact that a large number of manufacturers had joined the industry, selling prices of electric lamps were falling in the U.S. and Europe. It was this price war that made more and more small manufacturers go out of the industry and large firms were being threatened as well. Another effect of this price war was that the quality of the lamps was deteriorating (Heerding, 1988, p. 16).

In response to these conditions, the American electric lamp manufacturers, led by General Electric (GE), formed the first cartel in the history of the electric lighting industry called The Incandescent Lamp Manufacturers Association. The participants in the cartel reached an agreement on fixed selling prices and divided production quotas. At that time, however, the European market appeared to be different because of the differentiation in national culture and economic interests. Although some small electric lamp manufacturers continuously tried for several years to realise a similar kind of agreement to the American cartel, the leading manufacturers, like AEG and Siemens and Halske, were more interested in free competition because they were more diversified, and high volume sales of lamps with lower prices could enhance their other business sectors.

Despite the initial reluctance of the big manufacturers, the European electric lamp cartel was finally formed. In September 1903, under the agreement of the AEG, Siemens–Halske and Philips, later joined by 13 smaller European factories, the VVG (Verkaufsstelle Vereinigter Gluhlampenfabriken)—the Sales Organisation of the United Incandescent Lamp Factories—was established with its headquarters in Berlin. The agreement of this cartel specified that: (a) each of the component factories had a fixed percentage in the sale of lamps made by the central organisation, as indicated in Table 3.6; (b) members of the cartel were expected to provide the organisation with lamps at the standard price of 23 pfennigs; (c) profits made by the cartel were to be shared by the members in accordance with a carefully framed schedule; (d) the light bulb life-span was also stipulated by the cartel agreement.

The establishment of the VVG cartel was a major event in the history of the European electric lighting industry at that time. However, problems remained unresolved. According to the study by Heerding (1988), VVG only represented German, Dutch, and central European producers; the French electric lamp industry refused to join this German-dominated organisation and the British manufacturers chose to form their own organisation. Subsequently, the British electric lamp industry, under the leadership of General Electric Company Ltd (GEC), established the British Carbon Lamp Association as a counterpart to the VVG in 1905 (Heerding, 1988, pp. 20–21). Why did the French and British producers choose to stay outside of the VVG at the early stage of 'cartel movement'? The reason was sociopolitical:

> A striking aspect of the British and French attitudes is that they contain a hint of political bias against their German rivals (Heerding, 1988, p. 92).

Table 3.6. Quotas Allotted to the Companies in the Cartel, 1903.

Company	%
AEG	22.633
Siemens & Halske	22.633
Vereinigte	11.316
Philips & Co.	11.307
'Watt'	7.134
Pintsch	6.579
Kremenezky	6.010
Goossens & Pope	3.196
Gelnhausen	2.040
'Constantia'	1.927
Glühlampenfabrik Zug	1.626
Bayerische Glühlampenfabrik	1.259
Total	97.660

Note: The remaining 2.34% was held in reserve for new participants in the future.
Source: Heerding (1988), p. 93.

This social–political approach helps to clarify this issue. It is obvious that neither the British nor the French electric lamp manufacturers were prepared to join a European combine dominated by the Germans at that time. Indeed national rivalries at a European level continued to be an important part of the explanation of the history of the electronics industry. It was not only cartelisation at the early stage of the European lighting industry, but also the later development of industrial collaboration amongst European organisations in the consumer electronics industry, which substantially characterised by this national rivalry.[19]

VVG, the first cartel in the history of European electric lamp manufacturing, existed for about 10 years until 1914, when the First World War happened. During this period, managers, like Anton Philips from Philips company, contributed very important management work in turn with key figures from other European firms in achieving the objectives of the cartel; on the other hand, production and sales of electric light bulbs by Philips as well as the company's organisational structure was inevitably influenced by the VVG.

The relationship between the European cartel, VVG, and Philips seems to have been very complicated. However, the influence of the former, either positive or negative, on the latter can be summarised by the following aspects.

Firstly, the cartel allocated specific sales/production figures for its members and in addition fixed the sales price for carbon filament lamps. Therefore, price competition between Philips and other members was eased or eliminated. According to Heerding, from the point of view of the producers, the positive aspect of doing so is that the cartel action "made an extremely useful contribution to the much-needed standardisation of lamps and their quality, and to the creation of regular trade practice".[20] However, such claims must define the meaning of the term 'regular trade practice'. In reality, any reduction or elimination of price competition could only be realised at the expense of the consumer. Therefore, the effect of cartelisation may be 'positive' on the side of manufacturers but was undoubtedly negative on the consumers side.

Secondly, under the above circumstance, Philips had to seek alternative ways of competing with the other members of the cartel. For instance, Philips was able to increase its quotas or share of the existing market only by means of taking over the small manufacturers (like Constantia) who were also members of the VVG. To expand its business, Philips had to create new market areas beyond the European continent (like Britain and its colonies). As a matter of fact, competition beyond price policy amongst the members of the cartel was never completely avoided:

> Admittedly, price competition between the members had been eliminated, but in its place came other forms of rivalry. These mainly took the form of intensified marketing activity and continuous modernization of manufacturing processes (Heerding, 1988, p. 123).

Thirdly, Philips was forced to commit more resources to R&D in order to develop new materials and produce new products. In this aspect, Philips acquired considerable success. Because the scope of the regulations of the VVG only concerned the area of carbon filament lamps, Philips was free to enter sales and production of other lamps based on new technologies. The launch by Philips of tungsten lamps, which were categorised as metal-filament lamps, proved to be a great success at that time.

Finally, the establishment of the European carbon-filament lamps cartel and its practice played a very important role in shaping the structure of the Philips company in its early stages. The company was forced to stretch its marketing activities beyond Europe by means of appointing local sales agents, which was a very important step in the internationalisation of Philips. Moreover, Philips was forced to diversify its product areas (from carbon-filament to metal-filament lamps) as well as rationalise its production process. Heerding (1988) gives a positive assessment of the influence of the VVG on Philips:

> In a technical sense, membership of the VVG gave Philips & Co. an opportunity for continuous rationalization of their factory—which was designed for a relatively high output—and, as a result, further reduction of manufacturing costs. There can be no doubt that membership of the cartel, far from weakening the company, made it stronger (p. 123).

The official historian of Philips describes cartelisation as one of the two most important factors which shaped the company's organisational structure.[21]

Philips and the 'Phoebus'

After the First World War, General Electric's dominant position in the world lighting industry began to come under increasing challenge from several fast growing companies, in particular Philips. In order to protect its vested interests, mainly the vast American consumer market, GE, through its wholly-owned subsidiary, the International General Electric (IGE), made great efforts to bring about a new international market structure in the form of cartelisation. After strong promotion by GE and Osram, the Convention for the Development and Progress of the International Incandescent Electric Lamp Industry was signed by the world's leading electric lamp manufacturers including Philips of Holland on 24 December, 1924. The signing of this Convention finalised the formation of Phoebus S. A. Compagnie Industrielle pour la Développement de lÉclairage (Phoebus Limited for the Perfecting of Electric Lighting), the cartel of world electric lamp manufacturers. Under the original agreement of its members, Phoebus was due to expire by 1934, but later it was extended to 1955. However, the outbreak of the Second World War put an immediate end to the cartel in 1939 (Bright, 1949, p. 307).

The main purpose of the cartel, as stated in the cartel Agreement, was

> ... the cooperation of all signatories for a national use of their production capacity of filament lamps, the maintenance of an equal high standard of quality, an economic distribution of their products, an increase of the yield of electric illumination and the development of the use of light, taking the interests of consumer into consideration.[22]

In their intensive study about the 'cartel movement', however, Stocking and Watkins have suggested that a very different set of motives lay behind the cartel:

> Although the industry was nearly half a century old when the cartel was born, the business had developed from the outset on a monopolistic rather than a competitive pattern. The cartel was the culmination, not the inauguration, of a program to avoid competition in the manufacture and sale of electric lamps (1946, p. 304).

Phoebus was organised with an Administrative Board and a Board of Arbitration as its permanent administrative agencies, and four divisions—accounting, sales, propaganda and technical—as its functioning departments. Phoebus' major policy issues were discussed and decided by the General Assembly of cartel members. Voting power in the General Assembly was weighted according to each member's capital stock, and Philips was ranked second next to Osram of Germany.[23] Phoebus' policy was mainly focused on the following aspects:

Firstly, under the Phoebus rules, world markets for electric lamps were divided into three categories: home market, overseas market, and common market. In accordance with this classification, American markets (including

Canada, South America, and the U.S. itself) were the territory of GE, which the European producers were not supposed to enter; European markets were the territory of the European producers and GE was supposed to remain outside there. Moreover, each big producer was also allocated with 'home markets' beyond its national border. For instance, Philips' home markets were the Benelux countries, Italy (shared by Osram and Tungsram), and Spain (shared by Osram and Compagnie des Lampes).

Secondly, Phoebus set export quotas for its members. These quotas were transferable amongst cartel members, but only after approval from the General Assembly had been obtained. Under the Phoebus rules, penalties were applicable to members whose sales exceeded the assigned quotas and these penalties were distributed afterwards amongst members whose sales levels were under their quotas (Stocking and Watkins, 1946, p. 337).

Thirdly, as far as Phoebus' price policy was concerned, the cartel did not directly fix prices for each type of electric lamp, but advised the associated national assemblies on the level of prices for local markets. Thus Phoebus was able to control prices in an indirect way.

It was inevitable that the increased transnational activities undertaken by major firms would cause various international trade conflicts which could not be handled at a national government level or satisfactorily settled by national legal systems. Before the creation of the EEC, cartels or syndicates were mechanisms to coordinate the interests and reduce conflicts amongst producers in Europe as well as other parts of the world. Philips, in those days, missed no chance to participate and have its interest represented in the 'cartel movement'. Due to the fact that almost every cartel in the electric incandescent lighting industry was engaged in market division and price control, the producers' interests were secured at the expense of the consumers' interests. Moreover, free market competition was normally sacrificed because of the cartels' monopoly over production and trade practice.

In the case of Phoebus, the connection between the three biggest producers, GE, Osram and Philips, was very close and they signed cooperative agreements with each other. Consequently, Phoebus was mainly manoeuvred by these three companies and thus the world markets were divided amongst them and a few other manufacturers. A company official from Sylvania of the U.S. once described Phoebus as follows:

> ... the fact is that in the world at large the more important electrical interests, such as the G.E., Siemens [Osram] of Germany, Philips of Holland, etc., are closely bound together in a cartel with the result that they have entered into binding agreements, apportioning world markets between the respective companies.[24]

As a consequence of the world markets being divided and monopolised by the big companies by means of cartelisation, small companies within the cartel, like Sylvania, were driven into very disadvantaged market positions.

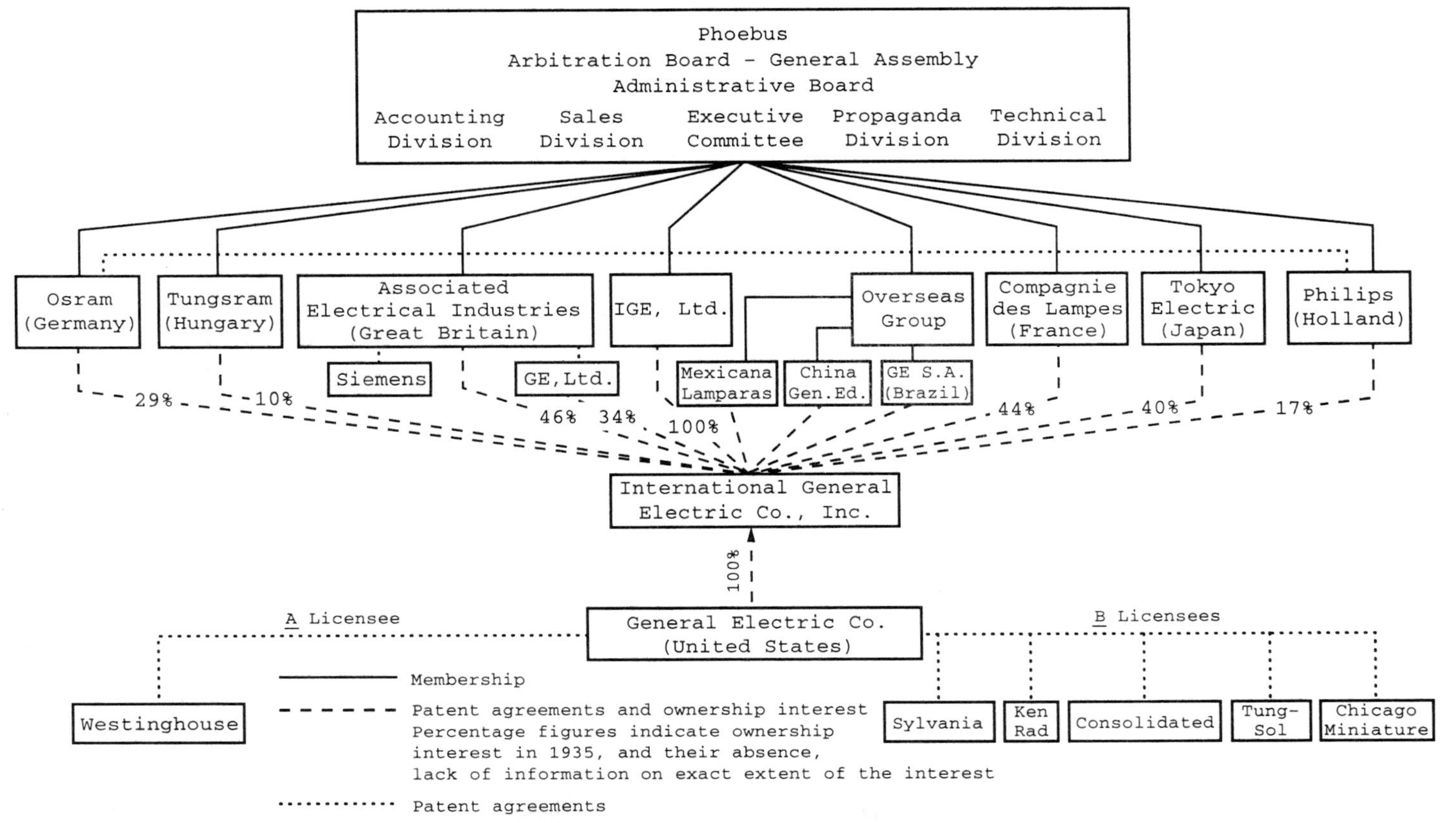

Source : Stocking and Watkins (1946), p. 334.

Fig. 3.1. International Incandescent Electric Lamp Cartel, 1935–1939.

The Largest Exporter

Compared with General Electric in the U.S. and AEG and Siemens and Halske (later these two German producers were merged into a single company Osram) in Europe, Philips was a latecomer and much more vulnerable in the world electric light bulb industry in the company's early days. However, the industry was surprised by the fact that Philips became the world's largest exporter of incandescent lamps by the year 1929, when Philips, the sole exporter of The Netherlands, realised an export value of lamps of $11,171,000 compared to total German exports of $8,927,000.[25] Philips kept this world export leader status until 1937 when the value of German exports exceeded that of The Netherlands. Philips was able to secure this position of leadership in exporting incandescent lamps in the following ways.

Firstly, Philips was established as an export-oriented company from the very beginning of its business practice due to the small size of its domestic market.

Secondly, Philips did not lose any chance to expand the production and export of tungsten filament lamps, which were not covered by the European Carbon-Filament Lamp Agreement, of which the company was a member.

Thirdly, Holland's neutrality during the First World War enabled Philips to expand its export of lamps to continental Europe and to other foreign markets at enormous profit. Bright (1949) argues,

> During World War I N.V. Philips increased its relative importance in the international lamp industry. Markets which had customarily been supplied by Osram and other belligerents were seized by Philips, and the Dutch company retained a considerable proportion of that trade after the end of the war (p. 403).

The point that Philips was able to acquire a considerable expansion of its foreign markets largely at the expense of the German industry has also been mentioned in Stocking and Watkins (1946).[26]

POSTWAR RECONSTRUCTION AND FURTHER EXPANSION

World War II not only led to the end of the operation of the international electric lamp cartel, the Phoebus, but it was also a destructive catastrophe for most of its European members, particularly N.V. Philips of Holland. According to the assessment of Anton Philips, 25% (967,000 square feet) of factory space and 72% (233,000 square feet) of the office space had been lost.[27] Many important auxiliary buildings such as the polyclinic, the trade school and the telephone exchange had been completely destroyed.

After the war years, Philips immediately entered another period of rapid expansion. Due to insufficient labour and factory space in Eindhoven,

Table 3.7. Philips' Postwar Growth, 1948–1957.

Year	Sales		Profit after tax		Current assets	Number of employees
	Amount	Increase	Amount	% of sales		
1948	651	31%	12	1.8%	614	81.000
1949	830	27%	16	1.9%	736	79.000
1950	982	18%	48	4.9%	821	90.000
1951	1,242	26%	63	5.1%	1,199	99.000
1952	1,384	11%	64	4.6%	1,276	95.000
1953	1,601	16%	88	5.5%	1,378	106.000
1954	1,936	21%	119	6.2%	1,481	123.000
1955	2,280	18%	148	6.5%	1,916	143.000
1956	2,686	18%	158	5.9%	2,135	152.000
1957	3,177	18%	187	5.9%	2,442	160.000

Notes: (1) All amounts in millions of guilders; (2) number of employees was accounted at the end of the year.
Source: Based on *Philips Annual Reports*.

Philips' production activities were spread to nearby cities, both in Holland and Belgium. This expansion was mainly concentrated on labour-intensive production processes and activities in which Philips had already experienced. Table 3.7 presents general information describing Philips' postwar business 'boom'.

Philips' postwar business boom happened not only in The Netherlands but also abroad. The rapid increase of Philips' factories and sales organisations was a convincing illustration of the company's expansionist policy during the postwar era. Table 3.8 compares the 1940 and 1950 figures concerning the increase of factories and sales organisations outside Holland.

According to Table 3.8, the number of Philips' production factories and sales organisations abroad both increased by about 70% during the period from 1940 to 1950. Meanwhile, Philips increased its domestic production capacity as well. According to the company's own records, Philips opened new production factories in eleven new locations outside Eindhoven in The Netherlands.[28]

Table 3.8. Philips' Factories and Sales Organisations Abroad, 1940, 1950.

Year	Factories	Sales Companies/Agents
1940	89	90
1950	152	1523

Source: Based on Philips (1991): *Philips: A Century of Enterprise*, *op. cit.*

The main characteristics of Philips' postwar expansion were characterised by the following aspects.

Firstly, a new production structure involving the spreading of export process to underdeveloped regions was developed.

Secondly, this new production structure gave a new dimension to Philips' plant location policy. Compared with the prewar plant allocation policy,[29] Philips' postwar expansion policy was characterised by a new objective which was to relate the geographical spread of production activities to various labour markets.

Thirdly, the postwar expansion was also characterised by Philips' product diversification which was mainly based on Philips' scientific research and technical innovation. This point will be further elaborated later in this chapter.

TECHNICAL CHANGE AND DIVERSIFICATION

As mentioned at the beginning of this chapter, right from the outset Philips was a company based on R&D. The company's continued commitment to scientific research and technological innovation, which were mainly carried out by Philips' experimental laboratories and research centres, allowed the company to sustain a growing dynamism and enabled it to realise its expansionist policy at certain stages of its history. For instance, in Philips' early days, Gerard Philips succeeded in experimenting with tungsten as a new material to produce metal filament lamps which later replaced carbon filaments. Following this, new machinery had to be installed, and new staff recruited. The metal filaments demanded five times as much labour as carbon filaments, but the new lamp gave three times the amount of light for the same amount of power (Bouman, 1970, pp. 54–55). Having innovated this technology, Philips was assured of a very strong position in international competition in the lighting industry. Meanwhile, the fact that technical changes gave birth to new products had caused Philips to rapidly diversify its production from lighting to numerous consumer electronics and many other products.

R&D within Philips

In retrospect, R&D remained to be the driving force that underlay Philips' expansion process, and from the beginning the company was committed to carrying out R&D in-house.

Philips' Laboratories

Even before the establishment of Philips & Company, Gerard Philips had set up a research laboratory with financial support from his father. Having been joined by a group of chemists, mechanical engineers and electro-technical engineers, a chemical laboratory was built ten years after the company's establishment in order to overcome the technical difficulties facing the production of metal filament lamps.

Later in January 1914, a physics laboratory was set up by Philips under the supervision of Gerard Philips. The main purpose of this physics laboratory was to achieve the production of small size half-watt lamps suitable for ordinary rooms.[30] From then on, Philips' physics laboratory was intensively involved in research on X-ray, radio, gramophone, television, audio/video tape recorders, etc.

Research Centres

Eindhoven has always been the most important location for Philips' scientific research activities. However, Eindhoven was not the only place where the concern's technical breakthroughs occurred. The significance of establishing scientific research presence elsewhere was strongly emphasised by the Board of Management in Eindhoven as early as the 1950s:

> By not confining our research work to the Eindhoven centre but by cooperating in the establishment of laboratories in other highly industrialized countries, we are strengthening the ties between these industrial centres and the potential of scientific research workers in these countries.[31]

Philips' scientific research had been decentralised, and other laboratories established after the Second World War in a number of industrialised countries. In addition, several research centres were also established by the company during the period of postwar expansion. For instance, Philips built its research centres in Germany (Hamburg and Aachen), France and England, as well as in the U.S. during and after the war. The Institute for Perception Research was founded by Philips in collaboration with the Technological University of Eindhoven in 1958. Furthermore, Philips' Computer Centre was established in 1960 for central administrative tasks; its powerful facilities were available for R&D projects requiring mathematical analysis (Bouman, 1970, p. 252).

The main purpose of this decentralisation of scientific research activities, as explained by Philips itself, was to enable the company to 'give a scientific backing' to its activities in these countries; and at the same time Philips could also contribute towards the national research efforts.[32] Philips has long been recognised across the world as one of the most innovative companies with its own R&D capabilities.

Commercialisation of Technical Inventions

It is often argued that Philips was good at scientific research but lousy at marketing. The V2000 failure was a classical example to be examined in the next chapter. According to this criticism, the company's main weakness turned out to be its inability to successfully commercialise its own technical inventions.[33] This argument, according to the current research, may not be applicable to the company's history prior to the 1970s.

As shown above, Philips was established on the basis of its own scientific research, which was initially carried out by Gerard Philips. It is also true that Philips' expansion and product diversification was based on the company's technical breakthroughs. Documents show that senior managers in the early days had given attention to the crucial importance of linking together laboratory activities and corresponding marketing environments. Philips' success in a wide range of areas, such as carbon/metal filament light bulbs, medical/professional equipment, radio parts/sets, audio tape recorders/players, monochrome/colour television sets, etc., speak for themselves. Although collaborating with competitors remained an important aspect of Philips' corporate history, the company, by and large, designed and launched new products on the basis of its own scientific and engineering work. The following quotation is part of a press interview with Anton Philips, which indicates the attitude of a Philips manager towards the relationship between R&D and marketing:

> If our laboratories are offered the possibility of a new product, we discuss the points involved. I put the questions, Is it the sort of thing we can handle? What type of customer would it find? What profit would it make? Has it been exploited by others to an extent that might spoil our chances? Is its potential sale sufficient to interest us? Or is it an article of importance only to a small circle but likely to enlarge our sales of other products in wider demand?[34]

This view reflects the historical experience that technical change and innovation had made it possible for Philips to successfully realise its pre- and postwar expansionist policies. The vital importance of scientific research and its relationship to the company's expansion and diversification was again elaborated by the Board of Management in the 1960s:

> Scientific research is the cornerstone of our constant efforts to find new fields of activity and to broaden the existing ones. This research work, the foundation for which was laid in 1914 in The Netherlands, is also undertaken on a wide scale in the laboratories of enterprises associated with us abroad, among other countries in France, Germany and Great Britain.[35]

The rapid increase of Philips patents, as shown in Table 3.9, has well demonstrated the company's long-term commitment to R&D.

As we shall see from the following section Philips succeeded in diversifying its product range by building market success onto its R&D activities. The criticism referred to above seems to be justified only for the period from the 1970s onwards when Philips began to face serious competition from the Japanese, to be shown in Chapters 4 and 5.

Table 3.9. Patents Owned by Philips.

Year	Patent Number
1905	1
1930	3000
1932	4500
1934	6000

Source: Based on various documents.

Historical Review of Philips' Consumer Electronics Industry

Philips diversified its production activities from the lighting industry into a range of sectors. The consumer electronics industry, in particular, has became one of Philips' most heavily involved areas.

Radio

Experiment and production of commercial radio actually began in the last few years of the 19th century. In the early days, the various methods of generating and receiving radio waves were usually regarded as systems. The waves used by one system were usually different from those used by other systems. All the early systems were spark systems. Spark systems depended upon the oscillatory nature of the current when a condenser discharged across an air gap (Sturmey, 1958, p. 21).

During that time, the most important systems were those of Marconi, the Telefunken, Lodge-Muirhead and those designed by Nevil Maskelyne and J.G. Basillie in England, Lee de Forest and R.A. Fessenden in America. These systems were closely allied to each other and made use of methods covered by the Lodge patent of 1897 and the Marconi patents of 1896 and 1900 (Sturmey, 1958, p. 22).

The discovery of the thermionic valve by J.A. Fleming in 1904 made valve transmitters important and radio telephony practical during the First World War (1914–18). The newly invented valve transmitters were mainly used for communication between the ground and aircrafts during the war.

During the first quarter of this century, major efforts were made to get more power into the aerial in order to increase distance. The development of valves reduced some of the importance of the search for higher power, and represented a revolution in the history of radio, as well as in the consumer electronics industry.

Based on its own scientific research Philips successfully built its first radio valve, which it marketed in 1917, and then began to sell radio sets in Autumn 1927. This event signified an entirely new era of production in Philips' history. By supporting the state-owned broadcasting network in

The Netherlands, Philips promoted the demand for radio sets and entered into mass production. Between 1927 and 1929, radio sales increased dramatically, and the profits for these years was 52 million guilders.[36]

In exactly the same way as happened in the lighting industry, Philips maintained innovation as the most important impetus in the radio industry. It was innovation in manufacturing techniques that made it possible for Philips to reduce the price of valves and to popularise their use. By the mid-1930s, Philips patents were so extensive that the producer of a good quality radio set could not avoid using them. It is striking that Philips became one of the most powerful competitors of Marconi in the world market in the 1930s. After Philips acquired half the shares in Mullard of the U.K., radio set makers had to choose between the Marconi and the Philips–Mullard licences. In the 1940s, radio production became one of the most important areas within Philips' Appliances Industrial Group.

Apart from the constant efforts made by laboratory staff, Philips management reorganised the company's structure to stress the significance of the radio business. On 1 June 1925, Philips Radio was established as a legally independent firm.

Tape-Recorders

Tape-recorders were not mass-produced until the 1950s. They gradually gained acceptance by becoming smaller, lighter and easier to handle. Nowadays, tape recorders are commonplace due to the design and production of small portable recorders. Philips launched its first tape-recorder for the domestic market in 1952.

The tape cassette, originally launched under the name of 'pocket-cassette', was a Philips invention. In 1963, Philips introduced its compact cassette for audio and this was followed by the launch of the company's first cassette recorder as well as the pocket recorder in Vienna during the same year. Soon after, Philips licensed its invention to other companies and this new product became a worldwide standard. Immediately, the Japanese entered this field and became Philips' most powerful competitor. Tape recorders are now manufactured all over the world according to the Philips system.

Television

Scientists and engineers at Philips started their research work on television in 1925. Three years later the first Philips television tube was developed and the company used it to demonstrate its moving pictures on a television screen with 48 horizontal resolution lines.[37] After another decade's R&D work, Philips succeeded in producing its first commercial receiver with 405 horizontal lines in 1939. The success of the 405 line receiver was a very important event not only for Philips but also for Europe as a whole as far as monochrome television was concerned.

Unlike various domestic appliances and personal care products, consumer electronics products consist of two aspects: hardware and software. More specifically, radio sets and television receivers without the support of transmitted programmes were empty boxes. Therefore, Philips committed great efforts in experimenting and promoting TV programme transmissions since 1948. On 18 March 1948, the company's first test transmission of TV signals was carried out by using 567 lines for picture elements and an FM system for sound information. In 1950 Philips manufactured a transmitter which successfully sent out the first TV programme using 625 lines. This transmission system (625 lines) has been adopted by most European countries as a standard format. Having succeeded in experimental transmission, Philips started public transmission the following year.[38]

Almost in parallel with the development work and promotion campaign on its monochrome television, Philips was also involved in Colour TV (CTV) research. The company made its first public demonstration of Colour TV in 1959. Just as it did with monochrome TV, Philips made efforts in experimental transmission of CTV from Eindhoven for three years from 1964.

Sales of television sets in The Netherlands were not very encouraging at the beginning. In 1953 only 500 receivers were installed to receive the Dutch Television Foundation (NTS). This situation was improved very quickly. In 1960 there were more than 600,000 Dutch homes watching TV programmes.[39]

Video Cassette Recorder

Philips was one of the pioneers in developing video recorders since the principles of recording pictures and moving images in an electronic way were worked out in 1959. In the following ten years video recorder products were mainly designed and produced for the professional market because of their high cost and bulky size. As a frontrunner in this technological area, Philips introduced its first video tape recorder in 1964. However, it was not until 1970 that Philips succeeded in developing a cassette-based video recorder. An improved version of the Philips VCR consumer model became available on the market in 1972. Thereafter, 'VCR', short for 'video cassette recorder', has become the standard name for this category of products, although Philips' technology failed to become the world standard.

The fact that, being a leader at the precompetitive stage, Philips did not translate its technology into a market success caused wide concern within and outside Philips. It seemed that successful commercialisation of a technical invention may only be realised with a proper marketing strategy. Detailed discussions on the reasons why Philips' VCR technology failed and the implications of this failure will be presented in the next chapter.

Diversification and Organisational Structure

As discussed above, R&D was regarded as the most important area by Philips management, and this allowed the company to diversify its product range. Diversification brought changes in the firm's organisational structure.

As early as in 1927, Philips' production areas spread over the following allied industries:

- a glass works
- a tool plant
- a rare gas factory
- a paper and cardboard works
- a printing works
- a building firm.

In 1946, the traditional management style was reformed, namely, the Board of Directors was turned into a Board of Management and top level management responsibility started to be shared by a number of experts rather than being in the hands of a few people who were either a member of the Philips family or closely related to the family. In the same year, the Philips company was divided into several industrial groups, combining related products. These industrial groups were as follows:

- Lighting
- Appliances (later Radio, Gramophone and Television)
- Telecommunications (later Telecommunications and Defence Systems)
- X-ray and Medical Apparatus
- Allied Industries and Glass
- Pharmaceutical-chemical Products.

After the 1940s, some new industrial groups were founded within Philips:

- The music group in 1950
- ICOMA (Industrial Components and Materials) in 1950
- Domestic Appliances in 1957
- Computers in 1964
- ELCOMA (Electronic Components and Materials) as a combination of the interests of the main industrial groups Electron Valves and ICOMA in 1965.

Product diversification had enabled Philips to successfully pursue its expansionist policy. An outcome of the rapid diversification process was the dramatic increase in Philips' annual turnover—fivefold over a decade. In 1950, the turnover of the company was 982 million guilders (approximately £113 million), while, in the 1960s it changed dramatically as shown in Table 3.10.

Table 3.10. Philips Turnover, 1950–1965.

Year	Turnover	
	Guilders (million)	£ (million)
1950	928	113
1960	4,762	550
1961	4,936	567
1965	7,545	867

Source: Based on various documents.

Diversification of production, however, also had its negative effects. In the process of successful technical change from carbon filament to squirted-metal filament and then to drawn-metal filament, electric discharge lamps and fluorescent lamps, and from the lighting industry to X-ray products and then to radio, tape-recorder, television and video, Philips had to make frequent changes in its manufacturing processes, which involved substantial costs in writing off old plants and machinery (Bouman, 1970, p. 74). These losses can probably be regarded as the inevitable costs of technical progress. Philips may not have changed from a family business into a multinational company without making enormous investments in R&D and the process of diversification.

FORMATION OF THE PHILIPS MATRIX STRUCTURE

Having experienced the expansionary postwar boom, Philips became a fully-fledged multinational enterprise and was ready to take the lead in the European consumer electronics industry by the end of the 1960s. As far as the organisational structure was concerned, Philips was born with a matrix configuration. Once the company started to face fierce competition from the Japanese in the 1970s, this giant matrix became one of the major issues—the characteristic to be blamed for the company's inflexibility and bureaucracy, and was consequently identified as the main target of the long-lasting restructuring process in the 1980s and early 1990s.

The Dual Management Structure

As mentioned above, Philips was established with a unique dual structure in which commercial management and engineering management were along two parallel hierarchies. This characteristic of the Philips organisation has had a profound impact on managerial practice throughout most of the company's history. Even if the parallel hierarchies once had their positive function (for instance, the two brothers' 'race' had promoted the company's expansion), the negative aspect of such a structure may not be ignored. For example, In the early days, the two brothers—Gerard Philips and Anton Philips—often came into conflict over the management of the company. This was followed later by the long-lasting conflicts between the National Organisations (NOs) and Product Divisions (PDs). The internal conflicts caused by the demarcation between the technical and the commercial domains were not always in accordance with the interests of the company as a whole. It was for this reason that Philips started a profound and expensive restructuring process.

Figure 3.2 briefly indicates the historical development of Philips' dual organisational structure. To understand Philips' organisational complexity, two layers of relations must be taken into account. The first layer of Philips' dual structure was the separate development and independent existence of the sales networks—sales agents, sales companies, and National Organisations—and the various industrial or production side of organisation—the origin of divisions, industrial groups, and Product Divisions.

The sales agents, the first generation of Philips' international sales network, came into existence in 1906. After the start of the First World War these local sales agents had to stop working for Philips, and the company changed its policy by setting up sales companies in foreign countries and employing its own people. There were three sales organisations in foreign countries between 1919 and 1927, while this was increased to nine countries in 1930. When these foreign sales companies were finally developed into National Organisations after the Second World War, each NO was managed by two managers who took on the commercial and production responsibility, respectively. In parallel, we see the origin of the PDs in the first ten years

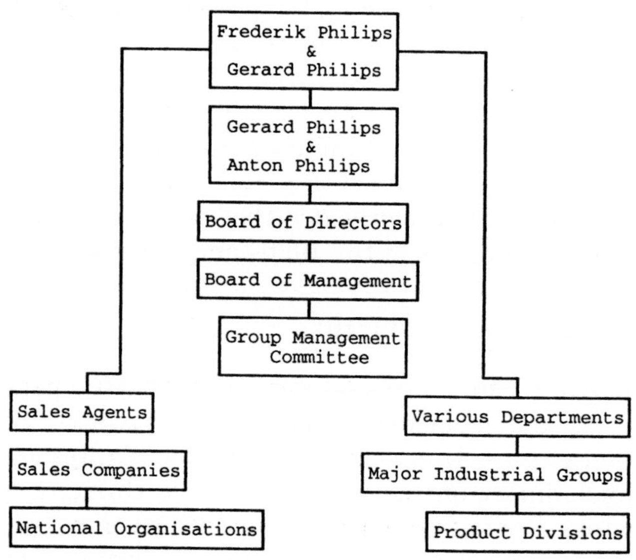

Fig. 3.2. The Origin and Evolution of Philips' Dual Management Structure.

after Philips was founded. Along with the development of product diversification, several industrial groups were formed in 1946, and these industrial groups were developed into Major Industrial Groups (HIGs) in 1948. Each HIG covered a number of similar products with respect to manufacture and/or application.

The second layer of the dual structure was the separation of commercial and engineering/production responsibilities not only at the top management level but also in the HIGs and PDs. As mentioned earlier, this dual decision-making system could be traced back to the very beginning of the company's history. The two Philips brothers, Gerard and Anton, were holding the engineering/production and commercial responsibilities, respectively. Later, each HIG was headed by both a technical and a commercial–economic director. The management of the PDs was also carried out along the same line since the end of the Second World War when Philips top managers, including Anton Philips, returned back from the U.S. to Eindhoven.[40] In Philips, this dual-headed management used to be regarded as division of labour and, it was a process involving what Philips called 'constructive friction'.[41] It was only in the early 1980s that this traditional management style was seriously challenged and subsequently reformed. A detailed discussion of this point is presented in Chapter 5.

Matrix: The 'Ghost' of the Philips Structure

Philips' organisational structure was usually seen as a 'matrix' with overlapping National Organisations and Product Divisions in addition to the twin hierarchies. On the one hand, the matrix structure of Philips was based on the long-term coexistence of NOs and PDs. On the other hand, the matrix was characterised by the separation of technical and commercial management.

Due to the fact that both the NOs and PDs were well established as horizontal and vertical pillars of Philips' corporate structure, it appeared extremely difficult for the company to reform the matrix structure. The matrix gradually took on the character of a 'ghost' to which Philips became deeply bonded for several decades. Therefore, to restructure or reform the corporate structure does not mean that the company could simply get rid of the 'ghost' but that it should regain its management efficiency and competitiveness by means of 'tilting the matrix'.

It is believed that the move to organise corporate operations by way of product divisions was started in 1921 by Du Pont and General Motors in the U.S. and this concept was adopted by the majority of American firms during the 1950s (Franko, 1976, pp. 198–199). As far as Philips was concerned, its management brought back the product-division concept from the U.S. and further adapted it into a unique corporate matrix structure:

> Philips and Ciba managers brought back to Europe the product-division concept, with its emphasis on assigning profit responsibility for operating decisions to managers of self-contained business units. ... Philips assigned joint divisional responsibility to coequal technical and commercial managers. Philips further adapted the divisional concept into the matrix structure ... by confronting the two coequal technical and commercial managers of product divisions with yet other coequal technical, commercial and financial managers responsible for all products within individual nations (Franko, 1976, p. 199).

An immediate result of the formation of the matrix structure within Philips was the complexity of the company's reporting system. As a matter of fact, "at every significant management level, every manager had at least two bosses".[42] Consequently, the second result was that various conflicts, within each PD and NO and between the PDs and NOs, were created. It was inevitable that constant corporate conflicts caused inflexibility and inefficiency.

When interviewed, a Philips company historian claimed that the separation of production and commercial responsibilities between the two brothers at the early stage of the company's history had nothing to do with the formation of the matrix structure and the various problems arising from this; on the contrary, the separation of responsibilities was 'a division of labour between them', and 'these two brothers worked together and

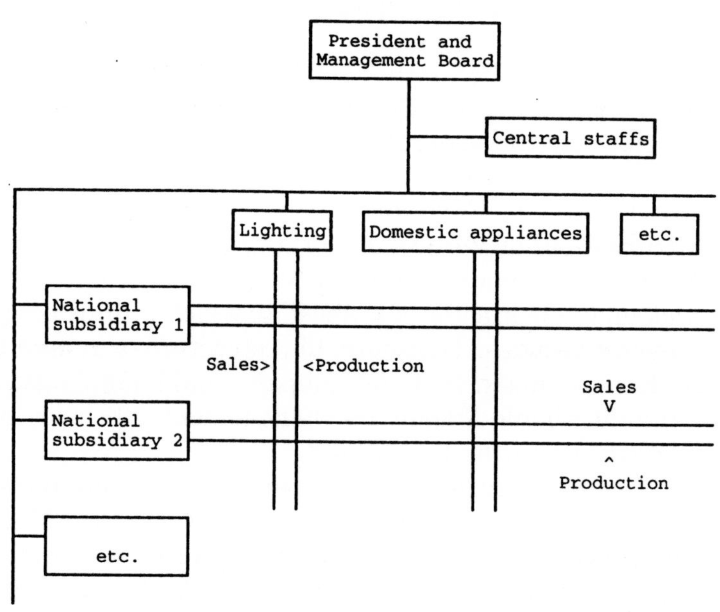

Source: Adapted from Franko (1976) p.196.

Fig. 3.3. The Philips Matrix.

cooperated very well'.[43] After the formation of the matrix structure, however, things began to change:

> Because of the division of the markets, the company structure was very much defined by the National Organisations [NOs]. We had that structure which was the famous Philips Matrix. When the divisional organisations [PDs] met the regional organisations [NOs], problems always rose. That's the problem of coordination. It is impossible to give instructions from Eindhoven to everywhere all over the world. In the seventies this problem became very hard for the company. ... Because of the fact that we were going through the postwar expansion boom, nobody could see the problem of the company's structure.[44]

Having failed to diagnose and overcome the problems related to the matrix structure at an early stage, Philips fell into a tremendously difficult situation in the 1980s and this, to some extent, contributed to the commercial failure of the V2000 system and the severe financial crisis at the beginning of the 1990s.

Philips' Structural Change and its Organisation

Philips' historical expansion was symbolised by several structural changes. The first happened in 1912 when Philips & Company was transformed from a simple family company into a more complicated financially regulated structure and renamed N.V. Philips' Gloeilampenfabrieken.

The second structural change occurred in the 1920s when Philips moved a large proportion of its production from Holland (mainly in Eindhoven) to other countries through the establishment of foreign ventures including sales offices, acquisition and new plants. It was in the 1920s that Philips developed its first international production structure.

In the 1930s, the worldwide economic crisis forced most countries to increase tariff barriers and tightened their foreign currency restrictions in order to protect their own economies. As an export-oriented company, Philips had to increase its workforce in foreign countries and reduce the number of its Dutch employees for the sake of avoiding the high tariffs. As a result, Philips' production was transferred abroad on a large scale. Philips formed its second international structure, which could be regarded as Philips' third structural change.

During the postwar expansion period, especially due to the creation and growth of the EEC, Philips restructured its plant allocation strategy and forced its original national organisations into a specialisation process in Europe. This specialisation provided Philips with the opportunity to explore economies of scale, and gave birth to its fourth production and managerial structure. The impact of the EEC on Philips will be further discussed in the following section.

In the late 1960s and early 1970s, Philips entered a more mature period with a very complicated production and managerial structure (Fig. 3.4).[45] During this period, Philips' organisation was very similar in many aspects to a national or European political structure (Bouman, 1970, p. 260). Within this structure, the Board of Management set out the main points of management policy under the supervision of the Supervisory Board. The Product divisions determined the company's policy on products and coordinated production activities. In the meantime, the National Organisations were given great autonomy to make maximum use of their local markets. In short, the structure of the Philips group during this period was a kind of loosely organised 'federation'. Figure 3.5. shows the Philips 'federation' of national organisations from many parts of the world. The term 'federation' was officially used by Frits Philips, an ex-president of Philips, to mark the retirement of his predecessor in March 1961,

> In his [P. Otten's] period of office, Philips expanded into a powerful industrial federation of large enterprises in the countries of Europe and in the major countries overseas.[46]

The official use of the term 'federation' to some extent explained Philips' positive position concerning the company's structure. However, this 'federalist' policy sometimes gives birth to badly controlled decentralisation, which could cause unexpected problems. Sometimes daughter companies might refuse to conduct its parent company's product policy, as shown by the case of Philips North America to be examined in the next chapter.

THE IMPACT OF THE CREATION OF THE EEC

In this section I will briefly discussion the major impact of the creation of the EEC on the performance of Philips in Europe.

Philips' Response to the Creation of the EEC

It seemed that the creation of the European Economic Community in 1958 complemented the interests of big European firms such as Philips. The Board of Management at Philips repeatedly expressed their positive attitudes towards the formation of the Community since the late 1950s when the EEC was emerging. Prior to the formation of the EEC, the Board of Management at Philips had foreseen the potential effect of the European Community upon the company's business development:

> ... our confidence in the future is partly derived from the gradual emergence of the European Community, within the framework of which we shall have the opportunity to distribute our production more efficiently on an international basis.[47]

Having been aware of the potential importance of economic integration in Europe to Philips' competitiveness, the company's official position was

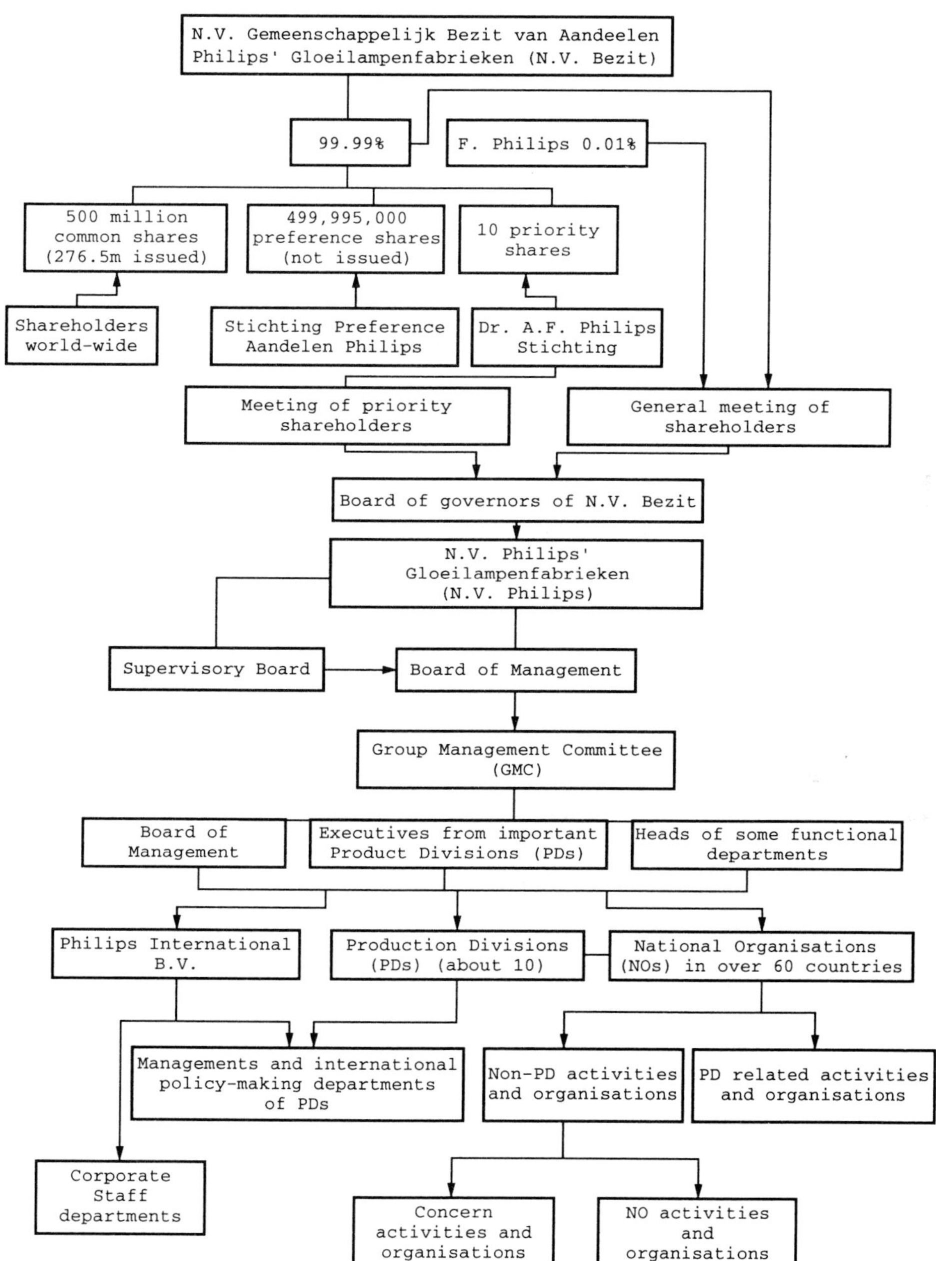

Fig. 3.4. The Structure of Philips as a Fully-Fledged MNE.

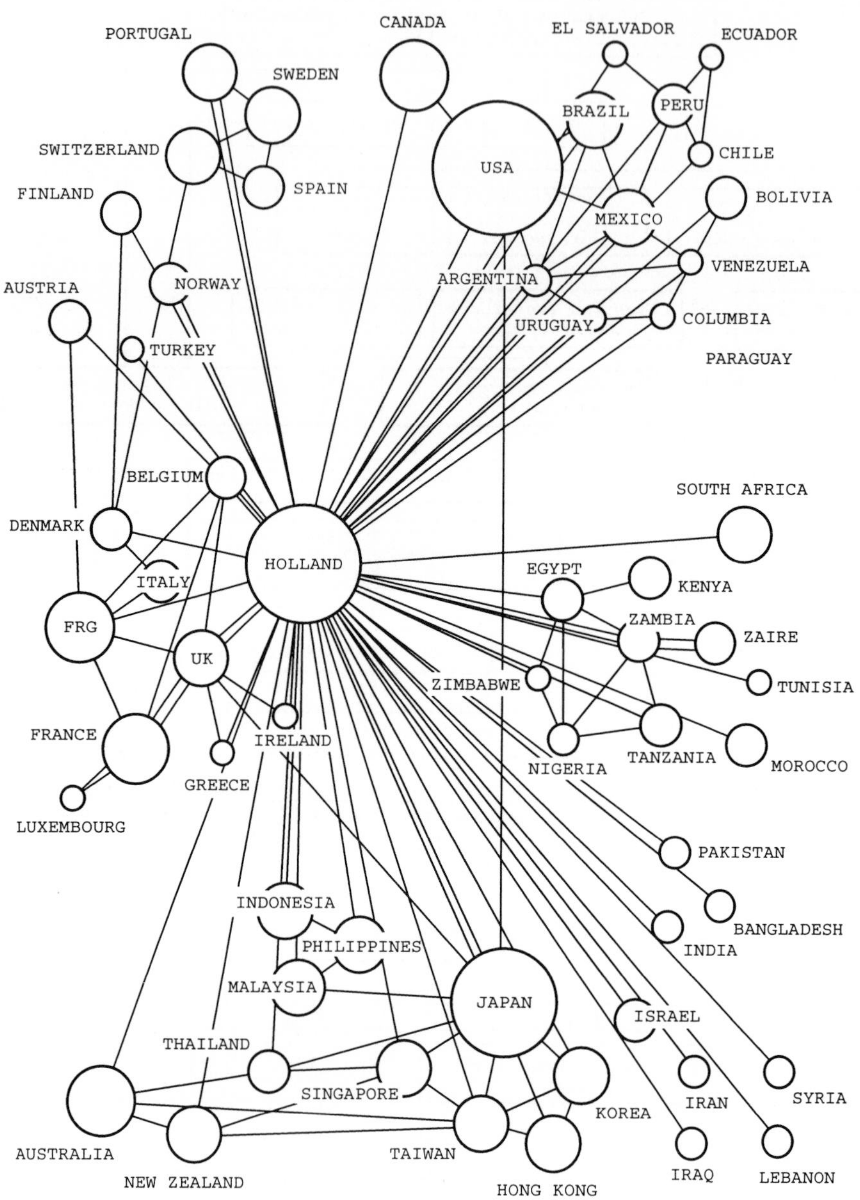

Source: Ghoshal and Bartlett (1993), p. 78.

Fig. 3.5. Organisational Units and Some of the Interlinkages within N.V. Philips.

defined to promote such a process towards possible forms of European integration including the European Community. This was confirmed in the company's own document:

> We shall be particularly concerned with inquiring into the way in which European industry can help to give substance to the existing integration plans—notably the EEC Treaty—and thus to contribute towards their success.[48]

Due to the building up of international competition in the consumer electronics industry, in particular from Japan, Philips and other European firms were quick to realise that the European Community might be able to play a more and more important role as far as international trade negotiations and regional industrial policy was concerned. In other words, the EEC, as a newly emerged political and economic structure, was regarded and interpreted by Philips as "an important means to improve the competitiveness of the European industries"; Philips saw the EEC as a 'comfortable structure' and, therefore, the company has always been interested in promoting the European integration process.[49]

To understand the relationship between European industrial organisations and the EEC, one must not forget the fact that European market integration was a very slow process, and for most of the time it was a political issue. Some Philips officials hold a critical view over this European integration process and its repercussion on industrial firms:

> The political process of the EEC is very slow. We are talking about the EEC which has been existing for more than 30 years. We are saying that the EEC will create a common market or single market after 1992. This means there is no common market at the moment. It is not a regular process. This makes it very difficult for companies to adapt because they don't know what they should adapt to and in which market. ... Even after 1992 there will be national divisions existing.[50]

Rationalisation and Economies of Scale

Most importantly, the European market integration made it necessary for Philips to reshape its European production policies and marketing strategies.

Until the end of the 1950s, Philips had already located production facilities in every EEC member country, and in each country it had been producing a wide range of finished products.

Responding to the creation of the EEC, Philips was quick to undertake a geographical specialisation process, which started in 1958 and has kept going ever since. Having realised the new circumstances, Philips started to redesign its European strategy—each European subsidiary was only allowed to specialise in a limited number of products, i.e. one Philips subsidiary could receive orders from other subsidiaries in Europe. This trend of specialisation has been enhanced along with the progress of the European integration process. It seemed to be an indispensable advantage for a multinational company located in the EEC that the common market presents an opportunity to transport products and components freely between member countries. Facing such a new competitive environment, Philips, as a whole, moved forward quickly in integrating its production and marketing activities in Europe. In the meantime, its national organisations in many other countries have been strengthened as well.

The impact of the European integration process on Philips' internal organisation was briefly commented on in the company's official document:

> The reorientation made necessary by economic integration in Europe affects the internal organisation of the Concern in many ways. The pattern of internal relationships is made more complicated, and this imposes high demands on organising skills and the readiness for international teamwork.[51]

New Opportunity for Industrial Collaboration

Another impact of the EEC on industrial firms in this region was that the European integration process provided more opportunities for industrial collaboration and mergers. Frits Philips, the then president of the company, showed his understanding and attitude towards such a new dimension of the European market in the following statement:

> Europe's economic integration inevitably led to numerous collaborations and merges. Many people view this development with suspicion, but I believe that only large combinations of European companies will be able to stand their ground against world concerns, and especially American and Japanese companies. It is rare for individual companies to possess sufficient strength for a contest in world markets (Philips, 1978, p. 217).

The acquiring of a majority shareholding of Pye, the largest exporter of radios and television sets in the U.K., in 1967 was a successful practice of Philips' ideal stated above for participating in European collaboration and mergers. In France, the French Philips organisation merged with a number of French electronic firms in 1968. Later, Grundig, the leading German consumer electronics manufacturer, gradually became Philips' daughter company. Philips' competitive position in Europe as well as in the world was, to some extent, enhanced by these arrangements.

It is worth mentioning that Philips became one of the most influential lobbyists within the EEC. In fighting against its non-European competitors, especially the Japanese and other East Asian companies, it was sometimes in Philips' interest to seek protectionist policies in the form of tariff and non-tariff measures from the European Community. This point is to be further investigated in the following chapters.

CONCLUSION

This chapter has examined the issue of how Philips grew from a Dutch family business into a large multinational enterprise and diversified its industrial production from lighting into a number of industrial sectors. What was the impetus behind this success? P.J. Bouman, the author of Anton Philips' biography, has summarised it in part as the following:

> The story ... and the impetus behind this 'industrial world federation' have been its technical quality, the triumph of applied science, the balance between centralization and decentralization from the organizational point of view, and the devotion of everyone in the company (Bouman, 1970, p. 255).

Amongst all the elements mentioned in the above quotation, the 'technical quality' and 'the triumph of applied science' might be the most important impetus during the whole historical process of Philips' expansion. Therefore, the criticism mentioned above that Philips' main weakness was its inability to successfully commercialise its technological inventions may not be justified by Philips' history as far as back we could trace, prior to the 1970s. However, the competitive environment in Western Europe changed in the 1970s and Philips was posed with new challenges from Japanese consumer electronics firms. Under this new circumstance the original Philips 'federation' with a matrix structure became increasingly unable to cope with changes. Moreover, it seemed that the long-lasting expansion and diversification process had stretched Philips too far on both the production and marketing side. It was this structural problem which handicapped Philips in its efforts to respond to intensified competition from the Japanese firms in the late 1970s and early 1980s.

The next chapter will examine the issue of how and why Philips failed to win the home VCR format battle against the Japanese.

NOTES

1. Note that 'EEC' is used throughout this chapter for the convenience of historical discussion. 'EU' is used elsewhere in this book.
2. On the occasion of the opening of factory at Roosendaal, reported in the *Philipskoerier*, 22 January 1949.
3. According to the same study, General Electric of the U.S. acquired 58 patents from Philips between the period of 1923 and 1939 under a bilateral patent-exchange agreement (pp. 310–311).
4. *Philipskoerier*, 22 January 1949.
5. Transnational Institute: TIE — Europe No. 12, p. 3.
6. Before the First World War almost all of the sales agents employed by Philips were exclusively local people. This was changed during the war, because lots of these agents and representatives stopped working for Philips during war time due to different national interests.

7. See Philips (1991) Philips: A Century of Enterprise (1891–1991), a Philips-designed CD-i disc, Eindhoven, September.
8. *Ibid.*
9. Philips Annual Report, 1930.
10. Bouman (1970).
11. *Ibid.*
12. According to Bouman (1970), Philips had its subsidiaries located in 38 different countries by 1938.
13. Note that Philips complained to the EU — successor of the EEC — about foreign companies of dumping in Europe — a practice of selling products at prices lower than their costs. This will be further discussed in the next few chapters.
14. Transnational Institute, TIE — Europe, No. 12, p. 3.
15. By the end of the 19th century and at the beginning of the 20th century, the AEG, Philips and Siemens–Halske were the three largest light bulb manufacturers in Europe.
16. International General Electric was founded in 1919 by General Electric as a wholly-owned sybsidiary to consolidate and supervise the latter's foreign business.
17. There was an exceptional occasion worth mentioning here. During the Second World War the Philips management was taken over by the Germans in those occupied countries including the Netherlands. German troops invaded the Netherlands on 10 May 1940 and the Germans inserted their military control over Philips in Eindhoven on the following day.
18. Interview with Philips Historian, Philips International B.V., Eindhoven, 28 October 1991.
19. For instance, the rivalry between Philips and Thomson prevented the two firms from joining forces for VCR in the early 1980s. More detailed discussion see the next chapter.
20. Heerding (1988), p. 122. By establishing the VVG, the possibility of monopoly by big producers, like AEG and Siemens and Halske, was, to some extent, avoided for the cartel forbid these companies to market their products at low prices. Therefore, VVG contributed to the creation of regular trade practices from the small producers' point of view.
21. Interview with Philips Historian, Philips International B.V., Eindhoven, 28 October 1991. The interviewee believes there were two important factors which created the structure of the Philips company, which he calls 'the industrial structure': the first factor was cartelisation prior to World War Two and the second one was the creation of the EEC at the end of the 1950s.
22. Cited in Bouman (1970), p. 126.
23. GE's interest in Phoebus was represented by acquiring stock-holding in each subscribed cartel members through its foreign business subsidiary IGE. For instance, IGE was holding 17% of Philips' stock under the two companies' bilateral agreement.
24. Cited in Stocking and Watkins (1946), p. 340, note 100.
25. Stocking and Watkins (1946), p. 318, note 42.
26. See p. 331, note 82.
27. Cited in Bouman (1970).

28. Philips (1991), Philips: A Century of Enterprise, *op. cit.*
29. Philips' prewar plant allocation policy was mainly oriented towards harmonising the geographical spread of its production and marketing activities.
30. Before setting up this physics laboratory, Philips had already been able to produce half-watt lamps of 300–1,500 watts used for street-lighting.
31. Philips Annual Report, 1956, p. 12.
32. See Philips Annual Report, 1964.
33. In analysing Philips company's structural problems, Hill (1982) has adopted the view of Mackintosh Consultants Ltd. to support his argument. He writes, "According to Peter Walker of Mackintosh Consultants Ltd., a firm that specializes in the electronics industry throughout Europe and in the U.S., Philips has always had problems converting first-class research capability into products that the market wants and that the company can make profits from" (Hill, 1982, p. 25).
34. *Niewwerotterdamsche Courant*, 17 December 1927, quoted in Bouman (1970).
35. Philips Annual Report, 1964, p. 10.
36. Figures are mainly from Philips (1991), Philips: A Century of Enterprise, *op. cit.*
37. In 1928 Philips achieved its first completed valve for television in its Scientific Laboratory in Eindhoven. This valve in combination with Paul Nipkow's disc and a photocell was used for the electrical transmission of a moving picture. See Philips (1991), Philips: A Century of Enterprise, *op. cit.*
38. *Ibid.*
39. *Ibid.*
40. The adoption of the concept of Product Division was very much influenced by American corporate culture.
41. Hill (1982), p. 20.
42. *Ibid.*
43. Interview with Philips Historian, Philips International B.V., Eindhoven, 28 October 1991.
44. *Ibid.*
45. By the 1960s Philips' national organisations were already operated in more than 60 countries.
46. Philips, F. (1960), 'On the retirement of Mr. P.F.S. Otten as president of the company', Philips Annual Report.
47. Philips Annual Report, 1957, p. 10.
48. Philips Annual Report, 1960, p. 71.
49. Interview with Philips Historian, Philips International B.V., Eindhoven, 28 October 1991.
50. *Ibid.*
51. Philips Annual Report, 1960, p. 12.

4

COMPETITION AND COLLABORATION IN THE EUROPEAN VCR INDUSTRY: PHILIPS AND THE V2000 SYSTEM

The previous chapter examined how R&D had characterised Philips' history from the founding of the company to the 1970s. It was argued that the tradition of strong in-house research had enabled Philips to diversify from the lighting industry to many other industrial sectors. On the management and organisational structure side, as we have seen, Philips had developed a unique dual-head management style and a very complicated matrix structure. Following an expansionist policy, Philips had made presence in over 60 countries by the 1960s, and its decentralised decision-making system had generated a corporate empire stretched as a loosely organised federation. Due to the radical change in the competitive environment in Europe since the 1970s, Philips' traditional expansionist strategy and decentralised organisational structure began to see challenges from competitors from East Asia led by the Japanese.

In this chapter, I will examine the issue of how and why Philips' VCR technologies, mainly the V2000 system, were defeated by the Japanese and, consequently, Philips fell from being a technology leader to the unexpected position of a Japanese VCR patent licensee. The major factors contributed to the failure of V2000 and the implications of this event to both Philips and the European industry will be analysed.

THE EMERGENCE OF THE EUROPEAN VCR INDUSTRY

The European VCR industry has been dominated for over a decade by a *de facto* world standard, the VHS system, developed by JVC. As the pioneer of the first consumer VCR system, Philips was forced by the market to buy the Japanese VHS technology in 1983. From then on, the long-standing VCR format battle was settled at the expense of European technology.

It is argued in this chapter that the failure of Philips' V2000 VCR system had a profound effect on the company's corporate strategies adopted in the 1980s and onward. Although Philips had been involved in a number of collaborative ventures with other firms in the world electric lighting industry, as shown in the previous chapter, Philips did not give priority to industrial collaboration in forming its strategies for new technologies and new products related to the consumer electronics industry. On the contrary, 'do it yourself' and trying to maintain a leadership in every aspect of technical change was a prevailing attitude within Philips. It was the difficulties experienced by Philips in promoting its own VCR technologies that made the company's management pay more attention to industrial collaboration with both European and non-European partners. This new strategy was subsequently reflected in the cases including high definition TV and multimedia technologies, which will be examined later in this book.

During the course of VCR format battle and afterwards, the EU stepped in and adopted a protectionist stance, such as antidumping policies, in order to secure a home market for the declining European manufacturers and their technologies. In addition to this, some EU member countries (e.g. France) also adopted very unpopular trade policies towards the usurpation of the European market by Japanese VCR manufacturers.

During the period from the early 1970s to the end of the 1980s, the home VCR industry had become the second largest product market after the CTV (Colour Television) market as far as the consumer electronics industry was concerned. It was reported that the market share of VCR products in 1989 accounts for $20.6bn, 18% of the total world consumer electronics markets ($117bn). In Europe, the VCR market share accounts for $7.1bn, 18% of the total European consumer electronics market and more than one third of the world VCR market. Having considered the size of the VCR market, it seems obvious then that Philips, as a leading manufacturer in the European as well as the world consumer electronics industry, would have suffered significantly from the failure of its V2000 technology.[1]

The Early Development of VCR Technology

The earliest efforts for developing video recording technology go back to the mid-1950s. In the U.S., Ampex became the first company to demonstrate a prototype magnetic video tape recorder, the Quadruplex format, in April 1956. The first video recording machine, called VR-1000, was sold only to broadcasting networks at the price of $75,000. Until 1959, when RCA launched another competitive model, Ampex's VR-1000 had remained the only machine designed for the professional market. In the two years after VR-1000's emergence, hundreds of units were sold to broadcasting organisations all over the world (Inglis, 1990, p. 356). Meanwhile, the Ampex Quadruplex format became the *de facto* professional standard in the broadcasting industry. When Ampex was enjoying such a success in launching the VR-1000 professional video, RCA's new initiative shocked the professional video world. Instead of following Ampex's model, RCA developed its own video recording machine which proved to be a big success in the next few years.

Some big Japanese companies, like Sony, Matsushita and Toshiba, also started to develop their own video recording technologies. Toshiba and Sony announced their independent systems in 1959 and 1961, respectively. But the Sony machine was never commercialised. In 1966, both Sony and Matsushita marketed their home video recorders but neither was accepted by the market. The major factors which led to the failure of these Japanese systems were the bulky size and high price of their products.

Meanwhile, the American and Japanese companies tried to join their efforts to develop video recording machines, but none of those ventures was successful. For example, Ampex and Sony signed an agreement in 1960 to establish a joint venture for producing both professional and consumer VCR products. This joint venture collapsed in 1961. In 1964, Ampex shifted to Toshiba and they formed another joint venture, called Toamco, for producing VCR products. Like others, this joint venture was also doomed.

In the professional market, the dominant position of Ampex and RCA was soon taken over by Sony after the company launched its U-Matic range in the 1960s.

Philips' Pioneering Work

Although several Japanese companies and Ampex of the U.S. had made great efforts in developing video recording technology, none of them succeeded in bring out a consumer model of videotape recorders onto the market. It was Philips that first introduced and launched a technically acceptable videotape recorder model for the consumer market at an affordable price level in 1972.[2] In the period 1972–78, Philips developed two incompatible VCR systems, i.e. Type A and Type B. Type A was exemplified by the N1500 machine which was capable of recording up to 1 hour with a tape speed of 9.82 metre/second. Type B was known to be the N1700 VCR machine, which had a technical capacity of recording/playing for a maximum up to 3 hours with a tape speed of 8.1 metre/second. Philips called the N1700 machine the Long Play system.

Under an information exchange agreement with Philips, Grundig developed the Super Video Recorder (SVR). Grundig's SVR system extended the recording/playback time to 4 hours and later achieved 5 hours in 1978. SVRs were marketed for only two years (1978–79). Table 4.1 is a list of major home video-recording systems, models and their makers in the world consumer market in the late 1970s.[3]

As indicated in Table 4.1, by the end of 1977 VHS was the biggest camp in which JVC was joined by another seven manufacturers. The second largest was Sony's Betamax system with five supporters. Apart from the technical information exchange agreement with Grundig, Philips was basically alone in promoting its VCR technology. This technology grouping left Philips in a relatively vulnerable position in competing with Japanese manufacturers.

Development of Japanese VCR Technology

As mentioned above, when the American companies were launching their video recording systems, some Japanese companies were also developing their own video technologies. In the wake of Toshiba's helical scan video recording technology in 1959, Sony introduced the world's first transistorised VTR in January 1961.[4] In August 1965, Sony launched its CV-2000 VTR onto the market at a price of $800–$1000. Customers of these machines were exclusively professional and institutional buyers.

Based on long term R&D investment in VCR technology and after many failures in launching home VCR machines, the Japanese companies finally reached the frontier of the home VCR market. In 1975, Sony launched its home VCR machines based on the Betamax format in the U.S. market. The Betamax machine was mainly designed as a 'time-shift' product for consumers' TV programme viewing. In the following year, 1976, JVC, a

Table 4.1. Video Recording Systems, Models and Makers, end 1977.

System	Maker/Brand	Model No.
VCR	A Philips	N1500 (N1502, N1512)
	B Philips	N1700
VHS	Mitsubishi	HS200
	Thorn (Ferguson)	Videostar 3292
	JVC	HR-3300
	Akai	VS-9300
	NordMende	Spectra Video Vision
	Hitachi	VT3000E (BS)
	National Panasonic	NV8600
	Saba	(T.B.A.)
Beta	Sony	Betamax
	Sanyo	Betacord
	Aiwa	Beta
	Pioneer (Shrio)	Beta
	Zenith	Beta
	Toshiba	Betavideo

Source: Adapted from *Electrical and Radio Trading*, 6 July 1978.

subsidiary of Matsushita, introduced another technologically independent home VCR system, the VHS format. From then on, various previously launched or announced VCR systems began to vanish gradually and a thoroughly Japanised world VCR industry was formed with the recognition of VHS as a *de facto* global standard in the early 1980s. Table 4.2 indicates Japan's VCR production during the period from the late 1970s to the early 1980s.

The development of Japanese VCR technology was an uneven process. The difficulties caused by software copyright disputes, rather than technological difficulties, brought this process into a very confusing situation, which partly contributed to the failure of Sony's Betamax format. By the end of 1976, Universal Pictures, a subsidiary of MCA, filed a lawsuit together with Walt Disney Productions against Sony. These two organisations were claiming that Sony's 'time-shift' advertising campaign for its Betamax VCR machines in the U.S. was breaching copyright regulations. Almost eight years later, the original suit was concluded by the U.S. Supreme Court in favour of Sony. If Sony had immediately lost the case in the U.S., the launch of the VHS by JVC, one year after the Betamax launch, would have been extremely difficult and, probably, the home VCR industry as a whole would have been killed off by nonmarket forces.

It is worth noting that, in the early days of the home VCR format battle, Sony appealed to the MITI (Ministry of International Trade and Industry), for help, and the latter tried to negotiate a settlement for the industry and favoured Sony's Betamax format during the discussions (Cusumano, Mylonadis and Rosenbloom, 1991). Ironically, it is the VHS system, rather than the government-favoured Betamax, that won the format battle.

Japanese Penetration and VCR Formats in Europe

Compared to the situation in Europe, the Japanese VCR industry was much more competitive. As shown in Table 4.1, there were more manufacturers in Japan than in Europe. This difference generated three immediate results.

Table 4.2. Japan's VCR Output (Million Units).

Year	Production	Shipments	Exports
1978	1.47	n.a.	0.79
1979	2.20	2.13	1.67
1980	4.44	4.41	3.44
1981	9.50	9.06	7.35
1981 Jan.–Sept.	6.18	5.94	4.83
1982 Jan.–Sept.	9.44	9.41	7.64

Source: *Financial Times*, 19 November 1982.

Firstly, VCR production soared in Japan, as indicated in Table 4.2, since the late 1970s, and the majority of VCR machines produced in Japan were exported to other countries.[5] Secondly, economies of scale, on which high volume production relied, quickly reduced the cost of unit production and selling price declined accordingly.[6] Thirdly, fierce competition among different brands forced Japanese manufacturers to improve product quality. These factors had effectively enabled Japanese firms to initially penetrate the European market by exporting large quantity and high quality of finished products.

The rapid increase of Japanese VCR exports to Europe was countered by heavy political pressures from both the EU authorities and European manufacturers. The EU's antidumping investigation and the subsequent VER (Voluntary Export Restraint) agreement signed between the EU and the Japanese government immediately made unlimited shipments of Japanese-made VCRs to Europe unrealistic. As required by the VER agreement, in 1982 and 1983, Japanese shipments of VCRs to Europe were not expected to go beyond 5 million units. Japanese manufacturers increased direct investment in Europe for the sake of diffusing trade frictions caused by exporting their VCR products. Japanese inward investment was encouraged by the EU and most national governments, particularly U.K. Table 4.3 shows the VCR production capability of Japanese manufacturers in Europe in the period 1982–1984.

In the early 1980s, Japanese penetration of Europe was not only in the area of VCR manufacturing but also in videotape cassette and component production. Table 4.4 gives some details of Japanese videotape cassette production penetration either directly or through local firms.

Due to the large quantity of Japanese-made VCR exports and increased direct investment, the European VCR market began to be dominated by leading Japanese VCR systems, firstly Beta and VHS then VHS alone since

Table 4.3. Japanese VCR Production in Europe.

Company or Partnership	Location	Production Per Month	Start of Production
Sony	Cologne, W. Germany	5,000	May 1982
J2T Video (JVC)	Berlin	20,000	May 1982
MB Video (Matsushita/ Bosch)	Osterode	2,500	Jan. 1983
Hitachi Europe	Bayern	5,000	Jan. 1984
J2T Video (JVC)	Newhaven, England	10,000	Oct. 1982
Sanyo	Suffolk	5,500	Sep. 1983
Mitsubishi	Haddington, Scotland	5,000	Mid. 1983
Mitsubishi	Livingston	5,000–10,000	Late 1984

Source: Adapted from JEI: *Journal of the Electronics Industry*, March 1983.

Table 4.4. Japanese Videotape Cassette Production in Europe.

Company or Partnership	Location	Production Per Month	Start of Production
Sony France S.A.	Dax, France	1 million	Mid. 1984
Maxwell U.K. (Hitachi)	Telford, England	750,000	Jan. 1984
JVC Magnetape Europe (JVC)	Mönchenglad Bach, W. Germany	250,000	Spring 1983

Source: Adapted from JEI: *Journal of the Electronics Industry*, March 1983.

the late 1970s. In 1981, the only remaining European-owned VCR system in operation was the newly launched V2000 format developed by Philips in cooperation with Grundig. In Britain, which had the highest rate of VCR household penetration amongst all the EU member countries, the total percentage of annual volume sales of VHS and Beta was 60% in 1978 and 85% in 1982, as shown in Table 4.5.

Although the market share of the V2000 format increased in the U.K. market in the first few year years after its launch, the figures it achieved were too small to enable Philips and Grundig to compete against the more established Japanese formats, in particular the VHS. Grant (1983) has analysed the market share by the three major VCR formats (VHS, Beta, and V2000) on the three biggest markets (France, W. Germany and the U.K.) in W. Europe.

Table 4.6 shows that, by the end of 1982, VHS obtained the overwhelming position in the three major European markets; while the V2000 accounted for only a very small market share; Beta's position was in between.

In 1983, when Philips and Grundig decided to buy JVC's VCR technology and manufacture VHS machines in Europe, the V2000 system was effectively dead and JVC's VHS format became the *de facto* world standard

Table 4.5. The UK VCR Market by System (% of Annual Sales).

System	1978	1979	1980	1981	1982
VHS	50	68	68	64	58
Beta	10	15	25	30	27
V2000	–	–	2	6	15
VCR/SVR	40	17	5	–	–

Source: *The Times*, 30 November 1982.

Table 4.6. VCR Market Share by System: France, W. Germany, and UK, 1982.

	VHS		Beta		V2000	
Country	Units	%	Units	%	Units	%
France	960,000	80	120,000	10	120,000	10
W. Germany	1,592,500	65	416,500	18	441,000	10
U.K.	1,512,500	55	852,500	31	385,000	18

Source: Based on Grant (1983).

for home VCR. From then on, Japanese VCR penetration in Europe has been carried out on an even larger scale mainly through direct investment either in the form of wholly-owned plants or joint ventures in collaboration with European partners. According to BIS Mackintosh Research (1990), VCR production capacity of Japanese based companies in Europe was 15% (310,000 units) in 1984 and it increased to 39% (3,615, 000 units) in 1989; while, the production capacity of European-owned companies declined from 85% (1,765,000 units) to 54% (5,080,000) in the same period.[7]

By January 1990, Japanese consumer electronics companies had already built 20 VCR plants employing 11,556 workers in Western Europe. Meanwhile, for the sake of localising their VCR production, the Japanese widely invested in VCR component production in European countries as well. A list of Japanese VCR component suppliers is presented in Table 4.8.

THE FAILURE OF THE V2000 SYSTEM

In the above section I have discussed the increased Japanese competition in the European VCR industry since the 1970s. Japanese competition has

Table 4.7. Growth in Ownership of European VCR Production Capacity, 1984, 1989.

	1984		1989	
Type of Ownership	Units (000)	Total (%)	Units (000)	Total (%)
European-owned Companies	1,765	85	5,080	54
Japanese-based Companies	310	15	3,615	39
Korean-based Companies	0	0	640	7
Total Non-European Ownership	310	15	4,255	46
Total Production	2,075	100	9,335	100

Source: Adapted from BIS Mackintosh (1990).

Table 4.8. Selected Japanese VCR Component Suppliers in Europe, May 1990.

Companies	Location	Components
Alps	Düsseldorf, Germany	VCR audio & Video heads
Amano	Belgium	VCR rubber components
Hosiden	U.K.	VCR components
JVC/J2T	Tonnerre, France	VCR deck assemblies
Matsushita	Peine, Germany	VCR deck assemblies
Mitsubishi	Livingston, Scotland	VCR deck assemblies
Sanyo	Nördlingen, Germany	VCR deck assemblies
Tunda	U.K.	VCR plastic parts

Source: Based on BIS Mackintosh (1990).

been regarded by some EU authorities and European firms as the major reason why indigenous European VCR technologies failed to be accepted by the consumers. However, it is my understanding that foreign competition was, at most, an external factor contributory to the failure of European technologies. I believe the failure of European VCR technologies, particularly the V2000 system, had much more complicated causes. In this section, I will investigate these reasons by analysing Philips' corporate strategies for the V2000 technology in comparison with JVC's strategic manoeuvring for VHS.

Time-Shifting Machine: The Launching of V2000

As indicated above, V2000 was the third home VCR system developed by Philips in cooperation with Grundig. Despite the technological superiority and huge investment it had cost, the V2000 VCR system failed to survive the fierce format battle both inside and outside Europe. V2000 was launched in 1981 and had a life-span of only three years. Although Philips and Grundig expected the V2000 system to become a world standard, or at least one of the leading systems worldwide, it never reached any other major consumer electronics market outside Europe (e.g. the U.S. and Japan).

Shortly after Philips brought the video-recording technology into the consumer market with its N1500 machine, the company realised that the N1700 series, the second generation of Philips' home VCR products, was losing its competitiveness in meeting the challenges from the new Japanese competitors. In comparison with Japanese VCR machines using the VHS and Betamax formats, the early Philips models (N1500 and N1700 series) were 'bulky, expensive, unreliable, and restricted to short recording times' (Cawson *et al.*, 1990, p. 226). With the intention of catching up with its Japanese competitors, Philips started laboratory work in collaboration with Grundig from 1976 to develop its third generation home VCR—the V2000 system. Five years later, in 1981, the V2000 system was introduced into the consumer market in Europe.

It is widely accepted that the V2000 system was technically more advanced than the VHS and Beta formats. Firstly, V2000 used two tracks on ½-inch magnetic tape instead of one track. The effectively doubled width of the magnetic tape enables the television programmes to be recorded along two separate tracks on the tape, and this allowed for the recording and playing-back time to be extended to eight hours—twice what the VHS and Betamax could do. Secondly, the spare track on the V2000 tape allowed extra commentaries to be added without erasing the existing audio track.[8]

Production and Market Share of V2000

By 1981, Philips had invested more than £100m on video production in Europe, of which £75m was spent on the company's production centre in Vienna mainly for automated VCR production on a large scale. In the Vienna production centre, Philips planned to produce 375,000 units on the V2000 format in 1981 and more than 750,000 in 1982.[9] At the same time, Philips was also planning to start a V2000 assembly plant at Krefeld in Germany and another one at Le Mans in France.

The only other European participant of the V2000 venture, Grundig, was operating daily production at its Nuremberg factory in southern Germany.

It was not long before Philips and Grundig found out that the marketing results of their V2000 machines were disappointing. Britain, as mentioned earlier, had the highest VCR household penetration proportion in Western Europe ever since home video recording machines were available on the consumer market. Philips' marketing ambitions for the V2000 machines, however, were never fulfilled. Table 4.9 shows the market shares in 1983 by the three major VCR formats in different regions.

When Philips was launching the V2000 system, it claimed that by the end of 1981 V2000 machines would take 15–16% of all video players sales in the U.K. market. Furthermore, the company predicted that 20% of the 900,000 machines expected to be sold in the U.K. in 1982 would be taken by its V2000 format.[10] In fact, the market share of the V2000 in the U.K.

Table 4.9. VCR Market Share by Format (%), 1983.

Country/Region	VHS	Beta	V2000
U.S.	75	25	–
Japan	70	30	–
Western Europe	66	23	11
U.K.	70	27	3

Source: Adapted from *Financial Times*, 4 January 1984.

in 1983, as indicated in Table 4.9, was only 3% which was much less than the market share Philips had expected for 1982! Table 4.9 also shows that, in Western Europe, the only market for the V2000 system, the average market share was 11% in 1983. This figure was far smaller than the 66% market share held by JVC's VHS system and 23% by Sony's Beta format.

The Issue of Collaboration and Industrial Support

Seeking partners and being supported by as many licensees as possible was becoming an important factor affecting a company's promotion of its new technologies or new products. The realisation of Philips' collaboration with Grundig helped its V2000 system to become another technologically competitive VCR format in Western Europe at the beginning of the 1980s; but the failure of Philips to find more partners, in particular major firms in the consumer electronics industry, contributed greatly to the V2000 disaster.[11]

To be sure, both JVC and Sony had already established wide-ranging partnerships with other companies (mainly Japanese firms) by the end of the 1970s, as indicated in Table 4.1. It seems that the company's new strategy of seeking partners for its V2000 system came too late to kill off the already well-established VHS. Compared with Philips and Sony, and of course Matsushita, JVC was very small in terms of size and it would have been extremely hard for the company to survive the VCR format battle alone. Having realised this disadvantage, JVC was quick to publicise its VHS technology and made it open to anybody who would possibly become a licensee during the late 1970s and onward.

JVC released its first model of the VHS system in 1976 and the company made extraordinary efforts ever since to promote its technology by means of issuing production licenses to consumer electronics firms throughout the world.[12]

Telefunken, a German consumer electronics manufacturer, and Thorn EMI of Britain did not have their own independent VCR production systems. Therefore, these two companies were both seeking opportunities to enter the VCR market by licensing either the Philips/Grundig V2000 technology or the Japanese VCR technologies since the late 1970s. However, Telefunken's negotiations with Philips and Grundig did not succeed. This led to the establishment of the J2T VCR joint venture—Telefunken, Thorn EMI and JVC joined forces to produce VCR machines based on the VHS format. This joint venture greatly enhanced the VHS system's competitive position against, firstly, Beta, then, V2000 in Europe. It seems that the establishment of the J2T joint venture served the purpose of building a 'Trojan Horse', through which JVC could easily penetrate the European VCR industry and the consumer market. Some commentators have expressed their concern about the failed efforts of European VCR contenders to collaborate:

> Had Thorn, Thomson and Telefunken chosen to go with V2000 following their very careful deliberations, the current reliance upon imports would be significantly reduced.[13]

When the V2000 system was launched at the beginning of the 1980s, Thomson appeared to be the ideal partner for Philips. In 1981 Philips announced that it would begin production of V2000 machines in France the following year under the assumption that V2000 could gain a major market share in Europe provided Thomson agreed to adopt the V2000 format for its VCR production (Cawson *et al.*, 1990). Thomson, however, shifted its alliance to the Japanese firms by aligning with JVC through Telefunken, which had been recently taken over by Thomson, to produce VCRs based on VHS. Thomson's decision to back the VHS format put Philips and its V2000 system in a very vulnerable position.

After Thomson took over Telefunken and joined the VHS camp, the previous joint venture J2T was strengthened and became a more effective channel through which JVC could present itself in Europe. As far as their national origins were concerned, Thorn EMI, Telefunken and Thomson were, respectively English, German and French companies. Therefore, the set-up of the J2T joint venture meant that JVC achieved 'inside' access to U.K., Germany, and France, which were the three major national markets in Western Europe. Needless to say, J2T formed a very important part of JVC's European strategy for promoting its VHS system.

It was widely accepted that collaboration between Philips and Thomson would have significant impact on the long-term development of the European consumer electronics industry. As the Mackintosh report has rightly argued above, the European VCR industry could have been substantially different had Thomson and other European firms joined the Philips/Grundig V2000 camp. However, the relationship between these two leading European manufacturers seems to have been very troubled, but there is hardly any research or published information on it. A Philips historian has revealed that the unhappy relationship between the two companies could be traced back to the early 1930s:

> Philips and Thomson worked together in a very large French incandescent lamp factory. Philips and Thomson made this joint venture in 1931. It was originally a 50/50 ownership. Sometime in the 1960s, when Philips took over the shares of Thomson and the venture became 100%-owned by the former, we saw a very odd relationship between these two companies.[14]

When asked about the reason why Thomson became a partner of JVC rather than joining forces with Philips, the Philips historian believed thats "it is possible that Thomson did not have any confidence in Philips".[15]

The difference between the corporate strategies adopted by Philips and JVC for their V2000 and VHS, respectively, may be summarised as follows: Philips and Grundig developed a better technology and marketed it themselves but JVC came to the market in cooperation with a wide range of industrial supporters and offered the consumer a more acceptable product. In short, interfirm collaboration played a significant role in the process of competition between competing technologies.

Victory for VHS

In 1983, when Philips and Grundig announced that they were to produce VHS machines in Europe for non-European markets, there were only nine VCR plants in Europe and three of them were Japanese-owned, according to Table 4.10; four years later, in 1987, the total number of VCR plants in Europe increased to 26, of which 14 were Japanese-owned.

When Philips was publicising its decision to adopt the VHS technology the company also expressed its intention to continue V2000 production; but few commentators believed this story:

> Philips was adamant that it would continue making V2000 for the European market, but the trade believes that the company will quietly drop V2000 after a period, as it did two earlier video formats.[16]

This suspicion was soon proved to be true in that both Philips and Grundig completely terminated their V2000 VCR production at the end of 1985. The disappearance of the V2000 system signalled a new era of a completely Japanised European VCR industry so long as the issue of technological standardisation was concerned. Although Sony's Beta system continued to have a small market share,[17] the dominant position of the VHS system was firmly established all over the world by the mid-1980s.

There were various reasons why VHS won the VCR format battle. JVC's corporate strategies in promoting its VHS system in the late 1970s and early 1980s may be understood from the following aspects.

Firstly, as discussed above, offering an open licence from the time of the VHS launch secured JVC a wide range of industrial alliances, in which many competitors became licensees of the JVC technology. This was confirmed in a recent report:

Table 4.10. VCR Plants in Europe.

Ownership	1983	1987
European-owned	3	6
Japanese-owned	3	14
Joint venture	3	4
Other	0	2
Total	9	26

Source: Adapted from *The Economist*, 11 February 1989.

> Takano hit on a clever strategy to sell the VHS format to other manufacturers. He saw Sony refusing to license Beta technology to other manufacturers and jealously guarding laboratory secrets and plans. So Takano showed his prototypes, humbly asked for frank comment and offered to license the technology to anyone capable of making it.[18]

Under Takano's initiative, JVC always kept its focus on expanding the VHS family of licensees. Earlier discussions showed that VHS was the biggest camp amongst the global VCR format contenders as far as the number of collaborating firms was concerned. In addition to those European manufacturers joined up through the J2T joint venture, several other European firms, such as Nordmende and Blaupunkt, had already entered licence agreements with JVC by the end of the 1970s. In the U.K., Thorn made an arrangement with JVC to market VHS machines under the Ferguson name from mid-1978.[19] The strategic importance of open licensing and industrial collaboration to the success of VHS over Betamax is also emphasised by Cusumano, Mylonadis and Rosenbloom (1991) in their study about the VCR format battle.

Secondly, capturing the home video rental market was crucial to the success of VHS. Prior to the commercial launch, JVC had made it no secret to the industry as well as the media of its future technological development for its home video system, and the company's openness had gained it great confidence from both hardware and software industries in VHS technology. Partly due to this, VHS was accepted not only by manufacturers but also by VCR rental companies. In Britain, the rental business proved vital to the success of VHS in entering homes in competition with other rival systems during the early days.

Thirdly, compared to other alternative systems, greater availability of prerecorded software on VHS tapes gained JVC an important competitive advantage. The success of the video rental business was an important influence on consumer purchases of VCR hardware; they preferred to buy machines for which there was the widest range of software available. The success of the VHS format has made the proposition that a successful launch in the consumer electronics industry largely depends on the synergy or combination of hardware and software widely accepted. Technical superiority is important; but support from the software industry is vital for a new system to survive. JVC did not offer the best technology to the home video market, but its success in VCR added a new dimension of strategic planning within the consumer electronics industry. Both the vanished V2000 and the defeated Beta formats suffered from the lack of software support because Philips and Sony paid more attention to the technological side of their systems. Initially, Matsushita, the parent company of JVC, was also involved in developing another VCR system itself; but the company dropped its own system in favour of its daughter company's VHS format. With the backing of Matsushita, JVC firmly grasped the film and music industry into its system.

Fourthly, a strong distribution team and an efficient distribution network was also indispensable. Having launched the VHS system and gained initial success in the marketplace, JVC committed itself firmly to satisfying its licensees' wishes, sometimes at expense to itself. The following story may partly confirm this point:

> Around 1980, JVC maintained shipments on an original equipment manufacturing basis to rivals short of capacity—even at the cost of suspending output for its own sales. JVC distributors became so strong due to the supply shortage that one JVC manager recalls being 'in charge of apologising'.[20]

Fifthly, conquering the American consumer market was critical for VHS in becoming the *de facto* global standard. The U.S. was the birthplace of magnetic video recording technology. Some American electrical/electronics companies, such as RCA and Ampex, had gained their leadership in the area of professional video recording systems before Sony launched its U-Matic in 1971. This was followed by Philips' launch of the first consumer VCR the following year, and the fierce format battle. Ironically, this format battle for the consumer VCR market excluded all American companies since not a single American firm was able to bring its traditional magnetic video recording technology into the consumer market. However, the U.S. remains one of the most important consumer electronics markets, where new gadgets are always appreciated.

Having abandoned its own system and adopted the daughter company's format, Matsushita started its campaign for VHS in the U.S. in 1977. As the largest consumer electronics manufacturer in the world, Matsushita's support was of great significance in determining the final VCR standard. In fact, immediately after Matsushita's decision to adopt the VHS format the American VCR market was substantially changed due to the company's unrivalled influence through its manufacturing capability and marketing network. First of all, Matsushita gained an OEM agreement to supply VHS machines to RCA. Following this, the majority of American manufacturers and distributors joined the VHS camp. In 1984, Zenith, Sony's OEM partner, changed its strategy and shifted to VHS at the expense of the Beta format. Obviously, the Matsushita action put both Sony's Beta and Philips' V2000 formats into very disadvantaged positions, and Zenith's shift concluded the VCR format battle with an overwhelming victory for VHS on the American market.

Finally, technical flexibility met the local conditions well. Immediately after JVC launched its VHS system, the company developed a PAL version of its VCR machines to meet the European market in 1977.[21] This technical flexibility demonstrated JVC's global vision, an important aspect of its corporate strategy over VHS. This was quite different from Philips' VCR production and sales in its own formats, which were almost exclusively confined to Europe.

As far as the technical parameters were concerned, it is widely believed that the VHS format was inferior to both of Sony's Beta and Philips' V2000 systems. Both Sony and Philips had gained good reputations for being technical innovators in the consumer electronics industry and they were more capable of handling technical product upgradings. From the late 1970s to early 1980s Sony pioneered many new technical improvements by adding appealing features to its Beta machines. Features like wireless remote control, Hi-Fi sound effect, multifunction players, etc., were all technically innovative and first introduced by Sony. However, shortly after Sony's introduction of such new features, JVC, as well as its parent company Matsushita, would quickly catch up by introducing its own version of the equivalent features.

The victory of the VHS format was demarcated, on the one hand, by the decision of Philips and Grundig to start VHS production outside Europe in 1983, not to mention the other European manufacturers' choice of the VHS technology; and on the other hand by some Japanese manufacturers' shift from the rival camp (Beta) to the VHS group. For instance, Toshiba had been one of the leading Japanese companies producing VCRs on the Beta format before the end of 1983. Following the announcement of Philips and Grundig, Toshiba decided to make and sell VHS machines.[22]

Why V2000 Failed

It might seem surprising that a technologically superior VCR system finally gave way to a technically less advanced system after a long-lasting format battle. However, it did happen. In addition to the major factors contributing to the success of VHS against Beta and V2000 discussed above, the following points shed more light specifically on the issue of why the Philips/Grundig V2000 format was squeezed out of the VCR market in the early 1980s.

Launching Time

When Philips and Grundig introduced their newly developed V2000 system in 1981, JVC's VHS format had already been well established in Europe as well as in other major markets. Because of the high penetration level of the VHS machines and the continually increased availability of various prerecorded software on VHS tapes, customers would not be easily tempted to replace their previous purchase with or, as a first-time buyer, purchase the new V2000 system at a higher price.

The timing of Philips' launch of its VCR formats was not favoured by the market. With its first product, Philips came into the consumer market too early.[23] In introducing its N1500 machines, Philips had to spend time and money to publicise the VCR technology and educate the customer. Philips' pioneering work in VCR definitely made it easier for the latecomers. Having tested the marketplace, Philips was not happy with its N1500 system and, therefore, brought out an improved version, the N1700, which was not compatible with the previous machine. On the other hand, Philips came into the market too late when it was jointly introducing the V2000 format with Grundig in Europe in 1980, because the VHS system had already become a fully-fledged format in the world markets.[24]

Price

The V2000 system was technically more advanced in comparison with the Japanese VCRs. Doubtless, more engineering work normally costs more money. Higher price levels, to some extent, made V2000 machines unpopular amongst the customers.[25] Some commentators believe that, "compared with Japanese VCRs, it [V2000] seemed overengineered and undermarketed".[26]

Immediately after the V2000 launch, Japanese VCR prices were substantially cut by 60% a year on the European market. Both Philips and Grundig believed that the Japanese were dumping their VCRs on the European consumer market and they filed a suit to the European Commission against the alleged unfair trading practice of Japanese companies. The lower price of Japanese VCRs might partly explain the sharp decline of V2000's market share in 1982 and the worsening situation thereafter.

Compatibility

It seems that the lack of compatibility between different generations of Philips' VCR products played a role in the failure of the V2000 system.

To be the first mover to start the consumer VCR industry, Philips came into the market with its own systems, the N1500 and N1700, which, surprisingly, were not compatible with each other. In other words, if you bought tapes for the N1500 you couldn't play them on the N1700, and *vice versa*. Moreover, the later V2000 machines were built on a completely new format with an 8-hour recording/playback capability, and it was not compatible with either of the previous Philips systems or with the Japanese formats. The lack of compatibility between V2000 and other Philips systems simply meant that previous purchasers had misinvested their money in buying Philips VCR products. One can imagine that it would have been extremely hard for Philips to rebuild consumer's confidence and persuade them to buy the new V2000 products. It was clear that most consumers did not want to buy a very expensive VCR machine which was unable to play other widely marketed tapes.

Prerecorded Software

It is widely accepted that, unlike the VHS system, the lack of prerecorded software was another major reason why the V2000 system failed. As mentioned above, the V2000 was designed mainly as a time-shift machine to tape programmes from a television channel without stopping the viewers watching another channel. Instead of promoting large-scale production of software, namely, prerecorded programmes, for its V2000 machines, Philips launched its consumer videodisc product, the LaserVision system, initially in Britain in 1982. Philips was hoping that consumers would buy the LaserVision machines for playing prerecorded software. It assumed that the V2000 system and the LaserVision player would be functioning as a perfect combination of off-air programme recording and prerecorded software playback. As a matter of fact, viewers were not convinced of this combination:

> Philips had adopted a strategy of promoting both V2000 and LaserVision as complementary purchases, in the belief that consumers would buy a VCR for recording off-air and a disc player to show prerecorded software. Thus in introducing the V2000 Philips underestimated the importance of software availability, and overestimated the willingness of consumers to spend heavily to achieve marginal increases in playback quality (Cawson *et al.,* 1990, pp. 226–227).

Contrary to Philips' expectation, LaserVision did not become a popularly used consumer product in Europe, although it is still being used for commercial and industrial purposes.[27]

A survey of VCR owners indicates that only 3% of total TV viewing time was spent watching shows that were recorded in advance using the VCR's programming feature, and "hardly any one used his or her VCR except to play rented movies".[28] This survey result completely disagrees with Philips' previous understanding about the role of home VCR.

On the other hand, Philips' declining VCR market share was a great disincentive to 'software' makers to record the best-selling films onto the V2000 tape cassettes. The lack of enthusiasm from the recording companies regarding the V2000 system contributed to the lack of suitable prerecorded programmes. This is a vicious cycle between the popularity of a hardware system and the level of enthusiasm software companies.

It is clear that the Philips management failed to understand the consumer market correctly as far as the V2000 system was concerned. In other words, the concept of 'VCR' conceived by Philips was substantially different from that of other companies, in particular, JVC. To Philips, 'VCR' was basically a time-shift machine. On the contrary, JVC firmly extended their understanding of the 'VCR' concept from time-shift to playback of prerecorded software, and, to some extent, the latter was considered more important. In commenting on this difference between Philips' and Japanese companies' understanding of consumer needs, a strategic planning manager at Philips recalls:

> We then decided that we would come in with a technically superior machine called the V2000 which was much better, a higher standard than the VHS, and also had eight hours recording—something could turn the tape over. Whereas we always say the VCR is a time-shift machine, and it enables you

to view later, the Japanese appreciated the value of prerecorded software. So they established themselves very firmly while we were still doing the V2000. When we came back with the V2000 we told the consumer that it was a technically superior machine, the consumer responded "so what?"[29]

Internal Organisational Conflicts and Managerial Inefficiency.

Although the headquarters in Eindhoven planned to launch an NTSC version of the V2000 system for the American market through its daughter company, Philips North America, the latter refused to commit itself to an uncertain system. Consequently, V2000 machines never reached the American market. It is obvious that, in promoting the V2000 system, Philips failed to achieve a concerted industrial alignment not only outside the company with other contenders but also with its own subsidiaries. The lack of strategic coordination between the headquarters and the subsidiaries was considered as one aspect of Philips' structural problem, and will be further discussed in the next chapter.

Repercussion of EC Public Policies

Another important factor which helps to explain the failure of the V2000 system lay in the public policies adopted by the EU in the late 1970s and early 1980s.

Unlike the European CTV sector in which national markets were strictly protected by the fragmented European standards and different national as well as European broadcasting regulations in the early years, the VCR industry was left mainly to the firms and subject to market competition from the early 1970s to the early 1980s. It is true that the EU undertook an anti-dumping investigation after Philips and Grundig as well as some other European firms complained about the Japanese dumping, and the VER agreement was signed between the EU and the Japanese government and adopted by the Japanese industry in the 1980s. However, all these procedures proved insufficient and were too late to prevent or scale down Japanese penetration of the European VCR market. High tariffs and export quotas might have been able to limit Japanese VCR exports to Europe, but the other side of the coin was that most Japanese VCR producers enhanced their presence in Europe through direct investment in VCR assemblies and VCR components production. From a long-term point of view, the latter was strategically more vital for Japanese companies in seizing a dominant position in foreign markets.

Apart from the EU policies mentioned above, the three biggest markets in Western Europe, U.K., France and W. Germany individually adopted very different public policies toward the presence of Japanese consumer electronics firms in their countries throughout the 1980s. Table 4.11 reflects

Table 4.11. Japanese VCR Investments in Britain, France and W. Germany, 1988.

Britain	France	W. Germany
JVC	Akai	Sony
Sanyo	Matsushita	JVC
Mitsubishi		Hitachi
Toshiba		Matsushita
Sharp		Sanyo
Hitachi		Toshiba
Orion		
NEC		
Funai		

Source: Adapted from Cawson *et al.* (1990).

the results of different national public policies in these three countries.

By 1988, according to Table 4.11, eight Japanese electronics companies had invested in the VCR industry in Britain, six in W. Germany and only two in France. These figures indicate the British government's more liberal attitude and the French government's protectionist policy towards Japanese direct investment in Europe during the decade of the 1980s. Between these two extremes was the moderate attitude of the German government.

Through direct investment, the Japanese companies widely established their VCR manufacturing/assembly plants and VCR component production factories in Western Europe. Although some Japanese direct investment was based in wholly Japanese-owned plants, a number of Japanese companies undertook their local VCR production through joint ventures with European partners. For instance, Matsushita/Bosch, JVC/Telefunken/Thorn-EMI/Thomson, and Funai/Amstrad were a few of these ventures.[30] To be sure, the majority of the workforce in the Japanese VCR plants were Europeans, and the production of many VCR plants were jointly owned and managed by Japanese and European companies. Thus, it seemed that neither the EU authorities nor any European national government could exclude the Japanese manufacturers from the European VCR industry without detriment to European interests. In fact, most European governments were in favour of having Japanese investment, with regard to partly solving their domestic employment problems. The issue of public policy *versus* the VCR industry will be discussed again later in this chapter.

IMPLICATIONS OF THE V2000 CASE

The failure of the V2000 format was a great set-back to Philips' financing, and turned out to be a great cultural shock for most of Philips staff at the time. But, on the other hand, this failure had many positive consequences for the company in the years to come. Having experienced the trauma of

V2000, Philips employees, in particular those at top management level, were forced to readapt their company's corporate strategy and improve the concern's corporate culture.

The Impact of V2000 Failure

Philips argued that changing VCR production from the V2000 format to VHS was not very difficult for Philips because the company had already invested large sums of money in its highly modern Vienna video factory for producing V2000 machines and most of the existing facilities in the Vienna factory could be turned to producing VHS machines:

> The money-losing production of the technically better, but more expensive Philips Video 2000 system is halted [in 1985]. Philips' knowledge and production capacity for videos allow the firm to produce VHS devices within a short period and to combat dropping sale prices by quality improvement.[31]

However, it is undeniable that shifting production from V2000 to VHS must have been a very costly process.

For Philips, once a pioneer in developing home video recording technology, shifting to making VHS machines in Europe was a loss of face. Although the failure of V2000 did not trigger a *domino* effect on the business operations of the Philips empire, it appeared to be a great shock to the company as a whole and brought about a profound impact on the European consumer electronics industry as well.

The heavy losses caused by the V2000 failure were far beyond the earlier expectation of the Philips management. When estimating Philips' losses in the VCR format battle, a financial analyst in Amsterdam said,

> Higher costs have been a factor in Philips' losing battle with Japanese manufacturers for the European VCR market, a battle that cost Philips $58 million in 1982 losses.[32]

In the space of 10 years, Philips group's consumer activities sharply declined from 47% to 23% of its total turnover in the decade 1975–1985.[33] This decline was largely due to the company's very costly but extremely unprofitable VCR business. Apart from the heavy investment in the Vienna VCR production centre as mentioned above and other intangible losses, it is also estimated that Philips spent nearly £3 million in the U.K. on a media advertising campaign with the intention of boosting the European market share of the V2000 format prior to Christmas 1982 (Grant, 1983, p. 21). Huge investments like this proved fruitless.

As a result of the V2000 failure and the almost simultaneous LaserVision disaster, a proposal for radical change was considered for implementation within the Group:

> The losses incurred on the V2000 led the main board in Eindhoven to consider in 1983 the possibility of withdrawing from consumer electronics altogether (Cawson *et al.*, 1990, p. 322).

If this option had been chosen, the whole structure of the Philips group would have been totally different from what it is today. Due to the interdependent relationship between the consumer electronics division and the company's other activities, the board of Philips group decided that the company should stay in the consumer electronics industry. In fact, consumer electronics has become one of the priority activities in the Philips group, and ironically in 1989 the most profitable Product Division.[34]

Japanese Identified as Main Competitors

At the beginning of the 1980s, Philips' strategic planners obviously felt the increasing threat imposed by the East Asian competitors, mainly the fully fledged Japanese multinationals in the consumer electronics industry. In particular, the insignificant market share of the V2000 system and its quick termination strongly suggested that the fierce Japanese challenge must be dealt with using proper long-term strategy if Philips and its European partners did not want to be excluded altogether from the European consumer electronics market. Having become aware of this challenge, part of Philips' long-term corporate strategy began to be focused on competition with the Japanese firms in the consumer electronics industry during Dekker's presidency beginning in 1982.

Other Aspects of Strategic Change at Philips

Having experienced the difficulties posed by the VCR format battle since the 1970s, senior Philips management realised that a single company could hardly survive contemporary global competition in high-tech areas by struggling alone. Instead, it felt that participating in industrial collaboration would be a better way of easing the plight created by the losses of the VCR format battle. Although Philips had been engaged in industrial collaboration before the V2000 disaster, the losses involved prompted the company to view industrial collaboration with both its European competitors and non-European rivals as a strategic necessity.

Having realised the extraordinary difficulties in launching and promoting the V2000 system without the collaboration of Japanese companies, Even before the collapse of V2000, the unsuccessful launches of the N1500 and N1700 formats had already pressed Philips to avoid direct competition with the Japanese. An immediate effect of this understanding was the decision by Philips management to launch its new digital audio format, the compact disc (CD) system in collaboration with Sony. It is acknowledged by Philips that without Sony's collaboration, its CD technology would not have become the world standard with such ease.[35] Philips' industrial collaboration with Sony and other firms in optical disc technologies will be further discussed in Chapter 8.

After the shock of the V2000 failure, Philips' corporate strategy radically changed. A senior managing director of Philips Data Systems was cited as saying:

> The Philips mentality has changed dramatically. Looking at the huge investments ahead, we would be fooling ourselves, thinking we could do it without cooperation.[36]

One important aspect of Philips' new corporate strategy since the early 1980s was setting up joint ventures with its major competitors in and outside Europe. During and after the 1980s Philips was involved in many joint ventures, of which the following were the most important:

- with AT&T in the area of telecommunication in 1984;
- with Control Data Corporation for developing video disc, CD-ROM and other optical memory technologies in 1986;
- with Du Pont de Nemours for developing chemical components used for compact discs in 1986;
- with Kyocera of Japan in the field of Home Interactive Systems (HIS) in 1986;
- Philips renewed its joint venture agreement with Matsushita in 1987;[37]
- More surprisingly, Philips and JVC announced an equally-owned joint venture agreement in 1990 and planned to start VHS VCR production in Malaysia in early 1991.

Another aspect of Philips' new corporate strategy was to promote technological and industrial collaboration amongst various European organisations. As one of the leading industrial organisations in Europe, Philips has been active in participating in several EU or pan-European programmes, such as ESPRIT, RACE, JESSI and Eureka, etc.

PUBLIC POLICY AND THE EUROPEAN VCR INDUSTRY

It is a widely held perception that, on the one hand, Japanese consumer electronics firms compete fiercely with each other both at home and abroad; on the other hand, however, a certain degree of government coordination and general economic/market planning, mainly through government agencies such as MITI and MPT (Ministry of Post of Telecommunications), is maintained as far as R&D for new technologies and key products are concerned. In contrast, such government performance was absent in Europe or North America for most of the time (Grant, 1983). In the case of the VCR industry, there was hardly any effort at a national or European level to coordinate the industry except those tough measures such as the anti-dumping policy implemented at a very late stage.

It is not my position to argue that the EU should have protected the European industry or to criticise the EU's policy measures taken against the Japanese in the 1980s. But it is true that, on the one hand, from the beginning the EU and the European national governments did not protect the indigenous European VCR industry against the competition of Japanese consumer electronics manufacturers, or they responded too late. On the other hand, the European protectionist measures adopted since the early 1980s forced Japanese manufacturers to change their European strategy from product export to direct investment in fear of the emergence of a 'fortress Europe'. As a result, the European VCR technology did not survive and the structure of the European VCR industry was completely changed since in the mid-1980s.

Japanese VCR Exports to Europe

Compared with Japanese VCR producers, the production capacity of European companies, like Philips and Grundig, was very small at the beginning of the 1980s. Table 4.12 is a list of the top 10 VCR producers in the year 1982 as estimated by Mackintosh International Ltd.

Amongst the top 10 VCR producers in 1982, as indicated in Table 4.12, only two were European-owned companies—Philips and Grundig—with a total estimated production capacity of only 1.0 million units, while the rest were all Japanese-owned. Claiming the majority of world VCR production output enabled the Japanese companies to export their products to foreign markets easily. By 1982, combined Japanese VCR production accounted for 94% of total world production (Cawson *et al.,* 1990, p. 255).

With the steady increase of Japanese VCR production, the volume of Japanese VCR exports to foreign countries was also rising from the beginning of the 1980s. No matter how big the domestic demand was, Japanese manufacturers would never be satisfied solely with their home market. On

Table 4.12. Worldwide VCR Production League (Estimates), 1982.

Company	Millions Units
Matsushita	3.5
JVC	2.63
Sony	1.83
Sanyo	1.5
Hitachi	1.3
Sharp	1.0
Philips	0.6
Toshiba	0.5
Grundig	0.4
Mitsubishi	0.4

Notes: (1) The above companies account for close to 95% of worldwide output; (2) JVC is a part-owned subsidiary of Matsushita.
Source: Adapted from Mackintosh (1985), "The European Consumer Electronics Industry".

the contrary, increasing exports became the most effective way for the Japanese to expand their VCR production in Japan. Apart from the U.S., Western Europe was the biggest market for VCR products outside Japan.

Prior to the boom of Japanese direct investment in Western Europe, as mentioned earlier, Japanese consumer electronics companies increasingly channelled large quantities of finished VCR products into the region. For instance, in 1981 Japanese companies exported 7.35 million VCRs worldwide, of which 2.85 million (38.8%) were to EU countries. In the following year, the total Japanese VCR export was 10.6 million units, of which 4.95 million (46.7%) were to the EU. Between 1981 and 1982 Japanese VCR exports increased 73.7% in volume.[38]

Firm Level Responses: The Philips/Grundig Lobbying

The boom of Japanese VCR production at home was sustained, to a great degree, by the increasing volume of exports to foreign markets. Despite export volumes of about 75–80% of total production, the remainder was still in excess of domestic demand. In June 1982 VCR stocks in Japan exceeded monthly shipments for the first time and subsequently savage price cutting was undertaken on the domestic market with some models down to under $250 for retail (Grant, 1983). This action was not confined to Japan. On the contrary, significant retail price reductions of Japanese VCRs followed suit on the European markets as a means of stimulating worldwide VCR demand. Of course, price reductions like this were undoubtedly beneficial to consumers and they responded to this Japanese action with a sharp increase of VCR purchases. European manufacturers, however, had a different reaction towards this price war. Philips and Grundig, the holders of the V2000 technology, were very much disadvantaged by this price cutting. Grundig, in the German market, had to cut prices by 30%[39] to be competitive against Japanese imports amid the intense VCR price war in Western Europe. Moreover, Grundig also made 25% of its V2000 assembly staff redundant. For Philips and Grundig, the newly launched V2000 format seemed unlikely to catch up with rival systems under this new market condition although they were spending heavily to publicise their technology all over Europe. Rather than concentrating on improving their VCR production and marketing activities, these two companies shifted their attention to the European Commission—in an effort to find a political solution to improve the declining V2000 system.

On 15 November 1982, Philips and Grundig went to Brussels with a file complaining about the dumping practice[40] of Japanese suppliers. Responding to the action taken by Philips and Grundig, the European Commission started an investigation into the alleged dumping activities of Japanese VCR firms. This official action signalled the start of a long-lasting but very controversial antidumping campaign against the Japanese and other East Asian exporters in the field of consumer electronics. As far as the EU regulations were concerned, manufacturers, like Philips and Grundig, were eligible to lobby the Commission and the Commission had to take action in accordance:

> Under the EU regulations, the European Commission is required to act on any complaint of dumping brought by Community producers of the product concerned, when they consider themselves to have been injured—or threatened—by dumped imports. The Commission must then determine whether there is sufficient evidence to justify initiating an investigation (National Consumer Council, 1990, p. 7).

As a matter of fact, as has been shown earlier, Europe's antidumping policies and the insistent lobbying of European manufacturers did not help much in preventing 'Japanisation' of the European VCR industry in the 1980s.

Shortly after preliminary talks between the two companies and the European Commission in Brussels, Philips and Grundig were joined by some other companies such as Siemens and ITT. The total number of lobbying group members increased to 11 within less than two months. All the members within the lobbying group were electronics manufacturers based in Europe. It did not take very long for a cease-fire to be called by the Japanese side in the Euro–Japanese VCR war. The Japanese industry accepted the VER (Voluntary Export Restraint) deal negotiated between the EU and the Japanese government. This point is explained in more detail in the following section.

Contrary to the expectation of Philips and Grundig, the EU's protectionist policies did not make much difference. The V2000 system was killed off by VHS, and Philips and Grundig became the licensees of VHS.

The European VCR price war was resumed by contenders from South Korea in the mid-1980s. Since 1985, some Korean companies including Samsung, Daewoo, and Lucky-Goldstar started selling their VCR products at prices 20–30% below the average market prices on the European market by means of direct export.[41] Once again, European manufacturers represented by Philips, Thorn EMI, and Thomson asked the European Commission to investigate the dumping activities of Korean companies. This time, some Japanese manufacturers in Europe joined the lobbying group against the Koreans. The reason was simple: Japanese manufacturers did not want their VCR prices to be challenged by the cheaper Korean products in a technologically Japanese-dominated European market. Ironically, a couple of Japanese companies, Funai and Orion, were also accused of being involved in dumping because these two firms were simply assembling parts imported from Japan in 'screwdriver plants' in Europe.[42] Obviously, the cheap Korean VCRs made the already intensified VCR price war more complicated.

The VER Agreement and EU Antidumping Policies

Having accepted the complaints and allegations of the European lobbyists against, firstly, Japanese and, then Korean VCR manufacturers, the EU authorities undertook a series of protectionist measures in favour of European firms.

The first significant measure was the signing of the VER agreement between the EU and the Japanese government. Commissioners Viscount Davignon and Wilhelm Haferkampf was sent by the European Commission to Tokyo to start talks with the Japanese government, represented by MITI, concerning the problems caused by Japanese VCR exports to Europe. In February 1983, an agreement was signed between the EU and MITI, under which Japanese VCR manufacturers agreed to collectively limit their VCR exports to the EU. As required by this voluntary export restraint agreement, the maximum annual VCR export from Japan to Europe should not exceed 4.55 million for 1983, compared to the total VCR shipment of 4.9 million units in 1982. This VER agreement was initially negotiated to be valid for a period of three years. Accordingly, the European Commission dropped its official investigation over the previously accused Japanese companies.

Shortly after the cease-fire of the Euro–Japanese dumping argument, Korean companies rapidly increased their VCR market share in the EU countries. According to EU official records, the VCR market share of Korean companies in the EU increased from 1.2% in 1985 to 6.1% in 1986 and 15.3% in 1987; this market share was achieved exclusively by direct export to the EU countries.[43] This provoked the European manufacturers (including Japanese firms producing in the EU) to file a second complaint of dumping in March 1987. In accordance with EU legislation, the European Commission singled out a list of Korean and Japanese companies which were believed to be involved with dumping in the EU market and provisional and definitive duties were imposed on them in 1988 (see Table 4.13).

EU antidumping actions were extended from VCRs to video cassette tapes and other key components. Companies affected in the latter areas of trade were mainly those from South Korea and Hong Kong.

Additionally, under EU's new regulations, all imported VCRs from non-EU countries were charged import tariffs at 14%.

Apparently, the purpose of the EU antidumping policies and procedures was to restore a 'fair' trading practice between the European firms and those from East Asian countries. However, the widely believed underlying purpose of these EU actions was to make sure that sufficient market space was available for the Philips/Grundig V2000 system.[44] Ironically, the big Japanese firms had all established their production capabilities inside Europe well before EU antidumping duties were actually imposed in 1988.

Another crucial factor regarding the VCR war was the consumer choice. It is believed that, apart from German buyers, most Europeans preferred to purchase Japanese brands:

> There is no avoiding the fact that European consumers like and generally prefer Japanese consumer electronics products, especially home video; to protect the Grundig and Philips V2000 range in the EEC is no guarantee that it will improve consumer acceptability either in Europe or in the rest of the world (Grant, 1983, p. 10).

Table 4.13. Rates of Antidumping and Price Undertakings Accepted on VCR Imports from Korea and Japan.

Exporting Company	Provisional Duties (%) 26.8.88	Definitive Duties (%) 26.8.88
Daewoo (Korea)	29.2	Undertaking accepted (17.8)
Goldstar (Korea)	26.4	Undertaking accepted (14.2)
Samsung (Korea)	25.2	Undertakings accepted (12.9)
Other Korean exporters	29.2	23.7
Funai (Japan)	18.0	Undertakings accepted (8.6)
Orion (Japan)	18.0	13.0
Average rate of duty	24.3	15.0
Estimated effect on selling prices	18.2	11.3

Note: Where price undertakings were accepted, it was assumed that the exporting company raised its cif import value by the amount of the dumping margin and this amount is shown in brackets.
Source: *Official Journal of the European Communities*, L240/5, 1988 and L57/55, 1989.

It is no secret that the European antidumping policies were implemented, to some extent, at the expense of the interests of the consumer because of the higher prices of VCRs. On the other hand, these actions failed to help maintain the V2000 format as a European technology. Figures in Table 4.15 indicate the costs paid by the consumers due to antidumping measures.

According to the National Consumer Council (1990) assessment, the 1983 VER agreement and the antidumping duties that followed together could possibly have caused an increase in VCR prices of about 26%.

The French Protectionism

As mentioned earlier in this chapter, the attitudes of national governments in the EU varied towards foreign VCR exports in the early 1980s. Amongst the twelve EU countries, the French government adopted the most restrictive policy on the import of foreign-made VCRs. Unlike the black and white TV and CTV sector in which the domestic market was mainly secured by technological barriers,[45] France did not have its own VCR manufacturing industry and all the VCRs had to be imported from foreign countries.

Aiming to reduce its trade deficit,[46] the French government implemented an extremely protectionist policy through the Finance Ministry concerning

Table 4.14. Rates of EC Antidumping Duty Imposed on Imported Video Cassettes from South Korea and Hong Kong.

Exporting company	Provisional rate of duty (%) December 1988	Definitive rate of duty (%) June 1989
Korean exporters		
Goldstar	10.8	2.9
Kolon Industries	7.6	2.0
Saehan Media	4.5	1.9
SKC	6.6	3.8
Hong Kong exporters		
ACME Cassette	59.3	9.3
Hanny Magnetics	59.3	21.9
Magnetic Enterprise	20.5	15.8
Magnetic Technology	59.3	21.9
Swilynn (HK)	8.1	0
Swire Magnetics	11.3	4.9
Wing Shing Cassette	59.3	Undertaking
Yee Keung Industrial	59.3	9.3
Casin	–	9.3
Unweighted average rate of duty	30.49	8.64

Source: *Official Journal of the European Communities*, L356/47, 24 December 1988 and L174/1 22 June 1989.

Table 4.15. Consumer Cost of EC Antidumping Measures Affecting Consumer Electronics Imports.

	Annual EC consumer cost (£ million)	Annual U.K. consumer cost (£ million)
VCRs:		
– 1983 VER agreement	175	80
– 1988 antidumping duties	174	48
Video cassettes	34	7
Total	383	135

Source: Adapted from National Consumer Council (1990).

VCR imports at the beginning of the 1980s. By the end of October 1982 the French government decided that all foreign-made VCRs coming into the country would have to pass through a 10-man customs office in the remote western town of Poitiers.[47] This procedure was universally interpreted as simply aimed at blocking or, at least, slowing down VCR imports although the French government officially claimed that this move was a simple reorganisation. Furthermore, under this new measure, all imported VCR products had to bear a label identifying their country of origin. Export documents accompanying goods had to be in French. In answering external criticism for this extreme protectionist policy, the French government explained its purpose as to encourage people to 'buy French'. Ironically, there had been no French domestic VCR manufacturing industry prior to the implementation of this policy! Additionally, some government officials argued for such a policy on the grounds that VCRs were not an indispensable item for the French people.[48] It is reported that after the French government announced its new import policy, more and more Japanese VCRs were blocked at the Poitiers customs point; and in January 1983 more VCRs were stolen from the growing pile than were permitted legitimate access to the French market (Grant, 1983, p. 9).

The Poitiers measures caused wide concern and controversy because the new regulations were not only curbing Japanese VCR exports to France, but also bringing difficulties to the already well-established intra-EU trade. Consequently, the European Commission urged the French government to provide more information to justify its Poitiers decision. Shortly after the VER agreement was reached by Japanese companies, the Poitiers measures were removed. It seems that the Poitiers action was a deliberate attempt on the part of the French government to provoke the EU to make radical policy changes towards Japanese VCR exports to Europe.

THE RESTRUCTURING OF THE EUROPEAN VCR INDUSTRY

The restructuring of the European VCR industry in the 1980s was mainly carried out through major direct investments by Japanese consumer electronics companies. On the European side, an important event is worth noting: the so-called Grundig affair—a planned but failed takeover of majority shareholding in Grundig by Thomson in the early 1980s.

Japanese FDI in the European VCR Industry

An immediate result of the European protectionist policy was the booming of Japanese direct investment in the European countries after 1982. Pursuing a new strategy, the Japanese consumer electronics multinationals started to build up a new image as 'good corporate citizens' through their European branches. From then on, the 'Made in Japan' VCRs in the European markets were gradually replaced with those made in Europe but still branded as Japanese. It is reported that, in the period 1983–1986, Japanese companies built more than a dozen VCR factories in Britain, West Germany, France and Spain.[49] Localised VCR production helped Japanese manufacturers to get round the protectionist measures imposed in international trade. Grant (1983) explains the strategic transition of the

Japanese consumer electronics firms to avoid new trade barriers:

> Capacity in Japan's consumer electronics plants is in excess of domestic demand, voracious though that is, and the Japanese have to export their products; throughout 1980 and 1981 VCR exports equalled 75–80% of production. Yet, because they export to all world markets and retain such flexibility on their production lines, Japan's consumer electronics companies have been able to meet the protectionist argument head on by generally agreeing limits to their exports and by switching production and/or assembly to locations in the consumer markets (p. 9).

In the case of Britain, VCR sales boomed to 3.6 million sets in 1982 from only 1.5 million the previous year and the rapid increase of VCR sales and household penetration rate was achieved in the post-1982 period as shown in Table 4.16. The figures of cumulative sales show a similar trend to increases in VCR annual output in Britain during the period 1982–1986. Although production and assembly of European firms increased after the 1983 VER agreement,[50] the booming sales and production of VCRs was still mainly explained by the success of Japanese firms in expanding their local production capacity in the U.K.

In the 1980s, the big Japanese consumer electronics companies in Britain started manufacturing VCRs to supply the British market and the continental European countries. In 1988, there were eight consumer electronics companies engaged in VCR production in Britain, of which seven were wholly-owned by Japanese companies and the other one was a joint venture between Amstrad of the U.K. and Funai of Japan, not to mention the Ferguson/-Thomson/JVC joint venture which had been producing VCRs in Newhaven before the sale of Ferguson to Thomson in 1987 (Cawson *et al.*, 1990). That is to say, the rapid increase of VCRs in Britain after 1982 was mainly promoted and supplied by the Japanese consumer electronics manufacturers in the country.

Table 4.16. Sales and Production of VCRs in the U.K., 1977–1986.

Year	Cumulative	Penetration sales ('000)	Production (%) ('000)
1977	20	–	–
1978	85	–	–
1979	200	1	–
1980	550	3	–
1981	1,520	7	–
1982	3,600	17	20
1983	5,780	27	180
1984	7,285	34	300
1985	8,885	40	400
1986	10,690	46	950

Source: BREMA; BIS-Mackintosh. Cited in Cawson *et al.* (1990).

By the mid-1980s, after the Japanese companies had well established their VCR production in Britain, they started to extend their direct investments to other European countries such as W. Germany. Britain used to be seen as a 'spring-board' for Japanese firms to export their locally-made VCRs to other European countries. Since the mid-1980s, it seemed more sensible for the Japanese to create more 'closeness' (both geographically and culturally) with other European national markets. This consideration brought about another wave of direct investment by the Japanese VCR manufacturers in W. Germany. Until the early 1980s, Sony and Sanyo were the only Japanese consumer electronics investors in W. Germany. Following these two companies, a number of Japanese consumer electronics companies including Matsushita, JVC, Hitachi, and Toshiba established their VCR production sites or links in the country.

As soon as the big Japanese consumer electronics multinationals had achieved their presence in Europe, the effectiveness of the European protectionist policies began to change. In other words, EU's policy measures, such as export quotas, heavy custom duties, antidumping investigations as well as the VER agreement, were no longer difficulties for big Japanese companies who were producing in Europe. On the contrary, these Japanese companies have benefited from those policies and sometimes joined the European lobbying groups demanding higher tariffs and antidumping procedures against those relatively smaller companies of South Korea and Japan. When commenting on the change in the stance of big Japanese companies, a Thorn adviser states,

> When the complaint was only against the Koreans, the Japanese companies were very eager to participate.[51]

It seems that EU antidumping procedures and other protectionist policies have only been continuously effective against those less established foreign consumer electronics companies from the Far East since the Japanese giants had all been comfortably accommodated by their European host governments from the mid-1980s.

In short, the death of the short-lived Philips/Grundig V2000 system unveiled an era characterised completely by the Japanised European VCR industry with VHS emerging as an unofficial but *de facto* standard.

The Grundig Affair

Grundig was a German company established in 1947 and one of the leading consumer electronics manufacturers in Europe. As partner with Philips in developing the V2000 system, Grundig offered 24.5% shareholding to Philips in 1979. Grundig's VCR sales increased from almost nothing in 1978 to 450,000 sets in 1980 and then rose rapidly to an estimated one million sets in 1982.[52] In the same year, Grundig stood as the only European-owned VCR manufacturer inside the EU.[53]

In the absence of protection from the German government, Grundig, together with its collaborator Philips, was being driven out of the VCR market by fierce Japanese competition, especially their 'cut-throat' VCR price war at the beginning of the 1980s. In the financial year 1980–81, Grundig had a major loss of DM 187m partly resulting from its unhealthy VCR business.

After six months' talks between the chief executives from Thomson and Grundig, an announcement was made by Thomson on 19 November 1982 that it planned to take over Grundig by assuming the majority interest—a 75.5% shareholding—in the latter. This news had a stunning impact because had the takeover plan been realised it would have led to the most far-reaching reorganisation of the European consumer electronics industry since the Second World War.[54] At the time the announcement was made, Grundig had just concluded an agreement with Telefunken which would allow Grundig 26% of control in the latter. In the meantime, Thomson had already taken over several smaller German firms including Saba, Nordmende and Dual.

Thomson's takeover bid for Grundig immediately caused wide concern and debate even at governmental level in Germany, and in the field of Franco–German relations. The dispute of the Grundig affair was concluded with a decision by the Bundeskartellamt (FCO—Federal Cartel Office) of W. Germany to reject the Thomson bid. The main reason for the FCO to reach this decision lay in the consideration that the interwoven business links between the three companies—Thomson, Grundig and Philips—would have great potential to be detrimental to competition.

It is believed that Philips did not give up its veto right[55] to interfere in the bidding process of the Thomson–Grundig affair. In other words, Philips did not want to see Grundig being taken over by Thomson, because Thomson showed no interest in joining the V2000 camp to compete with Japanese firms. The departure of Grundig from the V2000 group would have immediately put this besieged European VCR system at an end. As shown earlier in this chapter, the market share of V2000 had already been driven to an insignificant proportion in comparison to those of VHS and Beta in most European countries by that time. Meanwhile, Philips itself had never lost its interest in seeking majority control over Grundig's daily business operations.

Compared with Thomson, Philips' business links with Grundig were more long-standing. The commercial relationship between these two companies had made Grundig one of Philips' biggest TV tube buyers since the Second World War. It is believed that Philips had been gradually assuming more control over Grundig even before Thomson's takeover bid was announced:

> It has been clear for several years that Philips has wanted to acquire a larger stake in Grundig, with which it cooperated in production of the V2000 range.[56]

By the end of 1983, Philips, rather than Thomson, finally announced that it was to take control of Grundig from April 1984 and the latter was allowed to retain its independence as a company but the internal organisational structure would be changed in accordance with Philips' corporate strategy.

Euro–Japanese Collaboration

On the one hand since the beginning of the 1980s European companies, represented by Philips, had been seeking a large-scale European industrial collaboration to fight against the fierce competition from the East Asian consumer electronics giants, mainly the Japanese; but on the other hand most European firms in the consumer electronics industry were eager to join forces with their Japanese rivals. This contradictory corporate mentality of the European firms was embodied in most technology and product areas since the early 1980s, as far as the European consumer electronics industry is concerned. Traditional hostility between European firms sometimes left space for the Japanese companies to penetrate into the European markets by means of seeking access to or partnership with individual European firms. In doing so the Japanese companies could easily avoid the restraints imposed by the EU's protectionist policies. During the VCR format battle most of the major European consumer electronics manufacturers collaborated with their Japanese competitors in various ways at different times. Table 4.17 shows a number of Euro–Japanese industrial collaborations established in the VCR sector during the 1980s.

Table 4.17 indicates that the majority of the industrial collaborations established during the last decade were Euro–Japanese business arrangements.

Table 4.17. Industrial Collaborations in the European VCR Sector.

Firm	Firm	Nature of venture
Philips	Grundig	Collaboration on VCR technology
Thorn-EMI	Telefunken	Collaboration on VCR production (with JVC in J2T)
Thorn-EMI	JVC	Joint venture for VCR production (J2T)
Telefunken	JVC	Joint venture for VCR production (J2T)
Thomson	JVC	Licence agreement to manufacture VCR mechanisms; continued JT joint venture for VCR manufacture
Bosch	Matsushita	Joint venture for VCR production
Grundig	Matsushita	Collaboration on VCR components manufacture
Amstrad	Funai	Joint venture for VCR manufacture

Source: Adapted from Cawson *et al.* (1990).

CONCLUSION

In this chapter I have examined the development process of the European VCR industry, in which Philips experienced its rise (as an innovative technology leader) and fall (when its own systems failed to survive the competition posed by Japanese firms). Despite the fact that the EU authorities and some member states intervened in the VCR format battle at a late stage, the process of technical change and the competition between alternative systems was, by and large, determined by market forces. To summarise the study on the competition between the European VCR technologies, particularly the Philips/Grundig V2000 system, and the Japanese technologies, mainly JVC's VHS format, the following points are worth emphasising.

Firstly, corporate strategies played a decisive role in the VCR format battle. Earlier discussions in this chapter have shown that technical features might be important (e.g. Philips' N1500 and N1700 formats suffered from their bulky size, short recording/playing back time etc.), but it was not the technical parameters which decided the failure of the V2000 system and the victory of the VHS format. On the contrary, different ways of handling the technologies in the marketplace, namely corporate strategies of Philips and JVC, were the major factors for deciding the winner and loser.

On the European side, Philips made a number of mistakes in making its strategic decisions about its V2000 technology. First of all, the Philips management had put most of their attention on technical aspects and failed to make sufficient efforts in marketing the V2000 technology. This was in sharp contrast with JVC's strategy, which emphasised the importance of marketing. Another mistake lay in Philips' misunderstanding about the functionality of the VCR technology as a whole. Philips management perceived VCR as a time-shift machine which would be used for recording TV programmes when users were not able to watch them. Therefore, Philips engineers developed the V2000 systems with a recording time of up to 8 hours on a single tape. As a complementary system, the LaserVision video disc system was launched by Philips, almost simultaneously with V2000, to be used by consumers to play prerecorded software. By doing so, Philips did not make much effort to gain support from the software industry for the V2000 systems. Consequently, the LaserVision system did not take off on the European consumer market, and V2000 was unable to become popular partly due to the lack of prerecorded software in the shops.

Secondly, and closely related to the issue of corporate strategy, Philips' organisational structure had a significant impact on V2000. More specifically, because of the fact that Philips was like a 'loosely organised federation' with a decentralised management system, its daughter company Philips North America was able to choose not to cooperate with the headquarters in Eindhoven for marketing V2000 machines in the U.S.

Philips North America's decision effectively blocked the way for V2000 to get into the U.S. market.

Thirdly, the result of the VCR format battle clearly demonstrated the pivotal role of industrial collaboration in establishing a product standard in the market. Apart from wide range licensing of its VHS technology, JVC persuaded Matsushita to adopt the VHS system and entered joint venture agreements with a group of European companies, in particular Thomson (*via* the J2T venture). Matsushita was a world leader and Thomson was the second largest concern next to Philips in Europe as far as consumer electronics manufacturing was concerned. Having failed to get the support from Thomson and other European manufacturers, Philips and Grundig were left alone with its V2000.[57] The 'Grundig affair' showed the difficulties of industrial collaboration between major European firms, i.e. Philips and Thomson.

Fourthly, since the early 1980s EU public authorities adopted a series of protectionist measures, e.g. antidumping policies, negotiations of VER agreement with the Japanese government in order to improve the declining European VCR industry. In addition, the French government also applied special policy, i.e. the 'Poitiers affair' to curb VCR import from Japanese firms in the early 1980s. To avoid trade protectionism in Europe, Japanese firms shifted from exporting finished goods to direct investment to manufacture VCR machines and components in major EU countries such as the U.K. and Germany in the 1980s. Japanese direct investments further intensified the process of 'Japanisation' of the European VCR industry. Therefore, it was argued in this chapter that European public policies in the last decade failed to rescue the European V2000 technology and improve the competitiveness of European firms against foreign competition.

Finally, a protectionist trade policy of the EU was initiated in the early 1980s under strong pressure from leading European manufacturers such as Philips and Grundig. This chapter has shown that, when V2000 was faced with increased competitive pressure from alternative Japanese systems (VHS and Beta) in the marketplace in the early 1980s, Philips and Grundig shifted their attention to seeking political solutions. The first proposed antidumping investigation was announced by the European Commission after a series of industrial lobbying led by Philips and Grundig.

The EU antidumping measures were initially taken to protect indigenous technologies against foreign competition. But the nature of these measures have subsequently become more complicated. It was recently reported that Philips had lodged new complaints with the European Commission about VCR imports from South Korea and Singapore, where Korean firms (Daewoo, Goldstar and Samsung) and Thomson (in a joint venture with Toshiba) have manufacturing plants. In 1994, about 50% (or 5.4 million out of 10.5 million units) of VCR machines sold in the EU were imported.[58] It is known that Philips, together with its subsidiary Grundig, manufactures

VCRs in its German and Austrian plants; whilst Thomson has about 1.5 million VCRs supplied by its Singapore joint venture with Toshiba. To a large extent the recent Philips complaint was one against the commercial interests of Thomson, the French state-owned company. Ironically, both Philips and Thomson have been manufacturing their VCR machines with the same Japanese technology, i.e. the VHS format. In the early 1980s the French government was determined to keep the Japanese VCR machines outside its borders as shown by the Poitiers affair; this time, in the mid-1990s, if the Commission decides to apply antidumping measures, French-made VCRs may well become victims.

Having examined the failure of the V2000 system and the competitive pressure imposed by Japanese and other East Asian firms on the European VCR industry, the next chapter will turn to discussing public policy changes within the EU and Philips' strategic change and organisational restructuring in response to foreign competition in the consumer electronics industry since the 1980s.

NOTES

1. Figures in this paragraph are mainly from Mackintosh (1990), 'The Competitiveness of the European Consumer Electronics Industry'.
2. According to Grant (1983), the retailing price for Philips' first VCR machine, the N1500, in 1974 was equivalent to £1,200 in today's prices, and one tape with a maximum recording and playing time of one hour cost about £75. These figures slightly differ from those from other sources. For instance, Geddes and Bussey (1991) suggest that the Philips N1500 was launched in 1972 at £315, equivalent to well over £1,500 in 1990 (p. 412).
3. In an attempt to get into the home VCR market, Sanyo and Toshiba designed the V-Code system and Matsushita developed its VX format. Both systems failed to survive. See Cusumano, Mylonadis and Rosenbloom (1991).
4. Keen, B., 'Play It Again, Sony', *Science as Culture*, No.1, 1987.
5. According to Cusumano, Mylonadis and Rosenbloom (1991), Japanese firms exported 53% of the VCR machines they produced in the year 1977 and approximately 80% from 1979 onward.
6. It is believed that, at the beginning of the 1980s, the cost difference between the V2000 machines manufactured by Philips and Grundig and VCRs imported from Japan was as big as about 40% (Cawson *et al.*, 1990, p. 227).
7. Firms from S. Korea accounted for 7% of the total VCR production in Europe in 1989.
8. *Financial Times*, 1 October 1981.
9. *Ibid.*
10. *Ibid.*
11. According to Geddes and Bussey (1991), some two years before launching the V2000 system, Philips, in the hope of establishing an industry standard, had tried to persuade Thorn, Thomson, Telefunken and Grundig to adopt it and, in the end, only Grundig joined the V2000 camp (p. 414).
12. See *The Nikkei Weekly*, 1 February 1992.

13. Mackintosh International Ltd (1985), 'The European Consumer Electronics Industry', p. 79.
14. Interview with Philips Historian, Philips International B.V., Eindhoven, 28 October 1991.
15. *Ibid.*
16. See 'VCRs II: Will the Winner Please Step Forward', *Multinational Business* (U.K.), Issue No.1, 1984, p. 38.
17. Differing from the figures presented in Table 4.10, another source has suggested that, in 1983, there were 12 active VCR manufacturing plants altogether in the EEC, of which 2 were producing machines based on Sony's Beta system sharing about 8% of the total VCR output in the same year. See Mackintosh (1985), 'The European Consumer Electronics Industry'.
18. Fox, B., 'Shizuo Takano', *Electrical & Radio Trading*, 30 January 1992. Mr Takano was the head of JVC's video division and vice-president of the company. Takano was widely known in the industry as 'Mr VHS' or the 'father of VHS'.
19. See Geddes and Bussey (1991), p. 414.
20. *The Nikkei Weekly*, 1 February 1992.
21. The Japanese CTV system is NTSC.
22. *Financial Times*, 16 November 1983.
23. Interview with Senior Manager, Strategic Planning, Philips (U.K.), London, 11 July 1990.
24. *Ibid.*
25. See note 6 for the price difference between V2000 and VHS machines.
26. 'VCRs II: Will the Winner Please Step Forward', *Multinational Business* (U.K.), Issue No. 1, 1984, p. 38.
27. Thanks to Pioneer's efforts LaserVision has been successful in the consumer market in Japan. The system has also been well accepted as a consumer product in other areas such as Taiwan and Hong Kong over recent years.
28. *Business Week*, 29 April 1991, p. 37.
29. Interview with Senior Manager, Strategic Planning, Philips (U.K.), London, 11 July 1990.
30. Note that joint ventures between the Japanese and European firms formed an important part of JVC's strategy to penetrate the European VCR market. However, some other areas, such as CTV, in the consumer electronics industry did not see the same pattern. For instance, "it was not to be joint ventures but the establishment of wholly owned plants by Japanese companies (and by one Taiwanese company) that would transform the U.K. consumer electronics industry in the 1980s" (Geddes and Bussey, 1991, p. 405). Indeed, the only two joint ventures, Rank–Toshiba and GEC–Hitachi, established in Britain in 1978 failed to survive.
31. Philips (1991), *Philips: a century of enterprise*, a Philips-designed CD-i disc, Eindhoven.
32. 'Philips' High-Tech Crusade', *Business Week*, 18 July 1983, p. 100.
33. See *Philips Annual Reports*.
34. According to *Philips Annual Reports,* sales of consumer products as a share of total sales of the Philips group increased from 28% in 1985, when Philips finally dropped its V2000 production in Europe, to 41% in 1989. Income from consumer

products increased from only 1% to 48% of total income by product sectors in the same period.

35. Interview with Senior Manager, Strategic Planning, Philips (U.K.), London, 11 July 1990.
36. 'Philips' high-tech strategy: if you can't beat 'em, join 'em', *Business Week*, 28 May 1984.
37. The Philips/Matsushita joint venture was started in 1952 and the agreement was renewed in 1966 and in 1977, but Philips sold its holding to Matsushita in 1993.
38. Figures in this paragraph are mainly from Grant (1983).
39. According to *Financial Times* (16 November 1982), Grundig was forced to reduce the price of some of its top VCR models from about DM 2,900 (£679) to about DM 1,900 in 1981. By the end of the following year, the company started selling its VCRs at under DM 1,000 in a number of stores.
40. In accordance with GATT rules, EU legislation stated that dumping occurs when exporters sell at a price lower than the cost of production in their own domestic market.
41. 'VCR Wars: Korea threatens the cease-fire', *Business Week*, 13 April 1987.
42. The term 'screwdriver plant' refers to plants set up by foreign producers inside the Community member states to assemble finished goods using components and parts imported from their home countries. EU regulations state that products assembled by 'screwdriver plants' shall be subject to the same anti-dumping duties as products imported directly from the country in question (National Consumer Council, 1990, p. 11).
43. See *Official Journal of the European Communities*, L240/5, 31 August 1988.
44. According to National Consumer Council (1990), under the 1983 VER agreement, a quota for Japanese exporters was 4.55 million units and European VCR manufacturers were guaranteed a market of at least 1.2 million units for the same year.
45. In France, black and white TV signals were transmitted with an 809-line system for many years. The French CTV standard has been established as SECAM which is different from either the PAL system or the NTSC system since the 1960s.
46. By the end of 1982 France was facing a trade deficit of around 100 billion francs ($14 billion), of which 12 billion francs were caused by trade with Japan who was currently supplying 95% of the French VCR market. See *International Herald Tribune*, 28 October 1982.
47. *Ibid.*
48. *Ibid.*
49. 'VCR Wars: Korea threatens the cease-fire', *Business Week*, 13 April 1987.
50. After the signing of the VER agreement in February 1983, VCR production and assembly of European firms increased from 24% in 1984 to 29% in 1985, 34% in 1986, and 40% in the first half of 1987. See *Official Journal of the European Communities*, L240/5, 31 August 1988.
51. 'VCR Wars: Korea threatens the cease-fire', *Business Week*, 13 April 1987.
52. *Financial Times*, 20 November 1982.
53. Philips was operating VCR production in its Vienna factory in Austria.
54. *Financial Times*, 20 November 1982.

55. By having a 24.5% shareholding, Philips was eligible to maintain a veto right in deciding the selling of Grundig's other shares since 1979. With this veto power Philips was supposed to have a big say in Thomson' takeover of Grundig in 1982.
56. *Financial Times*, 20 December 1983.
57. Sony's Beta strategy was not better as far as the company's industrial collaboration was concerned.
58. See *Financial Times,* 19 April 1995.

5

EUROPEAN COMPETITIVENESS AND PHILIPS' RESTRUCTURING

The general situation of the European VCR industry, in which Philips failed to establish a world standard with its V2000 system due to fierce Japanese competition, was discussed in Chapter 4. The V2000 failure had a tremendous impact on Philips' future development. In particular, the V2000 experience prompted the company to restructure its internal organisation and readapt its corporate strategies.

The V2000 case also had a profound impact on the European consumer electronics industry. The EU authorities intervened in the industry by means of implementing protectionist policies, but these actions were too late to be effective, not to say whether they were appropriate or not.

The objective of this chapter is to examine in some detail the restructuring of Philips' organisation and the development of its new strategies. The first part of the chapter analyses the European consumer electronics industry and the EU's public policies towards the sector. This shows that Philips' difficulties should be understood within the more general problems of the industry, and the relative ineffectiveness of EU public policy to deal with the problem of competitive weakness at the level of major firms.

THE COMPETITIVE CHALLENGE OF JAPAN

During the period of the late 1970s to early 1980s, Japan posed a major competitive challenge in most sectors of electronics and, in particular, gained a dominant position in the world consumer electronics market.

In conquering the European consumer electronics market, major Japanese firms passed through three stages of activity: export of finished goods; direct investment; and, finally, 'Europeanisation' of their business on the local market. However, as argued by some Europeans and Americans, the Japanese home market remains 'impenetrable' to foreigners.

'Japanisation' of the European Market

Japanese consumer electronics companies first came to Europe with their finished products at the end of the 1960s. European firms were taken by surprise, as one Philips manager recalls:

> It was not till the end of the 1960s that the Japanese companies turned their attention to the European electronics market. It was about 1971 when I saw the Japanese produced portable radio appear on the European market for the first time, and I was quite shocked. The Japanese radio was much cheaper than the equivalent Philips product.[1]

The Japanese started their exports to European countries with small radio receivers, which were followed over the next two decades by audio equipment, small screen TV sets, VCRs and CD players. Gradually, the quality and design of Japanese products have been appreciated by consumers; and the Japanese made their products increasingly popular throughout the world with an aggressive pricing strategy.

Japanese exports to Europe were scaled down by the mid-1980s because of EU antidumping policies. In response to the increasingly upgraded trade protectionism in Europe, Japanese companies quickly shifted their strategy from exports to foreign direct investment (FDI).[2] This was the second stage of the 'Japanisation' process in Europe. On the one hand, many assembly plants and manufacturing factories were set up by Japanese firms (as shown in Table 5.1); on the other hand, Japanese firms also formed a number of joint ventures with local firms.

Table 5.1. Japanese-Owned Consumer Electronics Manufacturing Plants in Europe, January 1990.

Product area	No. of plants	No. of employees	Capitalisation (initial, $mn)
CTV	14	8,599	78.3
VCR	20	11,556	100.6
Camcorder	1	600	10.7
CD player	7	2,618	91.3
Audio	14	3,946	36.5
Total*	**46**	**20,454**	**264**

Note: * Note some overlap in plants by product.
Source: BIS Industry Research/Jetro, quoted in BIS Mackintosh Ltd. (1990), prepared for the Commission of the European Communities.

Some Japanese assembly plants were believed to be 'screwdriver plants'—which were simply assembling parts imported from their home market. These plants were later required by the EU to increase the proportion of local content in order that their output could freely circulate within the EU as EU-made products. One consequence of this was an increase in production in Europe by Japanese-owned components manufacturing plants.

Japanese competition, wherever, is not only a threat to the European industry but also a great concern to other parts of the industrialised world. Generally speaking, the history of electronics production has been characterised by the domination of different types of products since the 1960s. The 1960s was dominated by audio, the 1970s by video, the 1980s and 1990s by chips and computers, and the 21st century may be characterised by telecommunications. The Japanese emerged in the 1960s, and they gradually out-competed the Americans and Europeans with their cheap but high quality products in every decade. They won the VCR battle; they have become powerful chips producers; and they are gaining ground in the computer sector. It seems that it is not only Europe which has been Japanised; more broadly, the whole world is gradually becoming dominated by Japanese companies as far as the electronics industry is concerned.

'Europeanisation' of Japanese Firms

In more formal terms, 'Europeanisation' means the localisation and rationalisation of Japanese firms' European businesses. By the end of the 1980s, Japanese consumer electronics manufacturers had achieved a major presence in the biggest European national markets by means of export and direct investment. Initially, these Japanese investments, either in the form of wholly-owned subsidiaries or through joint ventures with local firms, were spread over Europe and there was hardly any coordination at a regional level, with each national subsidiary reporting back directly to Japan. These Japanese plants were generally smaller than comparable plants in Japan, and unable to achieve the same economies of scale.

The 1992 programme and the emerging single European market offers a lucrative business opportunity, not only for the Europeans, but also for the Japanese. In part to take advantage of the single market, European firms, such as Philips, are undergoing a radical restructuring process, and some believe that the Japanese firms will eventually take the same kind of action:

> The Japanese factories in Europe are going to suffer in the future, because they are all very small. In this country [U.K.], there are factories of Sony, Hitachi, Sanyo, Mitsubishi, etc. Some of these factories are only producing 100–200 thousand television sets per year. If you look at France, Germany, Spain, you'll see the same picture. So they will have to go through the rationalisation process that we did in the seventies.[3]

Cawson and Holmes (1991) comment on the recent rationalisation of Japanese firms as follows:

> ... recently there have been increasing signs of a move towards setting up integrated European companies. Sony led the way by centralising its European CTV manufacturing under a headquarters in Stuttgart, and by creating a 'European executive committee' to coordinate the development strategies of the subsidiaries and prevent duplication of effort. Others, including Hitachi, Matsushita and Pioneer, have followed by setting up Europe-wide organisations to coordinate their subsidiaries. ... there is an important organisational dimension which suggests that the Japanese will be more effective at 'Europeanisation' than the Europeans themselves (p. 176).

It seems that 'Europeanisation' of the existing Japanese business in Europe is another formidable push to further 'Japanisation' of the European consumer electronics industry.

The Impenetrable Japanese Market

It is widely believed in Western Europe that the competitiveness of the Japanese consumer electronics industry partly comes from the impenetrable Japanese home market. In other words, while a large number of Japanese multinational companies have been enjoying the benefits of exporting to and investing in other parts of the industrialised world, foreign companies are excluded from the Japanese market in many areas. As far as strategically important sectors, including the consumer electronics industry, are concerned, Japanese companies have well established their presence in both the U.S. and Europe, but neither American nor European firms have any substantial presence in Japan.[4]

In recent years, American semiconductor companies and European consumer electronics manufacturers have been complaining about the 'unfair trading practices' of Japanese companies. More specifically, American and European firms claim that they don't get equal treatment on the Japanese market because foreign firms face a variety of obstacles. Among others, the peculiar Japanese distribution system, the *Keiretsu* network as an alliance mechanism, and the alleged higher price level on the Japanese market are the most widely cited ones.

Firstly, in contrast to the U.S. and Western Europe, where independent retailing chains dominate the distribution of finished consumer goods, most big Japanese manufacturers have their own distribution network in which numerous retailing stores exclusively sell their parent company's products. This peculiar market structure is believed to be a very important factor in forging the Japanese business bloc, in which foreign goods face considerable difficulties in reaching the consumers.

Secondly, the *Keiretsu* are known as integrated and powerful industrial and trading groupings, in which various manufacturing companies and trading organisations are, either horizontally or vertically, connected to each other, normally with a big bank or large manufacturer at the centre of the industrial family. One of the most important characteristics of the *Keiretsu* system is that members within each industrial family offer each other share-holding. Although *Keiretsu* members compete with each other, the mutual share-holding system provides a mutual strengthening mechanism amongst members. It is believed that a foreign company would find it impossible to become a member of the *Keiretsu*.

Thirdly, it has been argued that the price level of Japanese electronics goods in Japan, particularly consumer electronics, is much higher than it is elsewhere in the industrialised countries. This price difference between Japan and other areas is believed by non-Japanese industrial organisations to be detrimental to the principles of free trade.[5] Table 5.2 has been picked out by Philips in its official position paper to demonstrate the unfair trading practice of Japanese electronics companies.[6]

In addition to the above three points, social and cultural factors are also believed to have contributed to the formation of the alleged impenetrable market. For instance, to set up a competitive business in a place, one needs to employ the 'top class' or the most capable staff, in particular managers, from the country concerned. However, it is extremely difficult for a European company such as Philips to get hold of the desired staff, because most Japanese are not willing to work for a foreign company operating in their own country.[7]

The 'impenetrable nature' of the Japanese market has triggered a new wave of protectionism in the U.S. and Western Europe since the early 1980s. International negotiations at a governmental level have been conducted to ease the tension of bilateral trade relations. The American–Japanese semiconductor agreement reached in 1986 (renewed in 1991) and the EU–MITI agreements on video cassette recorders in 1983 are two examples of this effort.

In criticising the alleged Japanese 'unfair trading practice', Overkleeft and Groosman (1988) argue that,

Table 5.2. Sales Prices of Japanese Branded Products.

1987 prices in U.S.$	Tokyo	Hong Kong	New York
Yashica camera	334	128	139
Casio calculator	41	6	17
Sony CD player	416	218	179
Brother typewriter	386	161	119
VHS Camcorder (JVC)	2,211	1,346	896

Source: *Wall Street Journal*, 21 April 1987, quoted in "Free Trade, Electronics and Europe", a Philips position paper, September 1989.

> It has proved that Japan is not really prepared, or not really able, to give Western companies access to its own market. Certainly not on a scale appropriate to a market of over one hundred and twenty million consumers. And absolutely not if that access might imply that foreign companies could get a lead on the Japanese market in those fields which have been defined by the Japanese government and companies as strategically important (pp. 78–79).

The above criticism is a clear indication of the European companies' discontentment about the current situation of the Euro–Japanese industrial and trade relationship.

PUBLIC POLICY AND THE EUROPEAN CONSUMER ELECTRONICS INDUSTRY

Generally speaking, prior to the beginning of the 1980s there was hardly any EU policy towards the consumer electronics industry. However, this 'hands-off' stance of the EU was substantially changed in the early 1980s, when major European consumer electronics manufacturers, such as Philips and Grundig, ran into trouble in competing with Japanese companies in the area of VCR. In response to the lobbying of European firms, the EU has implemented a series of interventionist measures, which were mainly aimed at scaling down foreign exports and protecting indigenous manufacturers. More specifically, EU antidumping policies and import quotas/VER agreements have been applied to sectors including VCR, CD players, and small-sized CTVs. More recently, stronger interventionist policies in forms of coordinating large-scale European R&D projects and financial subsidies at the EU-level have been implemented despite constant controversy. Among others, the EU HDTV policy and the European MEDIA programmes are prominent examples within the consumer electronics industry.

Debate on Free Market and Protectionism within the EU

It is no secret that official opinions within the EU have been divided between different policy domains as far as the relationship between competition policy and industrial policy is concerned. Whilst opinions on one side are in favour of free market and opposed to government subsidies to industries which are harmful to free competition, the other side believe that a competitive industry, such as consumer electronics, can only be ensured by strong interventionist and protectionist policies and, sometimes, proper EU or national government subsidies are necessary. The French government has been the leading advocate of protectionist measures, and of promoting leading firms as 'European champions' in the same way that 'national champions' were promoted in the 1960s and 1970s. The French Industry Minister, Dominique Strauss-Kahn, states that, broadly speaking, three kinds of businesses in which governments can legitimately act as guardians of national interests: defence; sectors where technological independence should be guaranteed; and vital supplies such as oil and nuclear energy. He also argues in favour of state funding for private companies for long-term and risky projects, such as making the next generation of semiconductors, and believes the EU gives too much power to its competition authorities.[8]

In the major consumer electronics product areas, the French government's protectionist stance within the EU has a long history. To protect its national economic, political, and cultural interests, the French government adopted its own SECAM colour television transmission system which was incompatible with both the PAL standard and the NTSC system. As discussed in Chapter 4, the French government implemented its notorious 'Poitiers' VCR imports policy which manifestly contradicted EU trade policy. During the course of debate over the European interim advanced television transmission policy, the French government, again, was the only one which strongly backed D2-MAC as the EU official standard for DBS (Direct Broadcast by Satellite) transmission. In fact, the first EU MAC Directive of 1986 and the 1991 Memorandum of Understanding as well as the renewed version of the MAC Directive for establishing a common European HDTV standard have been strongly influenced by the French government, who were joined by the major European consumer electronics manufacturers. More detailed discussion on this point is to follow in the next chapter.

It would, however, be insufficient to criticise European protectionism without referring to the international trade environment. Among the Triad economies, Japan, as argued by many Europeans and Americans, is still a substantially closed market to outsiders. Government intervention by means of either direct subsidies or national level coordination of key technological and product areas is not rare, not to mention the unique Japanese distribution system and *Keiretsu* networks. The American–Japanese semiconductor agreement and former President George Bush's visit, accompanied by a group of leading American industrialists, to Japan in January 1992 highlighted the increasingly enhanced protectionism and government intervention in the U.S. since the mid-1980s. Indeed, Europe's problem could not be solved by the EU alone; further negotiations at a global level, such as the GATT trade negotiations, would be necessary in order to liberalise international trade. It must be remembered, however, that GATT has no coercive power of its own and depends on national governments for its implementation.

EU Antidumping Policy

The first EU antidumping regulations came into existence in July 1976, and have been revised several times since. In accordance with EU regulations,

Community producers, either individually or in a group, are entitled to complain about the dumping of imported goods if they believe their interests are undermined. Under the requirement of the same regulations, the European Commission must respond to any complaint of EU producer(s) by means of undertaking investigations into the case, provided sufficient evidence is provided. The Commission has the power to impose provisional duties on the exporters of the product concerned for a period of up to four to six months; then, the Council of Ministers decides on whether to lift the provisional duties or impose definitive duties.[9]

To get round EU protectionist policies, foreign firms, mainly Japanese and Korean, quickly changed their strategy each time. One of the widely used tactics of East Asian electronics companies was to establish the so-called 'screwdriver plants'. There is no universally accepted definition for 'screwdriver plants', but the term is normally used to describe those plants set up inside the EU by foreign companies to assemble finished products using components and parts supplied by their home country.[10] To counter this problem, a 'screwdriver plant' regulation was added to the previous EU antidumping regulations under the decision of the Council of Ministers in June 1987.[11]

In addition to the above regulations, the EU has taken other measures such as tariffs and quotas to restrict imports into Europe. These measures have been applied to most product areas of consumer electronics since the beginning of the 1980s.[12]

Despite controversies over EU antidumping regulations, a number of antidumping cases involving consumer electronics products have been opened from the beginning of the 1980s, as listed in Table 5.3.

Very recently, European consumer electronics manufacturers, including Philips and Grundig, complained to the European Commission that some Japanese companies were dumping CD players indirectly into the European market by routing their imports *via* third countries like Singapore, Taiwan and Malaysia. Philips alleged that the price of the CD players being dumped into the European market was 40% lower than that of comparable models being sold in Japan.[13] In response to this Philips allegation, a surprised Japanese manufacturer declared that,

> We were flabbergasted. I think it is an example of a non-competitive industry unable to compete and therefore fighting in another way.[14]

Undoubtedly, argument and counterargument over the alleged 'dumping' practice have made it a more difficult task for the EU authorities to find a universally acceptable balance between competition policy and industrial policy. Philips and other European manufacturers complained about the unfair trading practice by the Japanese; but the Japanese claimed that European companies were 'fighting in another way'. It seems that while everybody in *principle* agrees to free market conditions or fair trading exercise, in *practice* each side of the trade conflict have been seeking or

Table 5.3. EC Antidumping Cases Involving Consumer Electronics Since 1983.

Year	Product	Exporting country	Action
1983	VCRs	Japan	Investigation closed
1987	CD players	Japan; Korea	Investigation opened
1987	VCRs	Korea; Japan	Investigation opened
1987	Video cassette tapes	Korea; Hong Kong	Investigation opened
1988	Video cassette tapes	Korea; Japan	Provisional duties imposed
1988	Small-screen CTVs	Korea; Hong Kong China	Investigation opened
1988	Video cassette tapes	Korea; Hong Kong	Provisional duties imposed
1989	VCRs	Korea; Japan	Definitive duties imposed
1989	Video cassette tapes	Korea; Hong Kong	Definitive duties imposed
1989	Small-screen CTVs	Korea	Provisional duties imposed
1989	CD players	Japan; Korea	Provisional duties imposed
1990	CD players	Japan; Korea	Definitive duties imposed

Source: Based on National Consumer Council (1990), *op. cit.*

utilising nonmarket forces to enhance their competitive advantages over the opposition. It appears that the process of establishing genuine free market competition is somehow resembling the search for the 'Holy Grill'! The European consumer electronics manufacturers have not been alone in 'fighting in another way' against Japanese exports. As early as 1959, the U.S. Electronics Association, backed by the U.S. government, filed allegations complaining that imports of Japanese portable transistor radios were threatening national security, and an overall quota was imposed on radio exports from Japan to the U.S. one and half years later (Gregory, 1986). In the 1960s and 1970s a strong coalition of manufacturers and labour was formed in the U.S. to lobby the Federal government to undertake antidumping measures to protect the American monochrome and colour TV industry, which was claimed to be under threat from Japanese exports. Although no antidumping duties were imposed on Japanese companies, a series of investigations was carried out, and the Japanese were forced to change their strategy. As happened later in Europe, Japanese manufacturers diverted their efforts from exporting to increasing direct investment in the U.S.

Antidumping policy is the latest phase of protectionism in the EU, following on from product-specific tariffs and voluntary export restraint agreements, as far as the European consumer electronics industry is concerned.

Bearing in mind the U.S. lessons and noting the 'Japanisation' trend in Europe, it remains to be seen whether the EU's antidumping policy is effective enough to protect its indigenous consumer electronics industry.

EU Coordination and Subsidies

Since the mid-1980s, EU policy-makers have become increasingly determined to take action to directly intervene in European industries. In the area of consumer electronics, the Commission has been deeply and widely involved by means of coordinating European R&D programmes and, sometimes, proposing financial subsidies.[15] Despite continuing disputes over the use of public money, financial subsidies from both the EU and national governments have been made available to certain new or 'strategic' technology sectors which are seeing strong international competition.

EU and the Hardware Industry

The Eureka 95 Project was launched in 1986 as an independent pan-European R&D collaboration aimed to develop a proprietary high definition television (HDTV) system in competition with the Japanese for a world standard. Gradually, EU95 became a *de facto* EU project because of the increased political and financial involvement at the Community level.

First of all, the Council of Ministers passed its MAC Directive to back the industry's initiative for establishing the HD-MAC system in 1986.[16] At the same time that the future of HD-MAC became increasingly gloomy, EU–industry relations have become more intense. Apart from the routine administration work undertaken by the Commission (particularly through DG-XIII and DG-III), both the Council of Ministers and the European Parliament have been frequently engaged in meetings and debates around the European HDTV policy and strategy.

Secondly, the EU also launched other Community projects under the umbrellas including RACE and ESPRIT to develop key technologies for HDTV.

Thirdly, the European Commission has set up a special European Economic Interest Grouping (EEIG), called *Vision 1250*, to oversee and coordinate efforts of various interested parties across the hardware industry, software industry, and broadcasting organisations to promote the development of European HDTV technology and advanced TV services. *Vision 1250* became fully operational in the first half of 1990. Detailed discussions about HDTV are to follow in Chapter 6.

EU and the Software Industry

In the audio-visual sector, or the consumer software industry, the EU has also committed considerable efforts to coordinate and promote progress in programme-making to support new technologies such as HDTV and interactive media. Investment in programme-making in these areas are believed to be more risky. The Commission exerts its influence in the software industry mainly through DG-X (Directorate-General Audio-visual, Information, Communication, Culture).

The overall Community audio-visual policy includes three major aspects: the establishment of the rules of the game (e.g. the Television without Frontiers Directive); promotion of the software industry; and the mastery of new technologies (particularly HDTV).

Recent EU audio-visual policies are in part embodied in the *Vision 1250 EEIG*, as mentioned above, and the *MEDIA Programme*.

As a Community programme, 'MEDIA' (Measures to Encourage the Development of the Audio-visual Industry) is one of those few areas which have been given priority in both the Council of Ministers and the European Parliament. The MEDIA Programme is designed in three phases.

Consultation of industry phase (1987). The first phase lasted a year and involved more than 2,000 representatives of the professions involved. It was accompanied by in-depth market studies in the production and distribution sectors.

Pilot experiment phase (1988–1990). On the basis of consultations and market studies, some ten pilot projects were set up in 1988, and more than 10,000 people and firms were involved from the Community.

Main phase (1991–1995). Starting from 1 January 1991, the MEDIA programme was intended to ensure that the achievements and guidelines emerging from the experiment phase bear fruit on a significant scale within a period of five years. Collaboration arrangements were established between MEDIA and the Audio-visual Eureka where and when necessary during this phase.[17]

Audio-visual Eureka is a sister or 'mirror' programme of the previous Eureka Programme, and aimed at promoting cooperation, development, and growth of the audio-visual industry on a pan-European base.[18] It is apparent that the 1985 Eureka Programme is mainly concerned with cooperation and projects for new technologies and the hardware industry; while, Audio-visual Eureka is individually concerned with the development of European audio-visual industry, namely the software sector. Audio-visual Eureka was launched under a joint declaration of ministers or representatives from 26 European states as well as the President of the European Commission at a meeting in Paris on 2 October 1989. With its secretariat located in Brussels, Audio-visual Eureka offers a similar host structure to *technological* Eureka to help professionals devise and execute transnational projects.

Generally speaking, Audio-visual Eureka and the Community audio-visual policy share a similar objective, i.e. to boost Europe's audio-visual production capacity. The relationship between the Community and Audio-visual Eureka is two-fold: the EU participates in Audio-visual Eureka through its MEDIA Programme, i.e. the Community is a partner of Audio-visual Eureka; in return, the Community's MEDIA Programme is extended to include new partners from countries within the Audio-visual Eureka framework.

EU Subsidies and Financial Incentives

Financial injection in forms of direct subsidies and economic incentives is the culmination of the development of EU industrial policy to intervene in the consumer electronics industry, which has already been under threat from Japan (the hardware sectors) and the U.S. (the software sector, particularly the film industry).

In an attempt to establish the increasingly controversial European HDTV system and encourage widescreen services, the European Commission proposed a budget of ECU 850m (£593m) to subsidise the industry over a period of five years from 1993 to 1997. Under the Commission's new Action Plan, industrial groups were expected to conclude bilateral contracts with the Commission and lay out their individual strategies if they wish to benefit from the Commission budget.[19] Financial incentives from the EU to the software industry excite less controversy. During the pilot phase of the MEDIA Programme the Community allocated ECU 13.5m as an initial outlay to attract private investors, professional organisations and various promotion bodies. For the Main Phase of the MEDIA Programme, the European Commission proposed a Community 'seed capital' of ECU 235m to be spread over five years from 1991 onwards.[20] The detailed 'seed capital' allocation plan is shown in Table 5.4.

Responses at the Industry Level

Government intervention in the consumer electronics industry has become more complex since the EU–Japan Business Round Table was initiated at the beginning of the 1980s.

The EU–Japan Business Round Table on consumer electronics was formed by representatives of leading consumer electronics manufacturers from the EU and Japan. Government officials with responsibilities for the consumer electronics industry from the EU and Japan were also invited as 'observers' to attend the annual meetings of the Round Table. The major intention of the Round Table meetings, as publicised by the two sides, was 'to have common orientations on technological and market developments in

Table 5.4. Indicative Breakdown of MEDIA Programme Costs (Million ECU), 1991–1995.

MEDIA Programme (The EC Action Plan) area	Proposed budget
Distribution mechanisms	**100**
– Distribution of films	40
– Distribution of video cassettes and TV broadcasting	20
– Support for multilingualism in TV programmes	10
– Development of markets for independent producers	30
Improvement of production conditions	**90**
– Development of preproduction	25
– Restructuring of the animated cartoon industry	25
– Promotion of the use of new European technologies in the production of programmes	25
– Contribution to the establishment of a "second market", using archive material	15
Stimulation of financial improvement	**15**
Others	**30**
Total	**235**

Source: Based on Commission of the European Communities, Directorate-General of Information, Communication, Culture (1990), *The European Community Policy in the Audio-visual Field: Legal and Political Texts*, Brussels-Luxembourg.

consumer electronics by means of an informal exchange of views'.[21] More specifically, the Round Table was primarily concerned with technical standardisation, and discussions at the Round Table meetings have been conducted in some key technological areas including DAT (Digital Audio Tape), HDTV, optical memory, erasable discs, home automation, etc. Of course, industrial cooperation between Japan and the EU was always one of the major topics discussed at the Round Table meetings.

Presumably, the Round Table has provided competitors from the EU and Japan with a mutual working environment to exchange information and views over key issues regarding the consumer electronics industry. The participation of officials from the EU and Japan would also possibly offer an opportunity for industrial organisations from both sides to present their views directly to government policy-makers. However, the EU–Japan Business Round Table, as a communication forum at an industrial level, failed to achieve the objectives as previously intended. In practice, trade conflicts between Europe and Japan have continued to exist as far as the consumer electronics industry is concerned.

In contrast to the intention of the bilateral EU–Japan Business Round Table, regional/national policy-making may be more likely biased by the lobbying of indigenous trade and industrial interest groups. Since its establishment in 1983, the European Association of Consumer Electronics Manufacturers (EACEM) has represented and served the interests of the European manufacturers. In 1984, EACEM suggested in its lobbying document presented to the European Commission that tariffs on VCRs, Hi-Fi equipment, and all new consumer durable products should be increased to 20% for a period of three years, after which they would be adjusted to 14%. To respond to the demands of EACEM, EU authorities agreed to a uniform rate of 14%, which was a significant increase for VCRs (8% prior to the increase).[22] More recent action taken by EACEM embodies the organisation's continued efforts to lobby the EU for protection. Immediately after the merger between Sky Television and BSB in the U.K., EACEM presented on the 11 December 1990 a position paper and industry proposal for action to the European Commission entitled "EACEM Position Paper on the Consequences of the Merger of British Satellite Broadcasting and Sky Television in the U.K. on the Future of European HDTV and Industry Proposal for Action on Broadcast Satellite Audiovisual Radiocommunication Services". The major objective of this Position Paper and Action Proposal was to pressurise the EU to implement tougher legal measures in order to protect the beleaguered European HDTV system.

In theory, EACEM is supposed to represent the common interests of all associated members from the industry. However, the demands of the leading members, such as Philips and Thomson within EACEM, are usually best and more effectively voiced under many circumstances.

Responses at the Firm Level

Conflicts between the European and Japanese manufacturers in the market-place have existed for many years, and neither industrial cooperation nor the EU–Japan Round Table could avoid them. It is widely observed that politically influential domestic firms tend to take the advantage of government intervention and industrial policy. Interplay of this kind between the two parties seems to suggest that public policy could easily be used by private (sometimes state) firms as a supplementary or alternative force to their corporate strategy in competing against their foreign counterparts.

Philips, on the one hand, has played a key role in slowing down or avoiding the process of 'Japanisation' in certain product areas in Europe. On the other hand, the company has shown consistent interest, and has made extensive efforts, in lobbying the European Commission for protectionist policies against East Asian, in particular Japanese, companies. In order to ensure an effective lobbying mechanism and more direct contacts, Philips maintains a unique 'diplomatic' relationship through its liaison office[23] in Brussels with the policy-making bodies of the EU:

> Compared to the limited resources of EACEM's resolving secretariat, Philips maintains a formidable lobbying organisation in Brussels. Its head, commonly referred to inside Philips as our 'Ambassador to Brussels', is in constant contact with Commission officials. Philips organises training seminars for EU officials to acquaint them with the problems of the electronics industry, and presents its own position papers on current issues.[24]

Philips' involvement in promoting the European antidumping policies in the VCR area has been discussed in Chapter 4. Together with other European manufacturers, Philips also made efforts to push the EU to take action and secure an independent European HDTV technology. More detailed discussion on Philips' role in the European HDTV campaign is to follow in Chapter 6.

EUROPEAN COMPETITIVENESS AND PHILIPS' PRESENCE IN EUROPE

As far as the consumer electronics industry is concerned, Europe's competitiveness is, to a great extent, contingent upon the technological and marketing strengths of a few European manufacturers, among which Philips, together with Thomson, have been playing a leading role. Despite the fact that Philips is a globally stretched multinational company, Europe remains the company's biggest and most important 'home base'.[25] In other words, the more Europe is 'Japanised', the less competitive Philips would be, and *vice versa.* Therefore, Philips' presence in Europe is vital to not only the company itself but also the European-owned consumer electronics industry as a whole.

Europe's Competitive Position

The electronics industry in Japan has already overtaken other major industries including automobile, chemicals and food and drink to become the largest industry. The same is expected to happen in Europe by the end of this century.[26] Within the global electronics industry currently valued at ECU 574bn ($689), consumer electronics represent 14% of the total; the global consumer electronics industry is currently dominated by about a dozen companies, of which eight are Japanese, three European, and two Korean.[27] The general characteristics of the European consumer electronics industry in the past 20 years are claimed to be competitive pressure, mistakes learnt and adjustment.[28] The competitiveness of an industry in a particular region is determined by various factors. The following aspects are chosen as major indicators of the European consumer electronics industry.

Ownership of Production

According to the BIS Mackintosh studies, as shown in Tables 5.5 and 5.6, of total production capacity by value ($11.8bn) of the consumer electronics industry in Europe in May 1990, European-owned ($8.5bn or 72%) was the highest compared with the Japanese ($2.8bn or 24%) and others-owned (less than 4% in total). However, the global picture for the same year was substantially different. Of the total global consumer electronics industry production value ($68.2bn), European-owned ($10.4bn or 15%) was the second highest next to Japanese-owned ($40.8bn or 60%). The difference between European-owned and Japanese-owned actual worldwide production value was very big ($30.4bn or 45%). If we look at the ownership of production for the Japanese consumer electronics industry, the difference was much bigger. In Japan, Japanese companies produce more than 95%, while the share of European and other firms together accounted for less than 5%, a very insignificant proportion.

Figures in Tables 5.5 and 5.6 also indicate that the European consumer electronics manufacturers were in a very vulnerable competitive position in comparison with their Japanese counterparts, although the former maintained a majority production share in their home base—Europe[29] and 30% of production ownership in the U.S. for the time being. As far as production is concerned, European companies have hardly made any significant presence in the Japanese consumer electronics market. On the contrary, having secured their home base, Japanese companies gained a majority production share (39%) in the U.S., and a considerable share in Europe (24%).

In the year 1988/89, as shown in Table 5.7, there were 167 manufacturing plants in the European consumer electronics industry, of which 46 were

Table 5.5. World Consumer Electronics Production by Country of Ownership (Actual Value $bn), May 1990.

	Region					
Ownership	Europe	Japan	Korea	U.S.A.	ROW*	**Total**
Europe-owned	8.5	Neg	–	1.6	0.3	**10.4**
Japan-owned	2.8	32.2	–	2	3.8	**40.8**
Korea-owned	0.3	–	7.6	0.6	0.1	**8.6**
U.S.A.-owned	–	–	–	1.2	0.8	**2**
Other-owned	0.1	–	–	0.2	6.1	**6.4**
Total	**11.8**	**32.2**	**7.6**	**5.4**	**11.1**	**68.2**

Note: * ROW = rest of the world.
Source: BIS Industry Research/EIAT/EIAK/EIA/EACEM, quoted in BIS Mackintosh Ltd. (1990), *op. cit.*

Table 5.6. Ownership of World Production for Consumer Electronics (% Value).

Production location	% of Total world production attributable to producer area	Japanese companies	European companies	USA companies	Korean companies	Others
World production	100%	60%	15%	3%	13%	9%
Japanese production	47%	>95%	←	<5%		→
European production	17%	24%	72%	0%	3%	1%
U.S.A. production	8%	39%	30%	20%	10%	1%
Korean production	11%	←	<15%	→	>85%	...
ROW* production	16%	←	45%		→	55%

Note: * ROW = rest of the world.
Source: BIS Research, quoted in BIS Mackintosh Ltd. (1990), *op. cit.*

Table 5.7. The European Consumer Electronics Industry (Number of Manufacturing Plants/Production Value), 1988/89.

Ownership	Number of plants	Production value ($bn)
Europe-owned	113	8.5
Japan-owned	46	2.8
Other Far East-owned	8	0.5
Total	**167**	**11.8**

Source: Based on BIS Mackintosh Ltd. (1990), *op. cit.*

Japanese-owned and another eight belonged to other East Asian companies. In the same year there were no wholly European-owned manufacturing plants in Japan, except Marantz, a Euro–Japanese joint venture, in which Philips was holding a 50% share in Japan.[30] The ownership of plants roughly matches the ownership of production in the European consumer electronics industry.

In each of the major traditional consumer electronics product sectors, European companies as a whole were also very much disadvantaged compared to the Japanese companies. Table 5.8 shows the ownership of companies from each region in product areas including CTV, VCR, camcorder, and audio in May 1990.

Table 5.8, with a 98% coverage of world camcorder production value, suggests that the world camcorder industry was completely dominated by the Japanese and Europe firms had no evident competitive strength in this area. In the VCR sector, European companies produced only 8% against 64% produced by Japanese companies of the total coverage (81%) of the world production value. It is worth mentioning that all European producers are licensees of the Japanese VCR technologies. In the audio sector, European-owned production value (16%) was also far behind the Japanese production capacity (40%) within the range of 64% of total coverage of the global production value. The position of European-owned production (23%) was slightly better in the CTV sector than in other product areas compared to the Japanese-owned production value (36%).

Market Share

Although the share of production ownership of European companies is much smaller than that of Japanese companies, Western Europe is the biggest consumer market compared to Japan and the U.S.

As indicated in Table 5.9, during the years 1988 and 1989, the market share of Western Europe in audio/video hardware products was the highest (34% and 33%, respectively) amongst the three major regions. Forecasts in the Mackintosh report suggest that the market development trend of the

Table 5.8. World Share of Manufacturing by Product, May 1990.

Profile companies	No. Main assembly plants	% World production value
CTV		
– European-owned	33	23%
– Japanese-owned	74	36%
– Other	21	10%
– Total coverage	128	69%
VCR		
– European-owned	10	8%
– Japanese-owned	42	64%
– Other	6	9%
– Total coverage	58	81%
Camcorder		
– European-owned	0	0
– Japanese-owned	9	98%
– Other	0	0
– Total coverage	9	98%
Audio		
– European-owned	19	16%
– Japanese-owned	82b	40%
– Other	2	8%
– Total coverage	103	64%

Source: BIS Mackintosh Ltd. (1990), *op. cit.*

Table 5.9. The World Market for Consumer Electronics Products (Actual Retail Value Market), 1988–2000.

	$Billion/% of total							
Region	1988		1989		1995		2000	
W. Europe	38,	34%	39,	33%	54,	30%	76,	28%
Japan	23,	21%	25,	21%	36,	20%	49,	18%
USA	23,	21%	25,	21%	32,	18%	46,	17%
ROW	27,	24%	28,	24%	58,	32%	101,	37%
Total	111,	100%	117,	100%	180,	100%	272,	100%

Notes: (1) Home audio/video hardware only—excludes magnetic and other media and nonconsumer products; (2) At 1988 $ exchange rates; (3) ROW = rest of the world.
Source: Based on BIS Mackintosh Ltd. (1990), *op. cit.*

concerned product areas will remain the same with Western Europe heading both Japan and the U.S. during the rest of this century.

Europe as a market is important to not only European but also Japanese manufacturers. Contrary to the successful market penetration of Japanese companies in Europe, European manufacturers, to a certain extent, have failed to explore the full potential and advantages of the European market since the early 1980s. Indeed, the market size of Western Europe does not necessarily suggest an optimistic future for the European consumer electronics industry. First of all, as discussed earlier in this chapter, Japanese companies hold an absolutely dominant position as far as audio/video production capacity and ownership is concerned. In other words, the European consumer electronics market is also supplied by Japanese manufacturers. Secondly, in the consumer electronic market a new generation of products, such as consumer multimedia, DCC (Digital Compact Cassette), Mini Disc, widescreen television, digital TV and digital video disc, etc., will become increasingly important during and after the second half of this decade. Experience suggests that the Japanese and American markets are critically important for the acceptance of new consumer electronics technologies. Therefore, a strong presence in the Japanese and the American markets for new products seems to be strategically vital for the European manufacturers, particularly those with a great innovative capacity, such as Philips, in competition with Japanese companies in the 1990s as well as the beginning of the next century.

Import and Export

Import and export figures are another indicator of the competitiveness of the European consumer electronics industry.

Industrial and market research reveals that, of the total value of the European audio/video home entertainment equipment market ($21.1bn) in 1988, over a half ($10.8bn or 51%) was supplied by imports, and 49% was locally produced (34% were local European-owned production, and 15% Japanese and others owned).[31]

The export side of the region sees a different picture. In 1988, total production value of the European consumer electronics industry, including all kinds of products, was $15.5bn, but only $2.4bn (16%) was exported to regions outside the EU and EFTA countries. The remaining $13.1bn (84%) was sold within the European market.[32]

Key Technologies

Market growth of conventional consumer electronics products, such as audio equipment, CTV, VCR, etc., has been seeing either a slow or declining pace. Therefore, further development of the industry and market will largely depend on the introduction of acceptable new products. The success of building a new consumer electronics product nowadays usually requires new components and new technologies such as semiconductors, LCD (Liquid Crystal Display), and miniaturisation technology, etc.

Semiconductors are crucial to the new generation of consumer products such as consumer multimedia systems and HDTV sets. In the beginning of the 1990s, Japanese companies became the dominant force in the world semiconductor industry.

As shown in Table 5.10, of the top 10 semiconductor manufacturers in 1990, six were Japanese, three American, and only one European (Philips). Intel and Motorola of the U.S. used to be the two leading semiconductor manufacturers for the computer industry. They were superseded by Japanese companies in the late 1980s.[33]

Philips' position in the world semiconductor industry has progressively weakened. After withdrawing from mainstream (SRAM—Static Random Access Memory) chips the company had partly pulled out of the European JESSI programme[34] (a programme which aimed to improve the competitiveness of the European semiconductor industry). It also divested its computer business in 1991. The controversial takeover of ICL, a European computer manufacturer, by Fujitsu in 1990 clearly indicated the increasingly aggressive strategy of Japanese companies towards Europe. While the 1980s saw a rapid expansion of Japanese consumer electronics assembly plants in Europe,

> [n]ow a new wave of investment is under way in large-scale semiconductor factories, led by suppliers including Fujitsu, Hitachi, Mitsubishi, NEC and Texas Instruments.[35]

It is becoming a reality that the development of the consumer electronics

Table 5.10. The Top Ten Semiconductors Manufacturers in the World.

Company	Country	Sales ($bn)
NEC	Japan	5.01
Toshiba	Japan	4.93
Hitachi	Japan	3.97
Motorola	US	3.32
Fujitsu	Japan	2.96
Texas Instruments	US	2.79
Mitsubishi	Japan	2.58
Intel	US	2.43
Matsushita	Japan	1.88
Philips	Netherlands	1.72

Source: Based on Dataquest, quoted in *Financial Times*, 25 July 1990.

industry will rely on a strong semiconductor industry, as the importance of semiconductor design grows in relation to increasingly complex products.

Another equally important high-tech area is display technologies, where LCD promises to replace the traditional cathode ray tube (CRT). It is widely believed that most of today's display devices, including TV screen, computer monitors, etc., will be replaced by flat panel screens. In particular, the large but thin and light flat screen may eventually make HDTV a desirable 'home cinema' for many families in the years to come.

As with the camcorder industry, LCD technology and production has been dominated by Japanese companies from the very beginning of the sector. Followed by a group of Japanese electronics companies (e.g. Hitachi, Toshiba, NEC, Mitsubishi and Matsushita), Sharp is the leading supplier of LCD products all over the world. Due to the huge planned investment of Japanese companies, as shown in Table 5.11, the LCD sector could be continuously dominated by the Japanese in the coming next few years.

In Europe, Philips is the major contender which has the potential to take on the Japanese domination in the area of LCD.[36] In an effort to catch up with its Japanese competitors, Philips recently opened a big LCD factory in Eindhoven. However, this action seems to be already quite late for Philips to win the LCD battle, because Japanese companies started LCD production about ten years earlier than Philips.[37]

Compared with European and American companies, Japanese electronics companies are also known for being good at miniaturisation. In this area, Sony has been the leader since its launch of the Walkman about 10 years ago. After the Walkman, Sony has introduced Discman and Data Discman

Table 5.11. Estimated Planned Investment in LCD, 1990–1992.

Company	Investment ($ mn)
Sharp	700
Sanyo	560
Matsushita	350
Hitachi	210
Hoshiden	140
Toshiba/IBM	130
Mitsubishi	70
NEC	70
Ovonic	26 (U.S.A.)
Total	**2256**

Source: BIS-CAP International, quoted in Commission of the European Communities, Directorate-General for Industry (1991), *Improving the Functioning of Consumer Electronics Markets.*

for CD-Audio discs and CD-ROM discs, respectively. To reduce the 12 cm CD disc to 8 cm in diagonal dimension, Sony has introduced its Mini Disc audio system to compete with Philips' DCC (Digital Compact Cassette) system.[38]

In order to get into the consumer electronics industry, Apple of the U.S. has already joined forces with Sharp, Sony and Toshiba to build the next generation of consumer products. Obviously, what Apple needs is not only the powerful marketing networks of the Japanese consumer electronics giants but also their technological advantages in LCD, miniaturisation and semiconductors.

Philips' Presence in Europe

In consumer electronics, Philips is the third largest manufacturer (after Matsushita and Sony) and the largest CTV producer as well as technology leader in CD-based optical disc products. Philips also holds the leading position in some traditional product areas such as electric lighting, shavers, dictation equipment, colour CRT, etc. Despite its presence in global markets, Philips retains a very strong European base.

Table 5.12 indicates that, in 1989, over a half of Philips' major global activities were based in Europe. In particular, 75% of Philips' R&D budget (which exceeds 4bn Dutch guilders per year) was allocated in Europe in 1989, and five of the company's six research laboratories, staffed with about 30,000 R&D personnel in total, were located in Europe.[39]

Philips is also a leader and active participant in various EU or European programmes, such as JESSI, ESPRIT, RACE, BRITE/EURAM, and Eureka, etc.

Philips' share in most EU and European projects, as shown in Table 5.13, was over 20% (e.g. 35% in BRITE projects and 26% in ESPRIT projects) in March 1990. During the period 1984–1990, Philips participated in 375 EU and European R&D projects, and contributed 6,222 man-year with a total investment budget of ECU 936m in various programmes. In these electronics-related technological programmes, Philips' leading position could hardly be matched by any other European firm. Therefore, Philips'

Table 5.12. Philips Activities in Europe (1989).

Activities	Euro-Presence
Sales	57%
Workforce	59%
Investment	60%
R&D	75%

Source: based on Philips: *Philips in Europe*, published by Corporate External Relations, Philips International B.V. (no date).

Table 5.13. Philips Participation in European Technology Programmes (Situation March 1990).

	ESPRIT 84/89	RACE MAIN 87/88	BRITE/ EURAM	OTHER PROGR.	TOTAL EEC SPONSORED	EUREKA
Number of projects with Philips participation	84	27	25	45	181	13
Projects terminated/ discontinued meanwhile	38	4	8	5	55	
RESOURCES						
Man-years all partners	7030	3600	437	360	11427	4150
Cost budget all partners (MECUs)	1069	555	47	53	1724	720
OF WHICH PHILIPS						
Man-years	1798	554	136	73	2561	1100
Cost budget (MECUs)	273	92	17	11	393	150
EEC/State contribution (MECUs)	137	46	8	6	197	60
Philips' share in projects (%)	26	16	35	20	22	21
Philips' share in programmes (%)	7.4	9.0	2.0	5.0	7.0	n.a

Source: Philips, *Philips in Europe, op. cit.*

commitment, to a certain extent, is vital to ensuring the success of most European programmes, and, eventually, the competitiveness of the European electronics (particularly consumer electronics) industry.

After the traditional audio/video products (radio, CTV, VCR, CD, etc.) have reached a high level of household penetration in the major world markets (Japan, U.S., and Western Europe), the competitiveness of the consumer electronics industry in a nation or region will be largely determined by the success of a new generation of products. Apart from the area of new TV systems, where Thomson, and Nokia are also heavily involved, Philips is the only European company with its own innovative products to compete with the Japanese and Americans. If Philips can make CD-i and DCC successful, both the company's position and the competitiveness of the European consumer electronics industry may be substantially improved.

Limits of EU Public Policy

To talk about Philips' active participation in various European programmes one must bear in mind the fact that Philips has been faced with trouble since early 1990, and consumer electronics, the company's biggest product division, has been severely affected by declining consumer demand during the long-lasting economic recession. Thomson, the second largest European consumer electronics manufacturer, has also been losing money in its consumer electronics activities. This is the reality which is most worrying to both the European industry and the EU. The search continues for a more effective policy solution to rescue the very vulnerable European consumer electronics industry.

As a matter of fact, the EU has implemented a series of protective policies and measures such as the above mentioned antidumping policy, coordinated European R&D programmes and direct subsidies. However, the level of competence of the European consumer electronics industry, as measured by its competitive position *vis-à-vis* Japan, has not been substantially changed since the early 1980s. In the meantime, in response to the European antidumping policy, many Japanese companies have substantially increased direct investment in Europe. This Japanese move might have been appreciated by the EU and some national government officials, but European manufacturers, particularly Philips, have been concerned with the steadily increased ownership by Japanese companies of production facilities in Europe. The decreasing European production ownership is certainly detrimental to the interest of indigenous European firms.

It might be possible that the EU will undertake certain legal changes in order to meet the international challenge. First of all, the EU may reconsider its unpopular technology policy, especially the coercive European HDTV policy[40] which was intended to give European manufacturers the opportunity to build a proprietary system for the next generation television. However, as the next chapter will show, the EU's HDTV policy failed to prevent HD-MAC from collapsing.

Secondly, it may be desirable to establish a Japanese-style distribution infrastructure in Europe. Having considered the impenetrable nature of the Japanese market, a European Commission official suggested that major European firms, such as Philips, might be allowed to set up their own distribution or retailing shops for those technologically complicated products like HDTV in the high streets in the future.[41] But, it is not guaranteed that the manufacturers' own distribution networks will be able to compete with the well-established retailing chains if the Community changes its competition policy.

Thirdly, so far Japanese companies have been excluded from EU R&D programmes and, in most cases, the pan-European programmes. If the Japanese companies come with expertise, know-how and capital, it might be advantageous to invite them to join the programmes. In return, the Japanese government may be obliged to further open the doors for European companies to participate in Japanese R&D programmes. In reality things may not be as easy as this.

To be sure, public policy is an important factor to the success of new technologies and, ultimately, the competitiveness of firms. However, I would argue, opportunities available in any business environment may favour only those companies whose internal organisational structure and competitive strategies are sound and effective.

STRUCTURAL CHANGE AND MANAGERIAL REFORM WITHIN PHILIPS

The optimal solution to the problem of competitiveness should, perhaps, come from inside a firm or industry. Apart from seeking an effective European solution at the EU level to tackle the increasing Japanese competition on the European consumer electronics market, Philips also committed great efforts to organisational restructuring and managerial reform from the 1970s. In other words, Philips' traditional corporate strategies and corporate culture no longer fitted in the new international competitive environment. More specifically, radical actions have been taken in order to overcome the drawbacks generated from the company's traditional technological diversification and geographical expansion policy, the dual management leadership, and the matrix configuration of the Product Divisions (PDs) and National Organisations (NOs), etc.

The major external forces which triggered the long-lasting structural reform inside the Philips group include the following.

- The formation of the EEC in the late 1950s and the subsequent European integration process, particularly the 1992 programme, offered a new market opportunity to European companies in terms of seeking economies of scale.

- In confronting the fierce Japanese competition, Philips and other European companies have clearly realised their many-fold competitive disadvantages—lack of efficiency in management, nonresponsiveness to the market, lower product quality, and lack of mass production, etc.
- From the beginning of the 1970s, production and trade have become increasingly globalised. Globalisation means that a company reaches the world market from one centre with a 'ONE' production approach—one TV centre, one computer centre, one switching equipment centre to serve markets all over the world.[42]

Rationalisation: Closures and Divestment

The first major aspect of Philips' corporate reform was the company's 'rationalisation' process. 'Rationalisation' *per se* does not necessarily mean that the previous corporate strategies and business activities were *irrational*. On the contrary, as far as Philips is concerned, rationalisation was necessitated by the accumulation of the long-standing historical development and the changing competitive environment in Europe since the late 1970s. On the one hand, the progress of the European common market provided manufacturers with new opportunities to explore economies of scale within Europe; on the other hand, the increasing Japanisation of some sectors, such as those of VCRs, camcorders and CD players, of the European consumer electronics industry left less and less performance room for Philips to pursue further profitability by simply following its traditional corporate strategies.

The meaning of Philips' rationalisation on the production side may be explained as follows.

Firstly, Philips dramatically reduced the number of production sites and now concentrates on only a limited number of international production centres.[43] Due to concentration and integration of production, a considerable number of factories were closed. Philips used to have 15 lamp manufacturing factories located in 15 different European countries, and 25 TV factories all over the world; in the U.K. alone, Philips had 4 TV factories in operation at the beginning of the 1970s, and it was not possible at all to explore the benefit of economies of scale, as far as TV production was concerned. It was hardly surprising at all that nonconcentrated production and market fragmentation in Europe substantially contributed to the total number of Philips plants at 512 as the highest point in the concern's history and this was reduced to 250 in the 1970s.[44] As a result of production concentration, the status of a number of factories was upgraded from national manufacturers to international production centres. The formation of international production centres means a complete transformation of the Philips company on the production side.

Secondly, Philips substantially narrowed down its product range. Nowadays, one production centre in a specific sector, such as lighting, can make products for different countries. In the past, each PD used to make many different product models to satisfy the demand of various markets. For instance, the Lighting Division used to make 2,000 types of lamps; after concentration the Division now only makes 1,000 types. In the past, whenever the customer wanted a specialised product, either lamp or radio or TV set, Philips always tried to make it. Philips used to lose money in trying to make everything the customers wanted, because a lot of small series of products were only made for very specialised markets.[45]

Thirdly, in addition to centralising its strategic decision-making power and concentration of international production, Philips has also undertaken an integration process to call together its R&D activities, production of components and assembly lines. Philips management believe this was another vital reform to ensure efficiency and quality:

> For the same product area, we used to have R&D in Eindhoven, but production of components in Germany. We spread ourselves all over the world. Meanwhile, we had various locations for production departments stretched from Eindhoven to Hamburg to Vienna. Managers from those places always came back complaining that they had problems in control of quality. As factory managers, they wanted all the things together for making the products. If you have better integrated production, you can better control the quality of your products. In trying to bring R&D, production of components, and assembly lines all together, we can be quicker in the market.[46]

Finally, as a globally stretched multinational company, Philips has no intention to solely confine its rationalisation reform to Europe; on the contrary, it is applicable at an international level. It is said that the ideal target the company is to achieve via its rationalisation process is that,

> ... we [Philips] have to have one-third of our production in America, one-third in Asia, and one-third in Europe.[47]

Obviously, there is still a long way ahead for Philips to achieve the above ideal of production allocation. For the time being, the largest proportion of the company's production and sales is allocated and carried out in Europe (see Table 5.12). Although Philips has made great steps in investing in many Asian areas such as mainland China, Taiwan, Hong Kong, Singapore, Malaysia, South Korea and Thailand in recent years, the company's sales in Asia accounts for only 8% of the total in 1989.[48] In particular, Philips' presence in Japan remains insignificant, considering the company's global business as a whole.

'Tilting the Matrix'

The historical aspect, namely, the formation of Philips' matrix structure was discussed in Chapter 3. In this section I continue the discussion and look at the company's long-standing commitment to and exhaustive effort in 'tilting the matrix'—the restructuring of the Philips organisation.

To tilt the matrix is not as simple as making the original columns and rows change positions. 'Tilting the matrix' is an expression within the Philips organisation, which implies that the Product Divisions demand and get a greater say in decision-making *vis-à-vis* the individual national markets. Therefore, if necessary the wishes of the national trading organisations are sometimes disregarded.[49]

From a historical point of view, Philips, as discussed in Chapter 3, had been following an expansionist policy until the 1970s. Partly because of the Second World War, Philips' diversification had brought about great structural and managerial problems. Due to the German occupation in Holland during the war time, most of the company's local organisations were left to do their own business. It seemed that there was no headquarters any longer in Eindhoven after the Germans army had inserted their military control to replace the previous normal business administration at Philips in Eindhoven. It is understandable that all the Philips NOs started running their own operation without having to report to Eindhoven under these circumstances. This unlimited autonomous decision-making power of the NOs had not been seriously challenged until 1987.

Before 1987 Philips had about 50 NOs and 14 PDs. Each PD was making products and selling to all of the NOs; while each NO had to buy from all of the PDs. PDs and NOs were the major elements (columns and rows) of the Philips matrix structure. Compared with the PDs, the NOs enjoyed greater autonomy. It is said that, within the very decentralised Philips group, about 95% of the decision-making power was held at the national level. At the headquarters in Eindhoven, the Board of Management (BOM) was only responsible for deciding on group management, investment, and the change of scope (whether or not the group should go to another new business). These three aspects were the sole 'property' of the BOM, and the rest of the corporate activities were completely decided by the NOs. Because the NOs were responsible for profitability, the PDs had to follow what the NOs said. Under normal circumstances, each NO issued orders of specific products to the PDs. Moreover, the NOs used to dictate the technical specifications of a product. As a result, for instance, the Lighting Division had to make special products for The Netherlands, special products for France, special products for Germany, etc. Other Divisions had to do the same. The tradition of designing and making an unrestricted range of products, to a great extent, was tolerated by the Philips management at the cost of economies of scales and, ultimately, undermined the competitive position of each Production Division in related industries. Indeed, each country has its own market environment, and adaptation to local customers and local preference was always necessary. However, the local preference of a country does not have to run into conflict with the principle of economies of scale. In the view of some senior Philips managers, this NO-dominated matrix structure caused tremendous inefficiency and bureaucracy within the Philips group.[50]

Within the matrix structure, conflicts always existed between the two parties—the technical side and commercial side during the last three decades. The commercial side always complained that the technical people had developed the wrong product; in response, the technical people said that they had developed an excellent product but the product failed to be accepted in the marketplace because the commercial people did not do their work properly. If a PD wanted to introduce a new product such as the Compact Disc system, it would be very difficult because certain horizontal marketing organisations might say 'No'. These NOs did not want the headquarters in Eindhoven to put all of its emphasis on Compact Disc; they have their own different priorities in their countries. Arguably, the NOs might want the BOM to put more emphasis on other products, such as CTVs or washing machines. This was the tension that has been concerning the Philips management.[51]

Due to the autonomous position held by the NOs, the administrative relationship between the headquarters and the national subsidiaries was not smooth at all. Sometimes, the BOM lost its central control over the NOs. An often cited example in this respect was the relationship between NAP (North America Philips) and the headquarters in Eindhoven. NAP remained almost an independent company during the period from the Second World War to the end of the 1980s. Strategic management decisions from Eindhoven were sometimes ignored or rejected by the subsidiary in the U.S., which some believed was the biggest problem once faced by Philips.[52] As mentioned in Chapter 4, NAP refused to accept instructions from its headquarters to market the V2000 system in the U.S. in the early 1980s. This was a typical case to indicate the challenges for effectively managing a multinational company with a federal organisational structure.

Philips was not alone. Many other big multinational companies, such as IBM, Unilever, etc., have also been faced with a difficult task: how to achieve the optimal form of coordination between the vertical PDs and horizontal lines of marketing organisations all over the world. We are not suggesting that Philips' solution is universally applicable to all the multinational companies with a matrix organisational structure. However, Philips' extraordinary experience in tilting its organisational matrix is certainly worthwhile considering by those international industrial organisations which are also being 'haunted' by a similar structural problem.

Within Philips the matrix structure problem was officially noticed more than two decades ago. In the later 1960s, some PDs started to try selling their products directly to the customers rather than through the NOs. The then Telecommunication Division successfully sold its products directly to customers like the PTT in The Netherlands without causing too much resistance from the local NO. Having obtained permission from the BOM, the then Minicomputers Division in France also attempted to do business in the same way in 1969. However, the ambition of the Minicomputers Division was not realised in most national countries. Wherever the Minicomputers Division people travelled to, the local NOs usually requested a commission charge over the products to be sold as an 'entrance fee' from the former; otherwise, people from the PD would not be allowed to do any business in those countries. Sometimes the NOs played various tricks to simply block the PD people from direct involvement in the local markets. People from the Minicomputers Division were 'nicely' asked by the NOs, "Would you please let me go to your customer?"[53] The resistance from the NOs was fed back to Eindhoven, and the BOM started formal discussions about the conflicts between the PDs and the NOs of the matrix in 1969. During the years 1969–1987 many discussions were carried out at the Board level but the matrix structure remained intact.

The bitter experience of some PD managers was also noticed by some experts in industrial organisational studies outside Philips. It is claimed that the 1960s was the time when the Philips matrix turned into mature, and conflicts, sometimes fights, between the PDs and NOs over decision-making power came to the surface.[54]

One year after taking over the presidency from Dr W. Dekker in 1986, Mr C. van der Klugt took the decision to radically change the traditional Philips matrix structure. The objective and result of this action includes the following aspects.

Firstly, most of the autonomy of the NOs has been taken away to Eindhoven, thus reducing the decision-making power from 95% to 5% at the NO level.

Secondly, PDs have been given more priority and autonomous decision-making power. All the NOs are going to be instructed by the PDs. Now, each PD decides on worldwide policy for products, technologies, manufacturing and investment.

Thirdly, regular consultation between the PDs and NOs has been established. Although the PDs have been given more decision-making power, they are required to have regular consultation with the marketing managers in the NOs. As far as consumer electronics is concerned, the PD and the management of the NOs from important countries including Germany, U.K., France, U.S., and Japan, together set the priorities of products and production; they together decide the global strategies for the development of consumer electronics.[55]

Finally, to achieve more flexibility, Philips adopted a 'Business Unit' (BU) approach during the restructuring process. The BUs are the cells of the Philips group; they are normally assigned specific projects to develop and market new products. For the time being the company has about 50 BUs all over the world. A very important function of the BU is that it can bring together people from different PDs to work for the same project. Every new product brought to the market is usually not from one part of the company;

on the contrary, it involves expertise from different angles of the company. For instance, CD-i started as a new project. To make CD-i a success, people from any part of the company could be called into the project whenever it was necessary.[56]

After the matrix structure had been tilted, the decision-making power was transferred from the NOs to the PDs. However, share or distribution of power at the PD level was not equal. In January 1987, the GMC (Group Management Committee) was established. Apart from the BOM members, the Chairmen of three PDs (Consumer Electronics, Lighting, and Components) were also members of the GMC. In other words, these three PDs have been given priority and hence regarded as the most important areas of the company, compared to other PDs.

In its essence, the BOM (five people) are responsible for setting the direction, major strategies, and the total mission of the Philips group; whilst the management of the PDs (eight in total at present) are respectively assigned to take the operational responsibilities of the company. Compared with the responsibilities of the BOM and that of the management of the PDs, the GMC (nine people) are concerned with major issues coming up from the operational level. More specifically, the GMC makes sure that:

- there are regular cross-PD contacts;
- there are no overlapping activities, and different parts of the company don't compete with each other internally;
- most importantly, the whole company moves as one machine in the same strategic direction.[57]

Although the restructuring decision was announced in 1987, and various radical actions have since been taken, Philips' organisational restructuring process is far from being completed.

Establishing Sole Leadership

As discussed in Chapter 3, the dual leadership, the separation of technical/production management and commercial management, was started by the Philips brothers—Gerard and Anton Philips—and this tradition had been maintained as an important characteristic of Philips' managerial culture. As required by the dual management structure, the BOM and each of the NOs and PDs was headed by a technical/production manager and a commercial manager in parallel. This tradition was not substantially changed until the late 1980s.

One of the effective challenges to the dual management structure came from the creation of the Business Units. To a certain extent, the functionality of the BU is to bridge the PDs and NOs, which used to be separate and independent from each other. Within each BU one single manager, i.e. the BU director, is solely responsible for the overall management. The creation of the BUs and the establishment of sole leadership since the 1980s is described by Overkleeft and Groosman (1988) as a 'cultural shock' to Philips people:

> It becomes clear that this was tantamount to a 'cultural shock' in Philips when we see that the policy of two-man management—technical side by side with commercial—maintained since the days of the Philips brothers has now been almost entirely replaced by one-man management (p. 58).

The position reserved for the technical manager/director in the BOM was abolished in the late 1980s.[58] In other words, the dual managerial responsibilities were taken over by a single manager at each level of the company during the last decade. When asked about the change of Philips' managerial reform and the new structure, a senior Philips manager gave his view:

> We are in an accountable business now. With one man in charge of the Product Division, there is no dual structure any longer. The dual structure went back to the early 1980s. After that time we went to single-headed management. That is to say, only one manager is responsible for the world business now.[59]

Further pursuit of the sole management responsibility at the top level of Philips company has been realised since Jan Timmer took over the presidency in July 1990. Timmer is believed to be personally responsible for a list of 15 key 'president projects', which includes the company's three major new efforts in consumer electronics (HDTV, CD-i and DCC) as well as high-resolution monitors, cellular radio and active matrix LCD.[60]

The Myth of the Philips Culture

There is wide coverage and discussion in the media about Philips' corporate culture. However, most writing has failed to give a detailed account of the nature of the Philips culture. Instead, most writers use the Philips culture concept to refer to a typical multinational corporation whose major characteristics are bureaucratic, inflexible, inward-looking, federalist or decentralised, and, most particularly, the dual management style and the matrix structure. Evidence from interviews conducted within the company suggests that the concept of Philips culture is much more complicated. More specifically, the term 'Philips subculture' is worth exploring.

According to one source, there does not exist a single or unified Philips culture. Because of the fact that Philips businesses are located in many parts of the world and there are many Philips NOs, Philips culture is very much diversified. To put it in another way, there is a Philips culture in Eindhoven, there are also many 'Philips subcultures' in individual national countries.[61] In the meantime, Philips now has eight different PDs, which are totally different businesses. Each of these PDs has its own history or track record. Some PDs are very profitable and some less profitable or loss-makers. Because of the difference of business nature and the difference of their historical backgrounds, there are different corporate cultures in individual Product Divisions as well.[62] This diversified corporate culture, as acknowledged by the Philips management, has its drawbacks. Firstly, some Philips people from one location or one part of the concern tend to believe they are different from those from other regions or other parts of the concern. This is an important source of internal conflicts or 'turf fighting'. Secondly, the diversified corporate culture had created difficulties for management. At certain times the headquarters in Eindhoven had to give great autonomy to the NOs because of the independent nature of these subcultures. A senior Philips manager commented as follows:

> Philips culture is quite diversified. That means there have always been conflicts. People in the laboratories really think and believe they are different from those in the rest of Philips. Philips people from the Far East think and believe they are different from other Philips people in Europe. So we had to live on these different subcultures. That sometimes causes tensions and makes it difficult to manage from the corporate headquarters in Eindhoven.[63]

Recently the differences have become more pronounced in respect of the PDs rather than individual countries. PDs have been given much more power in decision-making; each of them has got global responsibilities. But the PDs will now have to appreciate and recognise the uniqueness of national culture in each individual country all over the world.

In Chapter 4, I mentioned the 'cultural shock' within Philips caused by the failure of the V2000 VCR system. It is another aspect of the Philips culture in which most of the staff believed that their company should be the leader of technology or Philips should be good at as many technical areas as possible. The decision to license Japanese VHS technology was not emotionally accepted by everybody from Philips. This cultural mentality is one of the aspects to be changed during the current restructuring process.

PHILIPS' CORPORATE STRATEGIES SINCE THE 1980S

Since the 1980s technological and industrial development has been increasingly characterised by interfirm collaboration either locally or at an international level. Philips has been advocating its new diplomatic terms such as 'coexistence', while Japanese industrialists recently started to talk about *kyosei*—symbiosis with their American and European counterparts.[64]

Philips management have adopted a new set of corporate strategies while radically restructuring their company's internal organisation in order to adapt themselves to the new international competitive environment. The most important aspect of Philips' new corporate strategy since the beginning of the last decade, I believe, is industrial collaboration for its new technologies and new products. In addition to Philips' participation in

European R&D programmes, as discussed earlier in this chapter, the current section further addresses Philips' industrial collaboration from the following aspects: the adoption of a 'make and buy' policy; setting up a wide range of joint ventures; and forming strategic alliances and partnerships with competitors.

The 'Make and Buy' Policy

As discussed earlier in this chapter, one of Philips' rationalisation measures taken in the last two decades included narrowing down the product range in each industrial sector. This policy was substantially different from the company's previous expansionist policy and diversification strategy before the 1970s.

The Philips management learned from its own mistakes (e.g. the failure of V2000) that making everything yourself may not be a feasible strategy in an age of cooperation. In many cases Philips lost money due to weaknesses in marketing. The adoption of a 'make and buy' policy by the Philips management reflects a significant change in the company's new competitive strategy in the 1980s:

> On the production side, we used to make everything ourselves: we were making glass, cardboard, we had our own transport system, etc. We asked ourselves: should we make everything? Shouldn't we rethink if we ought to be diversified? The answer is that we should not do everything ourselves; we should have a more active 'make and buy' policy.[65]

In addition to the V2000 failure, another factor suggesting a 'make and buy' policy seemed to be Philips' severe financial difficulties seen in 1990. Heavy losses eventually forced the Philips management to accept that,

> ... it can no longer afford to be in the forefront of developing every type of technology. Nor can it continue to develop and build in-house every significant component that it needs for its products.[66]

In pursuing the 'make and buy' policy, Philips has pulled out from several industrial sectors. For instance, the company stopped production of military equipment and defence systems; the white goods sector has been transferred to Whirlpool of General Electric; and the minicomputers business has been sold to Digital Equipment.

Philips' 'make and buy' policy meant that, rather than trying to do everything itself, it was prepared to make difficult decisions to buy in expertise: the number of R&D staff has been reduced by more than 30% to 28,017 in 1993, compared to 40,752 in 1989; and R&D spending was also cut down from Fl 4.6bn in 1989 to Fl 3.7bn in 1992.[67]

In a highly competitive environment, fewer and fewer companies attempt to be good at making everything. Sony has been known for its miniaturisation technology, JVC for its VCR technology, Pioneer for its audio equipment making, etc. What is Philips' corporate identity? Philips is known as the leader in the lighting industry, but, in 1991, changed its company name from *N.V. Philips' Gloeilampenfabrieken* to *Philips Electronics*. This implies that the company would be mainly competing in the electronics related areas, particularly consumer electronics. For the time being, the popular image of Philips is being a generalist rather than a specialist.[68] It is still too early to say whether Philips can change its image by pursuing a 'make and buy' policy.

Joint Ventures

During the period 1982–86, Philips had about 25 joint ventures, of which those with AT&T for telecommunications, DuPont for magnetic tapes and optical discs, and Siemens for chips were believed to be of great significance to Philips.[69] Since the second half of the 1980s, a number of new joint ventures have been set up by Philips with others. Some time after the termination of its V2000 system and becoming a licensee of the VHS technology, Philips established a joint venture with JVC for VCR production in Malaysia, turning a former enemy into a partner. To ensure success for its CD-i multimedia system, Philips has been trying to set up as many joint ventures as possible. Philips Interactive Media Systems has been developing a large number of CD-i titles through joint ventures and partnerships with independent developers. In the area of HDTV, Philips set up BTS jointly with Bosch of Germany for studio equipment.

However, as far as joint ventures are concerned, there is some scepticism due to the difficulties and problems in organising new businesses between parties. First of all, there are difficulties due to the difference of languages and national cultures of the partners. Secondly, to bring two or more companies together, the difference between corporate cultures can be a big barrier to successful management of a new business. Thirdly, to create effective cooperation, the partners need to overcome the difference between each other's administrative procedures such as accounting procedures, staff recruitment and human resource management, etc. Fourthly, the organisational structure of one company usually differs from that of others. Fifthly, to create a joint venture, resources (capital and human) of the parent companies involved would have to be integrated together. In that case, it is very difficult for the new business management to get rid of the counterproductive influence from the parent companies. Finally, although companies can set up a new business together, the new business is not necessarily successful or competitive.[70]

Some industrialists are more than sceptical; they have no confidence in joint ventures at all. Helmut Maucher, Chairman of Nestlé, recently stated that,

> I don't share the euphoria for alliances and joint ventures. First, very often they're an excuse, and an easy way out when people should do their own

> homework. Secondly, all joint ventures create additional difficulties—you share power and cultures, and decisions take longer.[71]

It seems that Philips has sometimes used joint ventures as a temporary measure to rationalise and restructure itself. Through a brief joint venture with Whirlpool, Philips pulled out from the white goods sector; in order to enhance its position in the consumer electronics industry, Philips acquired all the shares from Marantz, previously a 50%-owned joint venture in Japan. In a similar way, Philips increased its share from 50% to 75% in BTS, and from 25% to 80% in Grundig.

Strategic Alliances and Partnerships

Strategic alliances and partnerships are another aspect of Philips' new corporate strategy. This new strategy is best exemplified by the subtle relationship between Philips and Sony, two of the most innovative firms in the consumer electronics industry.

It is no secret that Philips and Sony have been fierce competitors for many years in the consumer electronics industry. This rivalry turned into a partnership at the beginning of the 1980s when the two companies jointly launched the Compact Disc system. Partly due to this alliance, CD gained wide support from the hardware industry manufacturers and the music industry hence became the world standard. From Philips' point of view, the major purpose to join forces with Sony was not because Philips needed Sony's technology but the most powerful backing of the Japanese industry, and the same reason also explains their further collaboration for CD-i:

> We have developed compact disc on our own. Why Sony came in? The only reason is that we needed a partner in Japan to have the Japanese industry behind us. That is why Sony came into compact disc [business]. Now, we are moving to CD-i, and it is basically the same. If we want to introduce a new product on the market, we need not only the support of the software industry but also the other companies.[72]

Another interesting but apparently confusing alliance between Philips and Sony was their agreement to support each other's new digital music technologies: DCC from Philips and Mini Disc (MD) from Sony. From a technological point of view, Philips' DCC (Digital Compact Cassette) is backward compatible with conventional audio cassettes. MD is a Sony innovation; it is a recordable digital optical disc with a diagonal measurement of only 8 cm (compared to 12 cm for an ordinary compact disc). DCC and MD were both intended as the world standard for the next generation of digital audio products. In contrast to the usual way that format battles happened in the history of the consumer electronics industry, Philips and Sony announced at the end of 1991 that they were going to support each other for DCC and MD. There might be untold stories behind the scene, but the obvious motivation for Philips and Sony to come up with this deal was that both companies wanted to gain competitive advantages for their own systems through alliance with the opposite. Philips believed that Sony's support would give DCC a better chance to be accepted in the marketplace.[73] Sony followed the same logic; otherwise, it would not have come into alliance with Philips. However, the dilemma was that there might be only one winner, either DCC or MD, in the future.[74] In describing Philips' new corporate strategy, one senior Philips manager commented:

> We are now working in cooperation with other companies. We are making friends with our competitors. If you make a list of Philips' strategic partnerships, that has increased tremendously during the last ten years [the 1980s]. Now we work together with Sony, Matsushita and lots of other Japanese firms. We also have many joint ventures like those in China.[75]

THE LAUNCH OF 'OPERATION CENTURION'

As mentioned above, it took the Philips management about two decades to decide on how to reorganise the company's matrix structure. Similarly, the advantages of the new organisational structure may not be realised in the short term. Since 1987 restructuring policy has been implemented in the company to tilt the 'rows' (NOs) and 'columns' (PDs) of the matrix. In the hope to further improve its efficiency, Philips launched 'Operation Centurion' at the beginning of the 1990s to radically change its internal managerial system and corporate culture.

'Operation Centurion' was intended as a radical reform programme, which involves almost every concern staff from top to bottom. The immediate spur to the launch of 'Operation Centurion' was Philips' 1990 financial crisis, which was deeply rooted in the company's long-term strategic mistakes.

Philips' Strategic Mistakes

As far as the development of consumer electronics is concerned, research for this book suggests that Philips has made three major mistakes in its long-term strategic manoeuvring since the 1970s.

Firstly, although Philips was already feeling the strong challenge from the Japanese by the early 1970s, the management of the company failed to make a correct judgement about the potential of Japanese consumer electronics companies during the following decade. At the beginning of the 1980s when Philips started to talk to the Japanese and try to take on the competition, it was already too late—Japanese companies had already become the leaders of the industry all over the world.

In retrospect, W. Dekker, the President of Philips from 1982 to 1986 and now the Chairman of the company's Supervisors' Board, was the first Philips CEO who correctly identified the Japanese companies as the major competitors in the electronics industry. Strategically, this was a new vision;

a European company such as Philips would have to take into account the Japanese industry and the Japanese market if it wished to survive competition in consumer electronics.

Philips led the way of innovation for a number of new consumer products: compact cassette, the first company to bring home VCR into the consumer market, the developer of the LaserVision system, and the inventor of CD. However, Philips failed to translate all of its new technical inventions into commercial successes. At the beginning of the 1980s, Philips got into serious trouble due to the V2000 failure. The picture on the Japanese side was completely the opposite. Having beaten both Betamax and V2000, JVC turned its VHS system into the *de facto* world standard for VCR. After Philips had dropped its LaserVision system, Pioneer turned it into a commercial success, firstly in Japan and then some other Asian areas followed by the U.S. While Philips made no substantial commercial return from CD players, Japanese companies, led by Sony, have made tremendous profits from them.[76]

Having been aware of the intensifying Japanese competition, the Philips management, headed by Dekker, strongly emphasised industrial collaboration and strategic alliances with competitors as one of the company's principal strategies since the early 1980s.

Secondly, as discussed in Chapter 4, Philips underestimated the indispensable role of the software industry, and, consequently, failed to create a synergy between hardware and software for several of its new products. The same mistake was made in launching the LaserVision system.[77] In describing the distribution of software as 'of fundamental importance for the success of hardware systems', Timmer, the current President of Philips, stated,

> Our somewhat unfortunate experience with LaserVision in Europe was due to the fact that coordinated distribution of software and hardware was, to a great extent, not properly organized. In Japan it did succeed.[78]

To a certain extent, consumer electronics, or more broadly electronics industry, is a software-driven sector. From the early history to the present time, every consumer electronics product is a combination of two essential aspects: hardware or the technology part and software or the content part. Without transmission of programmes radio and TV sets would be dead boxes; without music titles consumers would never be persuaded to buy CD players or other types of audio equipment. The success of VHS and the failure of V2000 were largely determined by the availability and the lack of prerecorded software, respectively. As the owner of CD patents, Philips' profit in this area mainly comes from software (through PolyGram) rather than hardware.

In the newly born consumer multimedia industry, the format battle between the contenders including CD-i, CDTV (now CD32), 3DO and other systems will be mainly decided by each system's software catalogue with a satisfactory quality level available on the market. Nonspecialists could hardly tell the technological difference between a CD-i player and a CDTV player.

For the next generation of digital audio products, both Philips and Sony have made considerable efforts in promoting their DCC and Mini Disc systems, respectively. It is very unlikely that both systems will survive the competition between themselves. Again, the winner will be the one who can get wide support from the music industry. The same can be said about the newly emergent Digital Video Disc (DVD) industry, in which the Multimedia CD system from Philips and Sony and the Super Density format from the SD Alliance are competing against each other.

Thirdly, in the last two decades, Philips usually started production from the top-end of the products and, consequently, failed to create a mass market and economies of scale. On the contrary, Japanese companies adopted a totally different strategy. They normally started from the low-end of products and competed with each other at their home market. By increasing production volume, the cost of the product was reduced. Following this logic the Japanese companies reached mass production and lowered their selling price. Having experienced cut-throat competition on their home market, Japanese companies improved their product quality, added features, and were also able to export to the U.S. and Europe at an aggressive price level, as demonstrated by the VCR case. In short, it is very likely that a company will lose the mass market if it focuses only on top-end production.[79]

'Operation Centurion'

On 3 May 1990 Philips disclosed its first quarter earnings, which plunged to Fl 6m (£1.9m) from Fl 223m in the same period one year earlier.[80] Many people from inside and outside the company were shocked by this news, because shortly before announcing the first quarter results Cor van der Klugt, the then President of Philips, told the shareholders that the company was heading for profits. Cor van der Klugt immediately resigned from the post of President and Jan Timmer became the new President. It was reported that the year 1990 saw a total record loss of Fl 4.24bn (£1.3bn),[81] a figure unprecedented in Philips' history of 99 years (1891–1990).

To pull the company out of the crisis and turn it into profitability, the management of Philips launched the 'Operation Centurion' programme in Autumn 1990. According to Prof. Prahalad, who was believed to be the architect of this radical restructuring programme, 'Operation Centurion' was focused on creating efficiency—efficiency in the management of current businesses as well as the efficiency with which new business is launched.[82] The Centurion process started with two objectives: (1) to close the performance gap—the accumulated deficiencies in quality, cost, cycle time for product development, response time to customers, productivity and logistic performance—when compared with the best global competitors; and (2) opportunity gap—the growth rates of Philips were much lower than that of the industry (the annual growth rate of the electronics industry was about 14% in the 1980s and is expected to remain the same rate in the 1990s.[83] More specifically, the Centurion involves the following aspects.

To substantially slim down the company. To achieve this objective, on the one hand, the number of total employees would be reduced, and on the other hand those nonprofitable business areas would be either closed down or sold off. The former was called 'cutting fat', and the latter was to 'cut the dead wood'.[84]

As far as the first aspect was concerned, the headquarters in Eindhoven was believed to be the biggest problem. Eindhoven is a well-known Philips town—there are about 60,000 employees working at the corporate centre, and about 200,000 people dependent upon Philips one way or another. The company's bureaucracy problem and inefficiency is partly explained by the huge Philips population, particularly the high proportion of those working in the managerial and administrative areas. However, to lay off employees at any time has always been a very difficult decision to make for the Philips management, because the company had been following a long-term employment policy. In other words, to lay off employees means you are making a 'very unpopular decision' in the company.[85] Having been nicknamed as 'Hurricane Gilbert', or not very nicely, the 'butcher of Eindhoven', Timmer was determined to take the 'unpopular decision'. In the first nine months of 1991, Philips cut 24,200 jobs—about 9% of the total employees.[86]

In respect to the second measure, the Centurion action would make sure that loss-making activities, noncore businesses, and long-term costly investments would either be sold off or liquidated. After the launch of the Centurion, the loss-making Information Systems Division was sold; the white goods business was also sold; the company discontinued its joint venture with AT&T for telecommunications and pulled out from some expensive JESSI projects for advanced chips. Now Philips has decided to concentrate its business mainly on consumer electronics.[87]

To raise individual accountability. The so-called 'old boys network' was believed to be a factor which blurs lines of responsibility within the Philips group.[88] Centurion was to make sure that every member of the Philips staff, particular the managers at different levels, should be aware of his or her own responsibility; and responsibilities will be clarified by a new contracting system:

> Now people are being made accountable. ... That is whereby all key issues per Division are itemised; all the suggested action plans that address those key issues are stated. ... problems have to be addressed and written down.

> Individuals are accountable for making those decisions ... Those plans become contracted with managers. In other words, they have to be delivered. If they have not been delivered, the contracts are there.[89]

To make a new 'mind set'. Centurion was set to call Philips people to pay more attention to those internal factors which have made the company's competitive position decline. Mr van der Klugt correctly identified the internal structural problem and started to tackle it from 1987. However, he failed to thoroughly fulfil his ambitious restructuring plan. In describing the new 'mind set', Prahalad suggests,

> Externalizing the reasons for competitive failure—unfair competition, high wage levels in Europe, and so on—must be replaced with an internal determination to 'win' in spite of such odds.[90]

Customer orientation. As part of the Centurion action, customer orientation was to turn Philips from a previously technology-orientated company into a market-oriented organisation. If this could be realised, it would be a great strategic as well as cultural change at Philips. To make the new 'customer-oriented' philosophy a common commitment of all the Philips staff, Timmer launched a 'Customer Day' on 7 January 1992. On 'Customer Day', Timmer gave a speech to address the issue of customer-orientation through a satellite television broadcast; and up to 100,000 employees listened to the speech at 1,000 Philips sites in 17 European countries.[91] The key message delivered on 'Customer Day' was that,

> It is not us but the customers who determine how we do things and every employee must be aware of this. Because everyone has customers, everyone has to play his part in satisfying those customers.[92]

Indeed, it is not Philips but the customers who make judgements about the company's corporate performance. The other side of the coin is that it is not the customer but Philips management who are ultimately responsible for the company's corporate strategy and the future of the company. 'Operation Centurion' has been launched, but Philips' corporate image remains blurred.

Can Philips Regain Competitiveness?

After its internal organisational revolution and the top-level strategic readaptation, could Philips as an electronics giant regain the competitiveness that it used to have in the international lighting industry? To be sure, it is not an easy task to make a convincing assessment of a global industrial empire like Philips. Instead of saying 'Yes' or 'No', I try to make a comparison between the Philips organisational structures before and after the company's radical restructuring started in 1987.

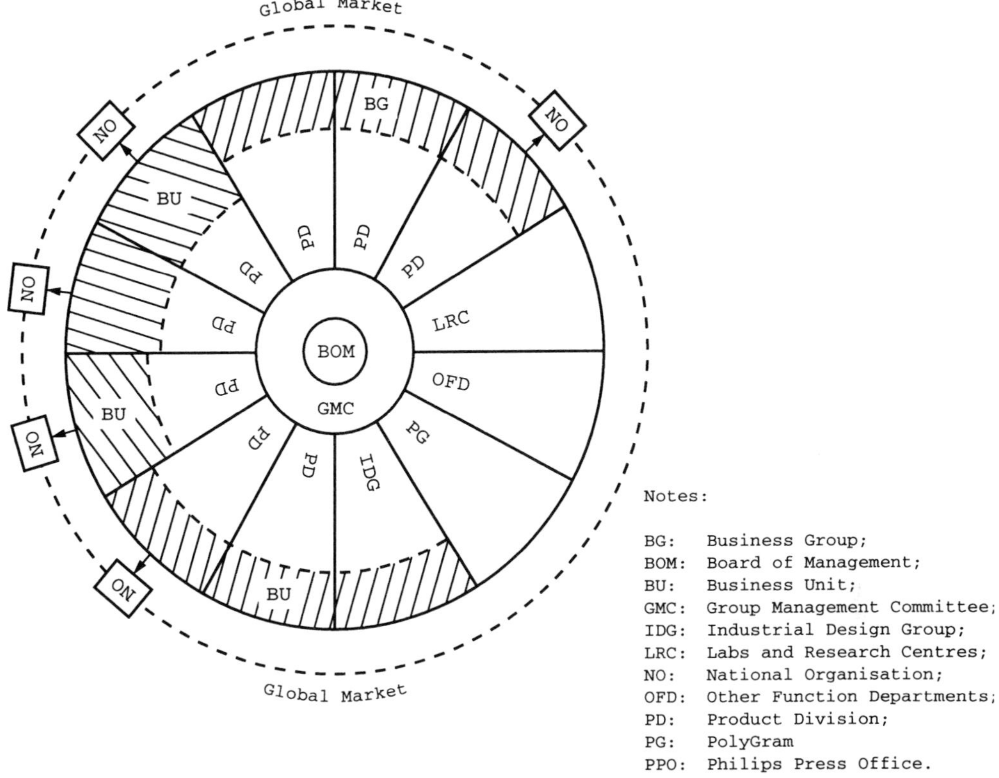

Fig. 5.1. Philips' New Organisational Structure after 1987.

Figure 5.1 represents Philips' organisational configuration after the 'van der Klugt action' taken in 1987. Figure 3.3 (The Philips Matrix), drawn by Franko (1976), is a 'sketch' of the traditional Philips organisational structure. Compared to the latter, Fig. 5.1 has unique characteristics.

Firstly, the dominant element in Fig. 5.1 is the PDs, which have direct access to the central power, but NOs have no direct access to it; in contrast, in Fig. 3.3, the columns (PDs) and the rows (NOs) have almost equal access to the Board of Management.

Secondly, in Fig. 5.1, PDs and NOs have no direct 'meeting points'; but in Fig. 3.3 there are many—PDs have to meet NOs for corporate activities (production and sales). This difference indicates that the internal conflicts or power struggle between the PDs and NOs have, at least in theory, disappeared after the 1987 restructuring action.

Thirdly, there is an organisational centre (Board of Management and the GMC) in Fig. 5.1; but this centre does not have a prominent position in Fig. 3.3. This indicates that after restructuring the global decision-making power has been centralised; the Board of Management directly controls the PDs again.

Fourthly, in Fig. 5.1 independent PDs, plus R&D activities and the Industrial Design Group, are bridged by the BUs, contingent upon the requirement of the company's new business development.

Finally, Fig. 5.1 has the Board of Management in its centre as the pivotal and, therefore, rotational and flexible. Opposite to this, Fig. 3.3 is squared and static. This difference reflects the aspiration that Philips has evolved into a new dynamic mechanism for everyday business operation after restructuring, in response to the fast forward process of globalisation of the world economy since the last decade.

It is obvious that the new Philips organisational configuration has many advantages over the traditional matrix structure. However, a new structure itself does not necessarily ensure high competitiveness and a successful future for the company. In its fifth year of operation, the 'Centurion' programme is not seeing an end to itself by the time of writing. On the contrary, Jan Timmer told Philips' shareholders that the 'Centurion' reform programme would continue, and 'restructuring is a life fact in our sector of industry'.[93]

CONCLUSION

In this chapter, I have examined the competitiveness of the European consumer electronics industry and Philips' presence in Europe against a background of intensifying competition from East Asian companies led by the Japanese since the late 1970s. It was argued that the VCR format battle and the failure of the V2000 format prompted not only a fundamental restructuring process and strategic readaptation within Philips, but also a substantial change in policy-making at the EU level towards the European consumer electronics industry since the mid-1980s.

It was shown that the EU has adopted a number of protectionist policies involving both the hardware sector and software sector in order to improve the competitiveness of the European industry, mainly the leading European firms. This corresponded to Philips' dual-track strategy: seeking political solutions to its business problems by means of lobbying and, internally, radically reforming its corporate structure.

Now that Philips has got rid of its inefficient matrix structure and dual managerial system and set a series of new corporate strategies centred on interfirm collaboration, would the company be more efficient and competitive in dealing with new technologies? Would EU's public policy necessarily ensure a success for the so-called strategic technologies? The next chapter will examine these issues through a case study on the development of the European HDTV system since 1986.

NOTES

1. Interview with Senior Manager of Strategic Planning, Philips (U.K.), London, 11 July 1990.

2. Japanese direct investment in the European consumer electronics industry started in 1974, when Sony opened its television assembly plant at Bridgend in Wales (Geddes and Bussey, 1991, p. 397); but the volume of investment remained relatively modest until the early 1980s when fears of growing protectionism in Europe reached a peak.
3. Interview with Senior Manager of Strategic Planning, Philips (U.K.), London, 11 July 1990.
4. Note that Marantz, originally a Japanese firm, has now become a Philips branch manufacturing consumer goods in Japan.
5. Because this price difference allows the Japanese firms a mechanism to accumulate capital resources which is not available to Western firms.
6. Note that European firms, such as Philips, used to complain that Japanese firms were dumping—selling products at a lower price than production cost—in Europe in the early 1980s, as discussed in the previous chapter.
7. Telephone interview with Senior Manager of Corporate External Relations, Philips International B.V., Eindhoven, 18 October 1991.
8. 'Advocate of evolving intervention', *Financial Times*, 8 June 1992.
9. See National Consumer Council (1990).
10. *Ibid.*
11. In the case of Canon photocopiers the Japanese complained to the GATT, and then the complaint was upheld.
12. In the case of VCR, as discussed in Chapter 3, the EU negotiated a VER (voluntary export restraint) agreement with MITI in 1983 to set quotas for Japanese import and reserve market for EU producers, i.e. Philips and Grundig.
13. *Financial Times*, 16 June 1992.
14. *Ibid.*
15. Any financial subsidy plan is subject to approval by the Council of Ministers.
16. The MAC-Directive was renewed in May 1992.
17. See Commission of the European Communities, *Action Programme: to Promote the Development of the European Audio-visual Industry—'MEDIA' 1991–1995*, Commission Communication to the Council, COM (90) 132 final, Brussels, 4 May 1990.
18. Interview with Secretariat Permanent, Audiovisual Eureka, Brussels, 21 November 1991.
19. *Financial Times*, 5 June 1992. The originally proposed sum of ECU 850m has now been reduced to ECU 228m.
20. Commission of the European Communities, *Action Programme: to Promote the Development of the European Audio-visual Industry—'MEDIA' 1991–1995, op. cit.*
21. Draft of Joint Press Release, the Sixth Meeting of Japan–EC Business Round Table on Consumer Electronics, Tokyo, 14 March 1989.
22. See Cawson, A. (1992), 'Interests, Groups and Public Policy-Making: The Case of the European Consumer Electronics Industry', in Greenwood, J. *et al.* (Eds.), *Organised Interests and the European Community*, Sage Publications, London.
23. The official name is 'Philips EC Liaison Office'.
24. Cawson, A. (1992), *op. cit.*
25. See figures shown in Table 5.12..
26. Directorate-General Internal Market and Industrial Affairs, Commission of the European Communities, *The European Consumer Electronics Industry in A*

Globally Competitive Environment, Brussels, 12 September 1990, unpublished document.

27. *Ibid.*
28. BIS Mackintosh Ltd (1990), *The Competitiveness of the European Consumer Electronics Industry*, prepared for the Commission of the European Communities.
29. Note that, within Europe, the U.K. is an exception in that a large proportion of output is carried out by Japanese-owned firms.
30. By the time of writing, Marantz has become a wholly Philips-owned company operating in Japan.
31. See BIS Mackintosh Ltd. (1990), *op. cit.*
32. *Ibid.*
33. However, U.S. firms regained the lead over Japan in the early 1990s, partly due to their greater strength in microprocessors, and partly due to the advances of South Korean firms in DRAMs.
34. JESSI (Joint European Submicron Silicon Initiative) started in 1989 and lasts for 8 years. The programme was joined by 29 research institutes and companies from 6 EU countries including Belgium, France, Germany, Italy, The Netherlands and the U.K. The initial investment of the JESSI Programme was $3120m. As a step to cut down the high cost of R&D activities during its restructuring process for recovery, Philips withdrew some of its involvement from JESSI in 1991.
35. de Jonquières, G. 'Shadows over the Sunrise Sector', *Financial Times*, 25 July 1990.
36. Note that, having identified LCD as one of the few important high-technologies, the S. Korean government recently pushed Samsung and Goldstar, the country's two leading electronics firms, to form an industrial alliance aimed at promoting the development of LCD technology.
37. Interview with Ex-Senior Manager of Corporate External Relations, Philips, Eindhoven, 25 November 1991. For more detailed discussion concerning Philips' LCD project, see Chapter 6.
38. Due to technical problems in production, Philips' DCC launch was postponed from September to the end of 1992, and the first shipment was only made to Germany, France and the U.K. This postponement caused scepticism about the future of this new digital audio technology.
39. Philips, *Philips in Europe*, published by Corporate External Relations, Philips International B.V. (No date of publishing), p. 4.
40. Note that, as discussed in Chapter 7, the EU has now substantially changed its HDTV policy from exclusively supporting the analogue HD-MAC to fully digital TV systems.
41. Interview with Principal Administrator, Directorate-General for Competition, Commission of the European Community, Brussels, 14 November 1991.
42. Interview with Ex-Director of Corporate External Relations, Philips, Eindhoven, 25 November 1991.
43. There were now about 100 production centres from Philips' eight Product Divisions by the end of 1991. Source: Interview with Senior Manager of Corporate External Relations, Philips International B.V., Eindhoven, 4 November 1991.

44. Interviews with Senior Manager of Strategic Planning, Philips (U.K.), London, 11 July 1990 and Ex-Director of Corporate External Relations, Philips, Eindhoven, 25 November 1991.
45. Information in this paragraph is mainly from an interview with the Senior Manager of Corporate External Relations, Philips International B.V., Eindhoven, 4 November 1991.
46. *Ibid.*
47. *Ibid.*
48. See *Philips in Europe, op. cit.*
49. See Overkleeft and Groosman (1988), p. 57; also interviews with Ex-Director of Corporate External Relations, Philips, Eindhoven, 25 November 1991; with Ex-Senior Manager, Philips Corporate Industrial Design Group, Eindhoven, 31 October 1991; with Philips expert in the Faculty of Organisation and Management, Eindhoven University of Technology, 29 October and 4 November 1991; and with Philips expert in Docent Marketing, Hogeschool, Eindhoven, 25 October 1991.
50. Information in this paragraph mainly comes from interviews with Ex-Director of Corporate External Relations, Philips, Eindhoven, 25 November 1991; with Senior Manager of Corporate External Relations, Philips International B.V., Eindhoven, 4 November 1991.
51. Information in this paragraph are from interviews with Ex-Senior Manager of Corporate External Relations, Philips, Eindhoven, 25 November 1991; with Senior Manager of Corporate External Relations, Philips International B.V., Eindhoven, 4 November 1991; and with Senior Manager of Philips International B.V., Eindhoven, 29 October 1991.
52. Interview with Ex-Senior Manager of Corporate External Relations, Philips, Eindhoven, 25 November 1991.
53. Information in this paragraph mainly comes from an interview with the Ex-Senior Manager of Corporate External Relations, Philips, Eindhoven, 25 November 1991. The interviewee himself was the then Manager in charge of the Mini-computers Division in Philips France, and personally experienced all the difficulties in most European countries hence repeatedly complained about the problems of the matrix structure to the Board of Management in Eindhoven.
54. Interviews with Philips expert in the Faculty of Management Studies, Eindhoven University of Technology, 31 October, 5 November 1991.
55. Interview with Senior Manager of Philips International B.V., Eindhoven, 29 October 1991.
56. Interviews with Ex-Senior Manager of Corporate External Relations, Philips, Eindhoven, 25 November 1991; with Senior Manager of Philips International B.V., Eindhoven, 29 October 1991.
57. Interview with Senior Manager of Philips International B.V., Eindhoven, 29 October 1991. Other sources suggest that, until recently, Philips was still faced with difficulties generated from 'divisional barriers', which effectively made 'technology transfer' within the company uneasy. See *Financial Times,* 11 May 1993. Cross-division power struggle has also been reported in Metze (1993) with more details.
58. Interview with Senior Manager of Strategic Planning, Philips (U.K.), London, 11 July 1990.

59. *Ibid.*
60. See *Financial Times,* 11 May 1993.
61. Interview with Senior Manager of Corporate External Relations, Philips International B.V., Eindhoven, 4 November 1991.
62. Interview with Senior Manager of Philips International B.V., Eindhoven, 29 October 1991.
63. *Ibid.*
64. See Philips Electronics Corporate External Relations, *Conditions for Coexistence and Cooperation in the Electronics Industry—Towards A Sustainable Relationship between the European Community and Japan: A View from Philips Electronics*, Eindhoven, September 1991; *Financial Times*, 30 June 1992.
65. Interview with Senior Manager of Corporate External Relations, Philips International B.V., Eindhoven, 4 November 1991.
66. *Financial Times,* 11 May 1993.
67. *Ibid.*
68. Interview with Manager of Philips Consumer Electronics, Croydon, 23 July 1991.
69. Interview with Ex-Senior Manager of Corporate External Relations, Philips, Eindhoven, 25 November 1991.
70. *Ibid.*
71. *Financial Times*, 17 July 1992.
72. Interview with Senior Manager of Philips International B.V., Eindhoven, 29 October 1991.
73. *Ibid.*
74. Despite marketing campaigns by Philips and Sony, both DCC and the Mini Disc have failed to become successful since their official launch in 1991. Given the shortening product life cycle and rapid technical change, the possibility can not be ruled out that both DCC and the Mini Disc could be superseded by an alternative digital technology.
75. Interview with Senior Manager of Corporate External Relations, Philips International B.V., Eindhoven, 4 November 1991.
76. Interview with Ex-Senior Manager of Corporate External Relations, Philips, Eindhoven, 25 November 1991.
77. Interview with Ex-Senior Manager of Corporate External Relations of Philips (Eindhoven, 25 November 1991) suggests that Philips made the same mistake in its computer business as well. Philips entered the computer industry at a quite early stage, but the company was not successful at all in either mainframes or minicomputers. The reason was that, strategically, Philips failed to understand the importance of software (including implications) for the computer industry.
78. Cited in *Philips News*, Vol. 21, No. 3, 2 March 1992.
79. Interview with Ex-Senior Manager of Corporate External Relations, Philips, Eindhoven, 25 November 1991.
80. *Financial Times*, 15 May 1990.
81. *Financial Times,* 25 February 1992.
82. Prahalad, C., 'Revitalizing Philips—the Centurion Process', *Philips News*, Vol. 20, No. 15, 16 December 1991.
83. *Ibid.*

84. Interview with Manager of Philips Consumer Electronics, Croydon, 23 July 1991.
85. *Ibid.*
86. *Financial Times*, 25 February 1992.
87. The Lighting Division, the company's traditional business, was still important and profitable. From a technological point of view, lighting is not closely related to consumer electronics and, therefore, the Lighting Division has been given a more independent legal status compared to other Divisions.
88. *Financial Times*, 15 May 1990.
89. Interview with Manager of Philips Consumer Electronics, Croydon, 23 July 1991.
90. Prahalad, C., 'Revitalizing Philips—the Centurion Process', *op. cit.*
91. *Philips News*, Vol. 21, No. 1, 20 January 1992.
92. *Ibid.*
93. Quoted in *Financial Times,* 3 May 1995.

6

REVITALISING THE EUROPEAN CONSUMER ELECTRONICS INDUSTRY: PHILIPS AND THE DEVELOPMENT OF HDTV

The last chapter has examined, on the one hand, the changed but growing political power of the EU in intervening the development of the European consumer electronics industry since the 1970s, and the organisational restructuring process and the changes in corporate strategies at Philips company on the other. What is common between these two aspects was the increased competitive pressure posed by East Asian firms, mainly the Japanese, on the European consumer electronics industry.

Through the case of HDTV development in Europe, this chapter attempts to further investigate the role of government, particularly the EU, in promoting strategic technologies. It is argued in this chapter that the European HDTV strategy was a combined effort between the EU and major European manufacturing firms to take on the Japanese competition and revitalise the European consumer electronics industry. I also argue that, while EU technology policy was biased towards protecting domestic firms on the home market, these firms became technology leaders without protection of foreign markets; i.e., the U.S.

INTRODUCTION

'The time for experimentation is over'.[1] This was a statement made in 1987 by a marketing manager at Sony of Canada Ltd. about high definition television (HDTV) development. The implication of this statement was that HDTV as a new generation of home consumer electronics product was ready for commercialisation. However, by the time of writing, HDTV is still something beyond the reach of ordinary consumers. In a technical sense, the battle between fully digital HDTV and analogue HDTV systems has not been concluded; the promised 'hang on the wall' flat panel HDTV screen, rather than the bulky and heavy HDTV set built with traditional CRT (cathode ray tube) technology, has not been made a technological reality anywhere in the world. Politically speaking, the collapse of the officially supported European HDTV system at the beginning of 1993 has marked the end of the EU's ambitious HDTV strategy.

Despite various uncertainties faced by HDTV development in major industrialised counties, it is a widely shared understanding that HDTV will mark the most important change in the consumer electronics industry since the introduction of Colour TV. Because of the complexity of technologies involved and the potential profitability as well as its professional applications, HDTV has been identified as one of the most potentially lucrative areas in the consumer electronics industry in the Triad economies—Japan, Europe, and the U.S. It seems that, to maintain a strong foothold in or re-enter the consumer electronics industry, none of these three leading industrial regions would like to miss the HDTV bandwagon or become a technological follower during the HDTV race. At the industry level, HDTV is believed too important to be ignored by any electronics firm which wishes to play a leading role in the future electronics, particularly consumer electronics, industry.

After more than 15 years' R&D work, NHK of Japan successfully defined the first HDTV system in the world, called Hi-Vision. Backed by the Japanese government and the Japanese consumer electronics industry and with support from CBS of the U.S., NHK proposed its HDTV system as a single global standard in 1986. The Japanese proposal presented to the CCIR (Consultative Committee for International Radio) was, however, blocked by European governments. Since then, HDTV has been added to the political agenda at an international, regional, and national government level. To take on the challenge of the Japanese, the Europeans and, somewhat later, the Americans joined the HDTV race with their own strategies and technologies, which are substantially different from those of their Japanese counterpart.

As far as Europe is concerned, the HDTV scenario is two-fold. On the one hand, HDTV development has provided an unusual opportunity for European manufacturing and broadcasting organisations to form the largest cooperative alliance in the history of the consumer electronics industry in the form of a Eureka (European Research Cooperation Agency) project. In promoting the establishment of a proprietary European HDTV technology, Philips, together with Thomson and Nokia, has shown the greatest enthusiasm. On the other hand, amongst the three major HDTV development camps, Europe has seen most of the political controversies since 1986, when the first MAC Directive was adopted.

This chapter mainly focuses on discussing government–industry and intra-industry relations related to HDTV development in Europe. In order to present a comprehensive understanding of the European HDTV policy and strategic manoeuvring in the past few years, I also take into account government intervention and major technical changes in Japan and the U.S., although in less detail. In fact, without the strong push of the Japanese HDTV proposal the Europeans might not have proposed and defined their independent system. As the leading firm in the European HDTV project, Philips plays a vital role and, therefore, I will pay particular attention to the company's involvement. Moreover, this chapter will also briefly discuss some other related issues such as the potential impact of HDTV development on Europe's industrial competitiveness.

Defining HDTV

Compared with conventional CTV systems, HDTV technology aims to provide the consumer with cinema quality pictures (over 1,000 horizontal scanning lines) on a widescreen (with a 16×9 aspect ratio) installed in the home. For the time being, the Japanese have already established MUSE as their HDTV transmission standard, and the Europeans had proposed HD-MAC as their independent HDTV transmission system prior to 1993. Although starting much later, the Americans have no intention of becoming technological followers of Japanese and European companies in the area of HDTV development. On the contrary, the FCC (Federal Communications Commission) has been coordinating various HDTV efforts towards defining an independent standard for the U.S.

Replicating the difference between current CTV standards, the HD-MAC system was proposed to transmit pictures with 1250 horizontal scanning lines, which would double that of the European CTV transmission systems (625 lines on PAL/SECAM); while the MUSE system transmits 1125 lines, which is more than double that on the current NTSC system (525 lines) used in Japan and the U.S. for CTV transmission. Both the HD-MAC and MUSE transmission systems are analogue systems except for digital sound. However, the increased number of transmission lines offers much higher resolution close to cinema screen pictures, which are far superior to ordinary CTV quality. Due to the large amount of information needed for HDTV pictures, the bandwidths available in terrestrial transmission were insufficient,[2] and satellite transmission was chosen for the European and Japanese HDTV systems. Of the HDTV proposals in the U.S., all are now fully digital systems, and the future American HDTV standard will be a digital system which is incompatible with both HD-MAC and MUSE.

HD-MAC was an extension as the highest level of the European MAC family.[3] It was claimed that the introduction of HD-MAC transmission would not make the current CTV receivers obsolete because the former was backward compatible. However, this backward 'compatibility' was conditional rather than 'unconditional'—those who wish to receive programmes from MAC or HD-MAC transmission through satellite would have to install a MAC decoder and, inevitably, the promised quality of the new transmission would be degraded. One of the major criticisms of the Japanese HDTV proposal at the CCIR Conference in Dubrovnik in 1986 was that MUSE was not backward compatible with either NTSC or PAL/SECAM mainly because the line number of the MUSE system is not exactly the double of that on any current CTV system. This criticism may not be true any longer due to the development of new technologies on the Japanese side. Major Japanese consumer electronics firms, for instance Sony, have developed converters which are capable of converting MUSE signals downward into the NTSC receivers although the price of this add-on device is very high. Table 6.1 gives a list of different TV standards including HDTV systems all over the world.

It is not the purpose of this book to explore HDTV in very detailed technical terms, but a brief comparison of the general technical features of conventional CTV and HDTV might be helpful in understanding the superiority or advantages of the latter.

Firstly, HDTV offers high resolution. As mentioned above, a new HDTV standard will use double or around double the number of scanning lines (1125 for MUSE, 1250 for HD-MAC) compared to conventional TV standards (525 for NTSC, 625 for PAL/SECAM). By increasing the number of horizontal scanning lines the picture or image on the HDTV screen will carry about four times the amount of information. The difference of resolution level can also been seen by comparing the numbers of pixels required by a CTV and HDTV picture. In its essence a TV image is a kind of electronic picture. Any kind of electronic picture is made from a series of individual dots, or pixels. The image which appears on the TV screen is created by varying the colour and intensity of the pixels. A traditional CTV set has 336,000 pixels, while, a wider, movie-like HDTV screen will require about 2 million of that to create a sharper or higher quality picture.[4]

Table 6.1. International Television Standards.

System	Lines
NTSC (currently used in U.S. and Japan)	525
PAL/SECAM (currently used in Europe)	625
MAC (used for European satellite broadcasts)	625 (enhanced resolution)
MUSE (Japanese high definition standard)	1125
HD-MAC (proposed for Europe)	1250

Note: By the time of writing, HD-MAC had already collapsed and a full digital HDTV standard is being defined through a "grand alliance" under the auspices of the FCC in the U.S.
Source: *Financial Times*, 17 July 1990.

Secondly, colour enhancement is to be achieved on any HDTV system. Colour information (or chrominance) sent to an HDTV receiver will be increased by about ten-fold compared with that of any conventional TV system—NTSC or PAL/SECAM.

Thirdly, the widescreen aspect ratio of HDTV is expected to be the most appealing feature. In order to create more natural or 'real world' images for the viewers, the aspect ratio of the HDTV screen has been universally accepted as 16×9 compared with the 4×3 aspect ratio of conventional TV sets. It is said that the aspect ratio of 16×9 is more close to the natural vision range of human eyes, and also close to the aspect ratio of most modern cinema screens. Due do its wider aspect ratio, HDTV transmission will substantially reduce the work of converting feature films from a cinema screen format to TV broadcasting.

Finally, CD quality digital sound is indispensable. All HDTV systems, including the Japanese one, embody multichannel 'surround sound' through several loudspeakers.

Thus HDTV, with the above technical characteristics, has provided consumers with the possibility of installing a much better audio-visual system which will enable them to enjoy cinema-like entertainment experience in the home.[5] The developers of HDTV say that it promises another revolution in the history of consumer electronics after the introduction of colour TV transmission in the 1950s.[6] Moreover, HDTV systems have also opened a great opportunity for various professional applications outside the consumer electronics industry.

The HD-MAC System

HD-MAC was the European HDTV standard proposed in 1986 by several leading European consumer electronics groups including Philips, Thomson, Bosch and Thorn EMI with strong backing from the EU and some national governments. Originally, HD-MAC was conceived as a rival system to MUSE, a more fully developed Japanese HDTV standard; both systems were intended to become *the* world standard for the next generation of TV transmission. The scenario of HDTV competition has become increasingly complicated since the CCIR Conference in Dubrovnik. It seems that to persuade all interested parties to agree on a single world standard for HDTV transmission and production is beyond the realm of the CCIR because HDTV has already become a sensitive political issue rather than a pure technical problem at either a national or international level.

As the intermediate step towards HD-MAC, MAC (Multiplexed Analogue Components) uses Time Division Multiplexing and transmits the sound/data, chrominance and luminance signals consecutively and, hence, eliminates effects such as cross-colour and cross-luminance from the transmission.[7] The MAC transmission system was developed by Britain's IBA[8] for DBS (Direct Broadcasting by Satellite). MAC signals can only be received on existing TV sets if a MAC-to-PAL or MAC-to-SECAM converter is installed. The MAC transmission system uses the same number of horizontal scanning lines as that of PAL or SECAM, i.e. 625 lines. Although some people claimed that the quality of a MAC picture is much better than that of PAL/SECAM, nonprofessional viewers may find it hard to tell the difference between the two, particularly when the PAL/SECAM picture is also transmitted through satellite such as the BSkyB channels in the U.K.

With MAC, more precisely D2-MAC, as its intermediate stage, HD-MAC has been agreed as the European HDTV transmission system. Because of the 'family' relationship, the popularisation of MAC, as conceived by Philips and some EU policy-makers, would eventually ensure the adoption and success of HD-MAC. The other side of the coin is that if MAC fails to achieve a wide penetration in European homes HD-MAC would be still-born.[9] Due to the fact that the number of horizontal scanning lines used on the HD-MAC system is exactly double that used by conventional European CTV transmission systems including PAL/SECAM and MAC, it would be technically easier to convert broadcasting signals from HD-MAC to PAL/SECAM in Europe than from MUSE to NTSC in Japan.

As leading players in Europe, Philips and Thomson promised to bring HD-MAC receivers to European viewers by 1995. To realise this ambition Philips and other hardware manufacturers would need the wide cooperation of the European broadcasting organisations and the full backing of EU authorities and the member states. However, in part thanks to the rapid progress in digital compression technology in the U.S. and Europe and also because of the failed attempt by the European Commission to get the 12 national governments to agree on a financial subsidy package of ECU 850m, HD-MAC came to a halt as major European manufacturers, such as Philips, suspended plans for mass production of HDTV sets to the European standard due to lack of EU funding for high definition programme-making for the European system in early 1993. This point will be further examined later in this chapter.

HDTV Development in Japan

Japan took a lead in the sense of HDTV strategic planning compared to their European and American counterparts. From a historical point of view, Japan was the first country to start HDTV R&D through its well-coordinated public and private organisations.

In Japan, MUSE (Multiple Subnyquist Sampling Encoding) is the only fully established HDTV transmission technology used for Hi-Vision, the entire high definition television system specified by NHK (Nippon Hoso Kyokai or Japan Broadcasting Corporation). NHK, the Japanese state-owned broadcasting organisation, started R&D on HDTV as early as 1968. Five years later, in 1973, NHK decided that HDTV should be a new television system and that it should not be compatible with any of the existing TV systems. This initiative received warm response from the Japanese government and Japanese consumer electronics manufacturers, particularly Sony. Having adopted 1125 horizontal scanning lines, the original MUSE proposal for transmission assumed a 27MHz bandwidth delivered by DBS. Now that compression technologies have improved dramatically it is possible to squeeze the MUSE signals required by HDTV pictures into a 6–8 MHz bandwidth. Therefore, it became possible to transmit HDTV programmes terrestrially using MUSE.[10]

In 1986, the Japanese HDTV proposal was rejected, mainly by European governments, as a world standard at the Dubrovnik CCIR Conference. The Japanese failed to persuaded the Europeans to accept the Hi-Vision package. This did not stop the Japanese going ahead in their own country. In the same year, MUSE transmission tests started on a Japanese direct broadcasting satellite. To accompany the tests, Sony, Toshiba, Matsushita and some other companies have built prototype HDTV sets aimed to receive MUSE signals.

While the Europeans were still arguing on renewing their MAC Directive and the Americans were waiting for the FCC to make a final choice amongst different proposals, the Japanese started experimental HDTV transmission through satellite with a service of one hour per day, and this was extended to a full Hi-Vision broadcast channel of eight hours per day on 25 November 1991.[11]

The enormous progress of the Japanese HDTV technology, as other key technologies in that country, may not have been achieved without government promotion. Various industrial efforts related to Hi-Vision have been strongly backed by the Japanese government in terms of financial support and high-level coordination. Regarding the development of the Hi-Vision system, Japanese government agencies including MITI (Ministry of Trade and Industry) and MPT (Ministry of Posts and Telecommunications) have played a significant role. Of course, as a state-owned broadcasting organisation, NHK's spending on R&D for HDTV over the years since the end of the 1960s and the launch of the pilot MUSE transmission service has also involved government budget. Table 6.2 contains several MITI and MPT programmes for Hi-Vision.

In addition to those MPT-sponsored HDTV programmes, the MPT also set up a steering committee to promote HDTV in Japan. During the occasion of the 1988 Seoul Olympic games, this steering committee chose 81 sites in Japan to show live events on HDTV screens, and 3.72 million people were attracted by the high quality TV pictures all over the country.[12]

In the end, a satisfactory return on the huge investment at both governmental and corporate levels can only be achieved by a successful commercialisation of HDTV products, particularly consumer equipment. In promoting the sales of HDTV receivers in the consumer market, a high price level has been one of the major bottle-necks. HDTV receivers were introduced in retail stores at an average price level of ¥3.5–4.5 million (approximately $30,000) in Japan. Obviously, HDTV sets at this price level could hardly become a popular consumer product anywhere in the world. As a matter of fact, the public in Japan have hardly paid any attention to the new generation of television technology since the launch of the 8 hours service at the end of 1991 partly because of high-priced hardware and the lack of attractive software.[13]

Recent reports indicate that consumers might not have to wait very long to see the HDTV price level being lowered to an affordable level. In late January 1992 Sharp Corp. of Japan announced that from 10 April in the same year it would be ready to market consumer HDTV sets, called 'Home 1125', for ¥1 million (or £5,000) with a built-in MUSE decoder and MUSE-NTSC converter; and a monthly production output after release was planned at 3,000 units.[14] Simply because of the inside converter, the user can record any HDTV programme with their ordinary VHS VCR machines, provided they don't mind losing the high picture quality.[15] Although the announcement did not provoke any open criticism from either NHK or the government, Sharp's so-called 'million-yen strategy' appeared to be quite controversial because its 'Home 1125' did not adhere to the full specifications of the original Hi-Vision system, which is the government-backed official HDTV package. Despite the fact that the Sharp set was built with an aspect ratio of 16×9 and 1125 horizontal scanning lines which are the two most important defining parameters of the Hi-Vision system, it failed to offer the same picture quality as that of the full high definition TV. More specifically, Sharp's 'trick' was that the company reduced the number of semiconductors required for making the MUSE decoder from a total of 40 to only 6; and in so doing, the quality of the model's still or stationary pictures has been sacrificed while the quality of moving pictures has been maintained at a true high definition level.[16] In any case, Sharp's move was bold and competitive, and it might speed up the cost-price decline process for full HDTV sets in Japan.[17]

HDTV Development in the U.S.

Independent R&D activities towards HDTV started at a much later stage in the U.S. compared to Japan and Europe. But this does not mean that the Americans had ignored the issue of HDTV competition. In 1977, the SMPTE (Society of Motion Picture and Television Engineers) set up a study group on HDTV. Two years after NHK's first HDTV demonstration at the annual conference of SMPTE in the U.S., the broadcasters' Advanced Television Systems Committee (ATSC) was formed in 1983 to explore the need for advanced television standards (Farrell and Shapiro, 1992, p. 15). In 1985, ATSC approved the parameters for 1125/60, the Japanese system, as an HDTV production standard (Brown *et al.*, 1992, p. 33). Having been convinced by the allied Japanese and American corporate lobbying (mainly

Table 6.2. Current Japanese Programmes for HDTV Total Spending for 1985–1997.

Programmes	Companies involved	Years for programme	Budget ($million)
Key Technology Centre Investment Programmes		1985–1993	**63.7**
— Giant Technology Corporation	17	1988–1993	20.7
— Advanced Image Technology Research Lab	3	1988–1993	25.2
— Graphics Communication Technologies Ltd.		1986–1990	17.8
Key Technology Centre Loan Programmes		1985–1993	**9.0**
MITI Programmes		1989–1992	**150.4**
— Hi-Vision Promotion Centre	51	1988–1992	22.6
— Hi-Vision Communications (leasing)	69	1989–1994	16.7
MPT Programmes		1989–1995	**183.4**
— Hi-Vision City Programme	10 cities	1989–1992	111.1
— Telecommunications Satellite Corp of Japan (transponder leasing)		1991–1995	55.6
— Japan Hi-Vision (lease Hi-Vision equipment)	40	1989	16.7
Satellite Construction and launching for HDTV			**1321.4**
— HDTV Satellites		1989-1991	580.7
— HDTV Satellite Corporation		1991-1997	740.7
TOTAL FOR ALL PROGRAMMES			**1727.9**

Source: *High Definition Systems in Japan* (JTEC report), quoted in ATM (*Advanced Television Markets*, Issue 1, November 1991.

Sony and CBS), the American federal government decided to support the Hi-Vision HDTV system at the CCIR Dubrovnik conference in May 1986.

This initial American response (at the industry and governmental levels) to the Japanese move might have been deeply rooted in the fact that all American companies except for Zenith, the only American-owned TV maker, had left the consumer electronics industry by the mid-1980s. In other words, there did not exist an American-owned consumer electronics industry, as that in Europe, crying for government protection.

However, the international scenario of HDTV development changed very quickly after the Japanese proposal for a world production standard was blocked by the European governments. In the following year, 1987, the Association of Maximum Service Telecasters and other broadcasting groups asked the FCC to act on HDTV and, in response, the FCC formed an Advisory Committee on Advanced Television Services (ACATS) to look into possible transmission standards (Farrell and Shapiro, 1992, p. 16).[18] With hopes of achieving a single, global standard blocked by the Europeans, the FCC then decided that the U.S. should build its own future HDTV system. In other words, instead of becoming a technology follower in the world HDTV arena, the American government opted to go its own way.

The American electronics and broadcasting companies' scramble for a proprietary HDTV standard was effectively triggered in September 1988 when the FCC announced that it would decide on a new transmission system by 1991.[19] The trend towards digitalisation has presented a window of opportunity for the Americans to re-enter the consumer electronics industry.[20] Most of the next generation consumer products including HDTV and consumer multimedia products depend substantially on computer technologies and semiconductors. Apart from their active involvement in multimedia technologies (e.g. Apple and Commodore), American companies, who have competitive advantages in digital compression technology, semiconductors (particularly microprocessors) and other aspects of the computer industry, have the potential to become influential players in the arena of HDTV. Whereas, the lack of technical and large-scale production bases for CTVs and VCRs as well video cameras could possibly turn the American contenders into components suppliers to the giant Japanese and European companies.[21] The work to build an entirely U.S.-owned HDTV industry in their own territory has started, and a multibillion dollar global HDTV market was undeniably lucrative.[22]

Since 1988, six proposals for new TV systems, of which five are HDTV, have been accepted by the FCC for consideration. The five HDTV systems, as shown in Table 6.3, were respectively proposed by three technologically powerful industrial alliances plus NHK competing with each other to become the official American HDTV standard. The three HDTV industrial groupings are:

- American Consortium (AC)—Zenith/AT&T;

Table 6.3. Contenders for the TV of the Future, U.S.A.

Entrant	System method	Scanning lines Transmission rate	Comments
General Instrument/MIT	Digital HDTV	1050 lines 30 times/second	GI started the digital transmission stampede
General Instrument/MIT/ Toshiba	Digital HDTV	787.5 lines 60 times/second	Alternative GI/MIT plan would show motion better
NHK	Analogue HDTV*	1125 lines 30 times/second	Only holdout for analogue HDTV transmission
Philips/Thomson/ Sarnoff/NBC	Analogue EDTV*	525 lines 60 times/second	Cheapest system but has lowest resolution
Philips/Thomson Sarnoff/Compre-Labs/NBC	Digital HDTV	1050 lines 30 times/second	Group says picture will hold up at edge of service area
Zenith/AT&T/	Digital	787.5 lines	Zenith chose AT&T to help with digital technology
Scientific Atlanta	HDTV	60 times/second	

Note: *Both the EDTV (Extended-definition television) developed by the ATRC consortium and NHK's MUSE system have now been withdrawn.
Source: Based on *Fortune*, 8 April 1991; Evans (1992); and Brown *et al.* (1992).

- American Alliance (AA)—General Instrument/MIT;
- Advanced Television Research Consortium (ATRC)—Philips/Thomson/ Sarnoff Research Centre/Compression Labs/NBC.

NHK worked alone in its attempt to extend its MUSE technology to the future American HDTV transmission, but NHK withdrew its proposal from the competition after the FCC announced that the future American HDTV would be a fully digital system. The ATRC group has also withdrawn its analogue EDTV system. Therefore, the remaining contenders for the future HDTV are the four fully digital systems developed by the above three camps. In the post-Dubrovnik period, HDTV development in the U.S., in terms of government policy and technical progress, had many distinctive characteristics, of which the major ones are as follows.

Firstly, the FCC ruled in 1989 that the future HDTV system should adopt a terrestrial transmission standard, although it would not retard the adoption of HDTV by other media (Farrell and Shapiro, 1992, p. 22). Despite the obscure attitude of the FCC in deciding the transmission method for HDTV, it is evident that this ruling was intended to protect the vested interest of the well-established terrestrial television broadcasters.

Secondly, it has been an unchanged policy of the FCC that the future HDTV standard in the U.S. must be either directly compatible with NTSC receivers or indirectly by means of 'simulcast'—all HDTV programmes should be transmitted by NTSC broadcasts alongside CTV transmission. If the future HDTV programmes are required to be either NTSC receiver compatible or transmitted simultaneously in both the new transmission standard and the current NTSC format during the transition period, NTSC set holders would not be excluded from watching new channels if they can't afford or are not willing to buy the HDTV receivers. In other words, the FCC approach has ensured a kind of backward compatibility as far as HDTV transmission and reception is concerned. From the consumer's point of view, NTSC receiver compatibility or 'simulcast' seems to be a more beneficial approach compared to the European or the Japanese solution.[23]

Thirdly, in 1988, the FCC's Office of Engineering and Technology found that most existing TV stations could be allocated with an additional 6 MHz channel without causing unacceptable technical problems in terrestrial transmission and, therefore, the FCC made a tentative decision that systems requiring more than 6 Mhz bandwidth, such as the then 9 MHz MUSE system, would not be authorised for terrestrial broadcast service.[24]

Fourthly, in 1990 the FCC stated that it was looking for a true HDTV system rather than initially an EDTV system (Farrell and Shapiro, 1992, p. 25). This ruling led to the withdrawal of the MAC-based EDTV proposal put forward by the ATSC consortium, in which Philips and Thomson are the leading members. The SuperNTSC system has also been withdrawn from the competition since then.

Moreover, significantly differing from the Japanese and European systems, the FCC has now decided that the future HDTV in the U.S. should be a fully digital system. This decision was effectively triggered by the technical breakthrough achieved by General Instruments in digital compression in 1990. As a result of this breakthrough, GI has made it a reality to digitally compress HDTV signals into the required 6 MHz bandwidth for terrestrial transmission. On 6 June 1990, GI and MIT jointly presented a fully digital HDTV proposal, called DigiCipher, to the FCC for test.[25] GI and MIT, and Toshiba since August 1992, have formed the American Alliance with two digital HDTV proposals in their DigiCipher bid. As shown in Table 6.3, the other two camps, the ATRC and AC groups, followed suit and switched their efforts to fully digital systems. The FCC has currently scheduled that, after a long transition period, NTSC broadcasts can be phased out by the year 2008 (Farrell and Shapiro, 1992, p. 20).

It is also believed that NHK has been involved in research activities on digital transmission through satellites but political considerations have stopped the state-owned broadcasting organisation from bring a technologically superior fully digital HDTV system into the U.S. while promoting an analogue MUSE system in its own country.[26] It seems to be only a matter of time until the Europeans and Japanese jump onto the digital bandwagon for HDTV.

Finally, the overall FCC policy on HDTV since the second half of 1986 can be summarised into two aspects. On the one hand, FCC decisions at different occasions have effectively ruled out the possibility for the Japanese and European systems, i.e. MUSE and HD-MAC, to be accepted in the U.S. As discussed above, the FCC required that the future HDTV system in the U.S. should be a terrestrially transmitted fully digital system with a bandwidth of 6 MHz or less. It is evident that neither MUSE nor HD-MAC could satisfy these technical parameters—both systems were intended for DBS rather than terrestrial transmission; both are hybrid systems (mainly analogue but with some digital elements); both systems require a transmission channel of more than 6 MHz.

On the other hand, the FCC has not completely excluded competition. On the contrary, outsiders, i.e. European and Japanese firms, were given the opportunity to participate in the bidding process with their own technologies. Consequently, the FCC policy has left room for market competition—FCC set the technical target such as the required bandwidth of 6 MHz or less and let the bidding companies do the rest of the work. This policy orientation was the major characteristic of the American HDTV competitive environment in which technical breakthroughs or improvements were fostered. Had the FCC adopted a similar policy approach to that of Japan and Europe, where only one technical proposal was chosen by policy-makers, the digital revolution for HDTV transmission may not have happened so soon. By fostering technical competition among bidding consortia, the U.S. government did not commit large sums of public money to subsidise private firms.[27] This is in sharp contrast to Japan and Europe, where public funds have been injected into HDTV projects.

At the time of writing, the three remaining HDTV consortia have entered a 'grand alliance' under the auspices of the FCC to pour their technical know-how together to create a fully digital HDTV system for the U.S. Most importantly, the digital revolution since 1990 and, in particular the recent 'grand alliance', has had a great impact on HDTV development in Europe. This point will be further discussed later in this chapter.

THE EUREKA 95 HDTV PROJECT

In this section, we look at the Eureka 95 HDTV consortium, through which the European HD-MAC system was developed.

Eureka, the coordinating agency of European advanced technology development, was set up in 1985 to stimulate cross-border collaboration in advanced technologies. Eureka currently involves about 19 European countries and 1,500 companies and educational institutions in more than 300 projects with a combined planned investment of $10bn.[28] The Eureka 95 HDTV (EU95) Project started in 1986 as a direct response to the

Japanese HDTV proposal, which had already gained support from the U.S.

At the May 1986 Plenary of the CCIR, a Japan/U.S. proposal for a 'revolutionary' global HDTV production standard failed to be accepted by European governments. Consequently, the Eureka HDTV project, EU95, was initiated by four leading European consumer electronics companies including Bosch, Philips, Thomson and Thorn EMI to develop and demonstrate an alternative 'compatible' or 'evolutionary', as claimed, system based on the MAC patent family of transmission standards adopted in Europe.[29] EU95 gained strong backing from the EU, and was widely supported by European institutions and organisations which brought the total participants to 43 in the year 1990.[30] To some extent, EU95 has become a *de facto* EU project because of the increased involvement by the European Commission.

To be sure, the European HDTV project has made tremendous technological progress, given the fact that European companies were beginning work in 1986, which was almost 20 years later than the Japanese. However, as far as EU public policy is concerned, supporting EU95 has caused tremendous political disagreement among different policy domains within the EU. The central point of the controversies was: while the hardware manufacturers have been enjoying EU financial subsidies, the satellite operators and TV broadcasting organisations would be unwillingly restrained by the EU MAC Directive, not to mention the fact that as an analogue system HD-MAC was increasingly challenged by the new development in digital HDTV technology which is becoming increasingly popular, particularly in the U.S. Due to their vested interests, those who were in favour of the European HDTV approach had all sorts of reasons to defend the official status of HD-MAC.

The Objectives of EU95

The major idea of EU95, as a European HDTV project, was to introduce high-quality TV pictures into the viewers' home. The approach adopted in this project to improve the TV system comprised the following three essential points.

Firstly, as mentioned above, the future European HDTV screen would adopt an aspect ratio of 16×9 compared to that of 4×3 in the PAL/SECAM format; and the horizontal scanning line number (625) currently used on the European CTV system would also be doubled, hence, a much higher level of picture resolution on the future HDTV will be achieved.

Secondly, the EU95 HDTV Project complied with a so-called 'evolutionary' principle. The European HDTV standard was said to provide for a sufficiently high level of performance. The EU95 consortium also promised to leave the door open to benefit from future technological improvements, avoiding for a long time any remodelling of the complete TV chain. New technological developments and their possible applications in the long run would be followed carefully.

The objective of the European industry was to provide at appropriate times and according to appropriate economic conditions the equipment required to operate HDTV services in Europe.

The Organisational Structure of EU95

As a Eureka project, EU95 was a very flexible R&D framework for the proposed European HDTV system; however, it was carried out through a well-organised structure which mainly includes the following elements.

Firstly, the EU95 HDTV Directorate. This Directorate was restricted to representatives of the major participants.[31] It directed the execution of the whole European HDTV project.

Secondly, the R&D Subproject Groups. There were 10 Subproject Groups at the beginning and this number was increased to 12 (groups 5 and 6 have been integrated and assigned the same task) as shown in Table 6.4. These Groups were open to the representatives of major and associate participants, who were responsible for R&D work on subprojects. Each proposed subproject could not become a part of the main project until it was accepted by the Directorate.

At the same level of the 12 Subproject Groups there was also a Chain Group (CG) as shown in Fig. 6.1. This Chain Group was first established

Table 6.4. EU95 Subproject Groups (Phase II: 1990–1992).

Group	Task	Project leader
1	Fundamental Picture & Sound	CCETT (F)
	Production Standards and Standard Convertors	THOMSON (F)
3	Studio Equipment	BTS (G)*
4	Transmission	IBA (UK)
5/6	HD-MAC System Standard	PHILIPS (NL)
7	Receivers	THOMSON (F)
8	Carriers	PHILIPS (NL)
9	Studio and Location Environment	RAI (I)
10	Bit Rate Reduction	THOMSON (F)
11	Sound	BBC (BBC)
12	Non-Broadcast Use of HDTV	BTS (G)

Note: * BTS (Broadcast Television Systems GmbH) is a joint venture between Philips and Bosch. The original 50/50 share ownership has now been changed to 75% for Philips and 25% for Bosch. BTS is a specialised developer and producer of HDTV studio equipment and HDTV VTR technologies.
Source: Based on Eureka HDTV Directorate: *Compatible High Definition Television System* (No date).

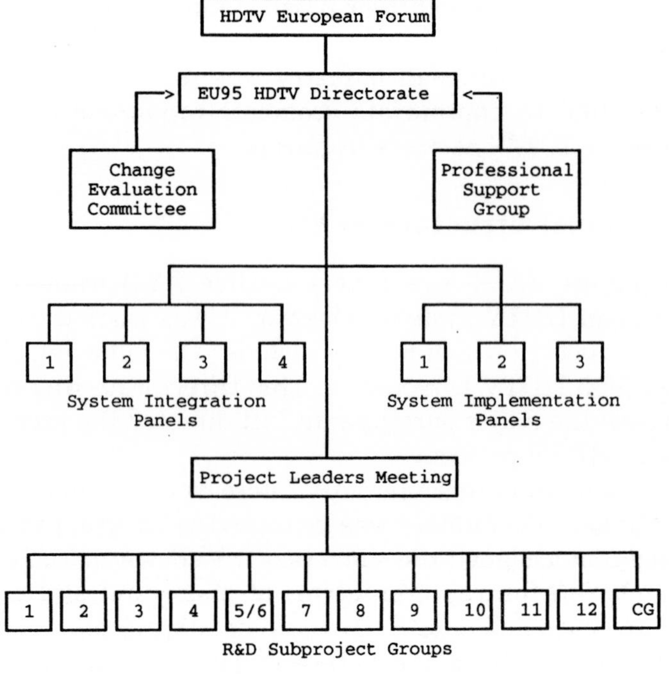

Fig. 6.1. Organisational Structure of the EU95 HDTV Project.

in 1987, and was responsible for configuring various HDTV equipment from different Subproject Groups for major demonstrations or exhibitions by EU95. In other words, the major function of the CG was to synchronise the separate parts of the HD-MAC system and present workable HDTV systems to the public during the period before final commercial launch.

Thirdly, within EU95 there were four system integration panels responsible for preparing the proper choice of the parameters underlying future TV systems. They dealt with the following aspects:

- technical aspects;
- economic aspects;
- regulation and standards;
- markets.

Apart from the above four System Integration Panels, there were another three System Implementation Panels which shared responsibilities to prepare future operation of the new systems. These three panels dealt with

- pilot experiments;
- government relations;
- relations with non-European efforts.

Finally, the function of the Change Evaluation Committee (CECO) is worth mentioning. The HD-MAC system specifications were concluded in July 1989. Afterwards, system enhancements and technical changes have been constantly made by the individual Subproject Groups. However, not all of these enhancements and changes could be accepted. The task of the CECO was to discuss various technical changes or modifications on top of the 1989 specifications, and decided on which enhancements and changes would be incorporated into the HD-MAC system. CECO membership included broadcasters and receiver makers such as Philips, Thomson, Nokia, CCETT, the BBC and IBA of the U.K.

EU95 Participants

As shown in Table 6.5, there were two categories of participants: major participants (list A) and associate participants (list B). List A contributed major R&D resources. List B were those institutions and organisations which took part in R&D projects and advisory participants, i.e. other European institutions and organisations which participated in the project in areas other than R&D. The total participants (including associate members) of EU95 accounted for 43 in 1990, and this was extended to 60 organisations from 12 countries by 1991.[32] As far as the number of participants is concerned, EU95 has brought technological and industrial collaboration to its largest scale in the history of the consumer electronics industry.

The figures in Table 6.5 are the original estimation by the time the EU95 proposal was revealed. After the launch of EU95, certain changes occurred. For instance, BTS was established as a joint venture between Philips and Bosch to develop HDTV studio equipment. A recent Eureka official report suggests that the total manpower effort of EU95 from 1986 to 1992, i.e. Phase I (1986–1990) and Phase II (1990–1992), equals around 5000 man/years.[33]

Table 6.5. EU95 Manpower Planning (1988–1990), man/year.

Category	Organisation	Year			Total
		1988	1989	1990	
	BOSCH	50	50	50	150
	PHILIPS	190	200	210	600
	THOMSON	165	184	176	525
List A		405	434	436	1,275
List B		214	135	93	442
A + B		619	569	529	1,717

Source: Eureka HDTV Project/Proposal, 16 May 1986.

EU95 Financing

Due to the wide range of participation and the huge organisational framework, the operation of EU95 as a high-tech project was highly costly. It is estimated that EU95 is the second largest single investment project in Europe after the Channel Tunnel, provided all of the commitments from both the firm level and governmental level turn into hard cash.[34]

Table 6.6 contains proposed figures to be invested mainly by the four leading manufacturers (Philips, Thomson, Thorn EMI and Bosch) during Phase I of the EU95 project. Recent report suggests that the total proposed budget for R&D through EU95 during the period 1986–92 (Phase I and Phase II) was ECU 625m.[35] This, of course, would mainly be contributed by the project participants and their national governments. As a matter of fact, the proposed budget for EU95 has been substantially superseded by actual investments. Some sources suggest that in total the Eureka 95 HDTV consortium had already spent £2 billion, of which half, i.e. £1 billion, was from the Community tax-payers to develop the HD-MAC system until the beginning of 1992.[36] In addition, the European Commission proposed an 'Action Plan' to provide ECU 850m to boost the progress and ensure the success of the European HDTV project.[37] More detailed discussion about the 'Action Plan' will follow later in this chapter.

Compatibility: 'Evolution' vs 'Revolution'

As discussed earlier in this chapter, 'compatibility' is a confusing term. Originally, this term was coined by the Eureka 95 consortium in its proposal in 1986 as far as HDTV is concerned. The Europeans argued that the MUSE transmission standard is incompatible with current CTV receivers; on the contrary, the European companies would build an alternative system which does not exclude conventional CTV users from receiving HDTV broadcast.

In Europe, there are two categories of CTV receivers—PAL/SECAM sets and MAC sets. The former receive traditional terrestrial broadcasts and the

Table 6.6. Estimated Costs of EU95 (Phase I: 1986–1990), Million ECU.

Project participants	Costs
Bosch	20
Philips	80
Thomson	70
Thorn EMI	13
Project overhead costs	9
Total	192

Source: Eureka HDTV Project/Proposal, 16 May 1986.

latter are designed for DBS programmes. In the future, HD-MAC transmission would be only compatible with MAC receivers including models with both 16×9 and 4×3 dimensions. Piet Bögels, a senior Philips manager and the president of the EU95 Directorate, states,

> The primary objective of the Eureka 95 project has always been downwards compatibility with the MAC/packet system to allow for easy introduction of High Definition Television.[38]

The above citation clearly indicates that it would be wrong to claim compatibility between HD-MAC transmission and the current PAL/SECAM receivers; as a matter of fact, the so-called 'downwards compatibility' of HD-MAC was only applicable to MAC receivers. To bridge the MAC or HD-MAC transmission and the PAL/SECAM sets users would have to buy an extra convertor. In that sense, the repeatedly emphasised superiority of HD-MAC to the Japanese Hi-Vision approach does not actually exist any longer, because the current CTV users in Japan can also get the new HDTV transmission provided they are willing to buy a MUSE-NTSC convertor which has already been technologically available now. In other words, there is no substantial difference between HD-MAC and MUSE with regard to downwards compatibility, except that Japanese convertors are more complex and more expensive.

It is argued that the extra device, i.e. the MAC-PAL/SECAM converter, would not affect the image of HD-MAC as being 'compatible' with PAL/SECAM; to receive PAL satellite broadcasts the viewer also needs to have a satellite dish and an extra box installed.[39]

The main issue is not the benefit of HD-MAC's backward compatibility with MAC receivers, but rather how many MAC receivers have been sold in Europe so far. Compared to the large population of PAL/SECAM sets, MAC receivers account for only a tiny fraction. MAC transmission in the EU countries has so far made little progress. For instance, among the 11 German-language channels that transmit in PAL from the two Astra satellites, only two also transmit in D2-MAC from TV-SAT, a German DBS satellite; and only a few thousand D2-MAC receivers have been installed to receive these DBS broadcasts.[40] In France, most of the D2-MAC channels transmitted through DBS are also available terrestrially. The situation in the U.K. is even worse. The takeover of the short-lived BSB, the only broadcaster using D-MAC, by Rupert Murdoch's Sky Television has effectively terminated the penetration of MAC transmission in the U.K. So far Scandinavia is the only region in Europe where MAC transmission is relatively successful.[41]

To set a stepping-stone for the future HD-MAC commercialisation, some organisations such as Philips and Thomson, joined by the European Commission, have been campaigning to promote the penetration of MAC receivers in the home. The manufacturers argue that because the number of MAC receivers and D2-MAC transmissions has been insignificant in the

Community, European manufacturers and some EU public authorities should put money to *create* such an 'interim' stage! In part to serve this purpose, the EU's MAC Directive was a coercive regulatory procedure which had made MAC an official standard for high-powered DBS service within the Community.

In a technical sense, the 'revolutionary' MUSE standard is not revolutionary any longer, if fully digital HDTV is the winning force in the current TV technology revolution; and the 'evolutionary' HD-MAC system is not evolutionary in its essence due to the fact that PAL/SECAM receivers would be made obsolete had the original EU strategy and the EU95 project turned HDTV service into a reality in Europe.

In short, the Eureka 95 HDTV project, as a strong European voice, was initially a direct response to the Japanese HDTV proposal. Having been joined by all the major European consumer electronics manufacturers and a number of important broadcasters and, perhaps more importantly, involved by the European national governments and the European Commission, EU95 became a regional political and economic alignment against foreign competition. Despite various controversies and disagreement, the central point of the collective efforts from the industry, governmental and EU levels embodied in EU95 was to sustain the declining European consumer electronics industry by means of establishing a proprietary new TV system. It might not have been a fair deal if the world had accepted the Japanese HDTV proposal without any competition in 1986; however, it seems that the EU95 action partly served the cause of building a 'fortress Europe', which is a long-standing fear of non-Europeans.

Phase II (1990–92) of the EU95 Project has now been concluded, and various prototype HDTV products have been built by the major participants such as Philips, Thomson and Nokia. Due to the efforts of EU95, some viewers have already experienced the high quality of HDTV pictures during some major events such as the International Broadcasting Conventions held in Brighton and Amsterdam in 1990 and 1992, the Winter Olympics in France and Summer Olympics in Barcelona in 1992 and many other exhibitions. In short, EU95 was a successful technological collaboration among the companies involved. However, HD-MAC, the European HDTV transmission standard, collapsed in early 1993, and this caused further uncertainties, both technically and politically, to HDTV development in Europe. This point will be further explored later in this chapter.

FORMATION OF THE HDTV TRIAD POWER

As discussed above, a Triad power for HDTV is being formed with Japan, Western Europe and the U.S. as the three major pillars. Although research and evaluation have been going on in countries such as the former Soviet Union and China, and R&D activities are being carried out in South Korea and Taiwan under the auspices of government agencies, no independent

HDTV systems have been officially established in areas outside the Triad so far. Most likely, some countries from the rest of the world will become licensees of any of those developed in the Triad power, as what happened with colour TV systems, if they wish and can afford to go to HDTV as well.[42]

Despite the fact that a worldwide HDTV industry has yet to be created, a certain level of initial cross-border market penetration has already been achieved within the Triad power. For the time being, this penetration does not seem to be balanced.

The most important characteristic of the HDTV Triad power relation is that the Japanese industry has adopted a global HDTV strategy. Under the guideline of this global strategy, Japanese companies have already made considerable penetration in both the U.S. and Europe in the past few years. In addition to the NHK HDTV proposal presented to the FCC, a number of Japanese companies such as Matsushita, Sony, Hitachi, and Toshiba have also been engaged in R&D activities related to HDTV manufacturing through their well-established local branches. These companies will be among the first to be ready to supply HDTV receivers to the American market as soon as the FCC has made its final decision. Meanwhile, the purchase of Columbia Pictures and MCA by Sony and Matsushita, respectively, was in part intended to enable the leading Japanese manufacturers to have immediate access to the American film libraries, which may easily be turned into HDTV software for broadcasting transmission or distribution on high definition video tapes. While some commentators were worried about the immediate financial risks in taking over and the managerial difficulties in running the American software businesses, the Japanese might have been calculating the potential profitability of being able to combine new technological breakthroughs with the software sector in the long-term.

Although the Japanese organisations have been excluded from R&D activities for the European HDTV system, Japanese penetration in Europe has seen big progress since the beginning of the 1990s. In 1990, one of the two HD-MAC CRT (Cathode Ray Tube) TV receivers installed at the BBC Television Centre in London for displaying live HDTV coverage of the World Soccer Cup Final was made by Sony.[43] This example shows that the Japanese have been technologically ready to produce HD-MAC receivers to supply the future European HDTV hardware markets.

Sony's European strategy is also followed by other Japanese consumer electronics manufacturers. For instance, to make itself ready to start mass production for the European HDTV market, Sanyo was quick to set up a European HDTV R&D project to make a presence within the HD-MAC territory.[44] Meanwhile, the Japanese have also brought *their* HDTV technology to Europe, and a number of software producers have already adopted the Japanese production standard. Table 6.7 shows the presence of the Hi-Vision (1125/60) production equipment used by European and Japanese software producers *inside* Europe.

Another characteristic of the HDTV Triad power is that the Europeans and Americans have no presence at all on the Japanese HDTV arena in Japan. In contrast to the fact that the Japanese consumer electronics companies have access to the MAC and HD-MAC patents, the Europeans appear to have no access to the Japanese Hi-Vision patents. It was reported that Philips' request for Hi-Vision licences was turned down.[45] This might cause another round of political confrontation and bilateral negotiations between EU and the Japanese government with respect to the development of HDTV in the future.

The general situation of the HDTV Triad power relations for the time being is also shown in Fig. 6.2 as far as mutual penetration is concerned.

In the future, a firm's competitive advantages in HDTV equipment manufacturing will be largely determined by its capability in supplying key components. In this respect, HDTV assembly might not be as important as the production of key components such as advanced semiconductors and flat panel display (large LCD flat screen rather than the current CRT monitor). The Japanese companies, who are leaders in LCD and advanced semiconductor technologies, have the potential to exploit the profitability by supplying key components to international HDTV assembly for the global consumer markets and nonbroadcast markets.

As mentioned earlier, R&D activities on HDTV are also being conducted outside but closely related to the Triad power. In particular, HDTV development in South Korea and Taiwan should not be ignored.

Table 6.7. 1125/60 HDTV Production Centres in Europe.

Name of Centre	OB Van	Editing	Camera	VTR	CRT/Proj	Others*
EUPHON International	+	+	+	+	+	+
VTTV	+	+	+	+	+	+
Sony UK	+	+	+	+	+	+
Sony France	–	–	–	+	+	–
HDTV Holland	+	+	+	+	+	+
National Film Geluidsienst	–	–	–	+	+	–
Mazzo Video Grootbeekd Proj.	–	–	–	–	+	–
Heuvelman	–	–	–	–	+	–
Steiner-Film	+	+	+	+	+	+
Eizoh-Tsuusin (Paris)	+	–	+	+	+	+
Technovideo SA	+	+	+	+	+	+

Note: * Swichers, audio mixers, paintbox, converters, etc.
Source: Based on MICO (Media International Corporation) Europe, quoted in Flynn, B.(1991), "Why European independents go Japanese", *Advanced Television Markets*, Issue 1, November.

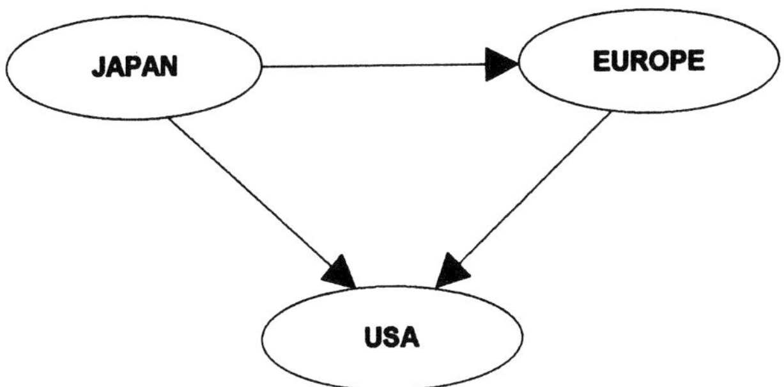

Fig. 6.2. HDTV Penetration in the Triad Power.

Under direct coordination of the South Korean government, an HDTV R&D and promotion consortium has been set up. This consortium has been joined by the manufacturing industry, public research institutes, and academic establishments in South Korea. As had been proposed, the consortium was to spend Korean Won 100bn, of which 40% was from government subsidy, by 1992. In February 1990, Samsung and Goldstar, the two leading Korean electronics firms, signed a contract with NHK of Japan to import HDTV signal processing technologies from the latter.[46] Given the proposal of the government-sponsored consortium, South Korea would be able to export HDTV sets by 1995.

Having been fully aware of the crucial importance of HDTV and its spill-over effect, the Taiwanese government has also been promoting R&D activities on new TV technologies through the Industrial Bureau[47] and the Research Institute of Industrial Technologies (RIIT) since the beginning of the 1990s. RIIT has proposed to invest NT$6.5bn (app. $260m), of which NT$6bn (or 92.3%) would come from the government budget, and employ 2,200 technical and engineering staff to develop HDTV based on the key parameters of the future American system during a four year period from 1992 to 1995.[48] To reinforce HDTV development in Taiwan, the Ministry of Transportation[49] has been preparing since the beginning of this decade to launch its own Quasi-DBS satellite in March 1995. This DBS satellite will be used to transmit HDTV signals.

Apart from domestic R&D activities, both Taiwan and South Korea have also established their own technical research laboratories in Princeton of the U.S. in order to catch up with HDTV development there. Currently, RIIT of Taiwan invests NT$300m (app. $16m) annually in its new HDTV labs in the U.S.

In Taiwan and South Korea there are two important factors which would enable these two areas to follow the HDTV development in the Triad power very closely. Firstly, both economies have a relatively competitive electronics industry and a tradition of strong government intervention, which is believed to be indispensable for organising and coordinating HDTV development.

Secondly, there are considerable sums of reserved hard currency which may be spent in part on HDTV-related activities by the government and the industry, particularly for buying key technologies from any region in the Triad.

EU POLICY AND STRATEGIC MANOEUVRING OVER HDTV

The general characteristics and development of public policy within the EU towards the consumer electronics industry have been discussed in Chapter 5. Here I will examine the role of EU policy-makers and the effectiveness of their policies and strategies in promoting HD-MAC, a widely believed strategic technology even before its inception.

Generally speaking, the pivotal role played by the EU authorities, similar to that of MITI and MPT in Japan and the FCC in the U.S., was their ruling which had made HD-MAC *the* legitimate HDTV standard in Europe before the system itself was technologically established. However, the policy was soon to run into trouble: on the one hand, a common European HDTV standard could only be viable should a consensus be reached by all parties involved; on the other hand, to harmonise the relationship among all the interested organisations from related industrial sectors has always been a difficult task. As a matter of fact, a common DBS transmission standard, firstly D2-MAC then upgraded to HD-MAC, was believed to favour equipment suppliers and set makers, i.e. it was impossible to implement a compulsory DBS directive without undermining the interests of independent broadcasting organisations and private satellite operators. This was one of the major reasons why the EU's policy orientation has been controversial. Another problem with the EU's HDTV policy lay in the fact that the importance of developing a fully digital system was not given proper consideration over the past few years when EU authorities were promoting MAC and HD-MAC, both hybrids of analogue and digital techniques.

The MAC Directive and Council Decision on HDTV

In Europe the major regulatory packages concerning HDTV were the MAC Directive[50] of 1986 and the Council Decision on HDTV of 1989. As mentioned above, the MAC Directive was mainly intended to back the Eureka HDTV consortium, which was launched as a response to the Japanese HDTV competition. The EU Council Decision on HDTV was designed as a complementary policy to the MAC Directive with regard to the implementation of HDTV services within the EU within due course.

The major aspects of the MAC Directive and the Council Decision on HDTV are summarised as follows.

Firstly, the MAC/packet family, including C-MAC, D-MAC, and D2-MAC, were given the official status of the standards for DBS in the channels defined by WARC BS-77 and RARC SAT-83, i.e. DBS broadcasts transmitted via high-powered satellites, across EU member countries.

Secondly, as an extension of the MAC/packet family, HD-MAC, which was being developed by the EU95 consortium, was adopted by the EU as the official transmission standard for any future nonfully digital HDTV service across the member countries. Meanwhile, HD-MAC with technical parameters such as the 1250 lines, 50 complete frames per second progressive scanning was adopted by the EU as the single world standard for production and exchange of HDTV programmes.[51] As discussed above, a single world standard for either HDTV transmission or programme production seemed to be out of tune already given the reality of regional fragmentation after the CCIR conference in 1986.

Thirdly, having been aware of the strategic importance of HDTV to the European consumer electronics industry and to the European audio-visual sector, the European HDTV proposal would be given full support by the EU and its member states in terms of legal administration and, as later revealed, financial subsidies.

Finally, in connection with the above point, it was evident that the interests of the consumer electronics manufacturers and the television and film industries, particularly that of the former, have been given top priority for consideration by the EU. In the MAC Directive, it is stated that,

> ... the implementation of common technical specifications leads to the creation of a large unified market, on which products will be freely exchanged without any technical barriers, which will be of great economic benefit for the European consumer electronics industry as regards its competitiveness.[52]

Being worded in this way, the Directive failed to give equal consideration to the interests of other concerned sectors including programme producers, independent TV broadcasters and private satellite operators. More surprisingly, the interests of the consumers and tax-payers were not properly considered—to be sure, they were not mentioned at all in the text of the Directive.

The economic benefits of the MAC and HD-MAC standards for the European consumer electronics industry and, consequently, ensuring the industry's competitiveness, as conceived in the Directive, seemed to be hypothetical. Doubtless, it would have been extremely disastrous for the major European manufacturing firms including Philips and Thomson if NHK's HDTV initiative was accepted as the world standard at the CCIR meeting in 1986.[53] However, it might be a wrong impression that the European HDTV regulation was a panacea to the declining European consumer electronics industry. In other words, the declared cause–effect relation between HD-MAC and the competitiveness of the European consumer electronics industry was hardly convincing. In commenting on the development of HD-MAC and the competitiveness of the European industry, an EU official said,

> Philips and Thomson would be severely disadvantaged by manufacturing on the Japanese standard. [However,] I don't think that these companies, who have incompetent management and make poor commercial decisions, can be saved by the European HDTV standard. On the contrary, the European HDTV standard will give them an opportunity, some competitive advantages, particularly in the early days of the market.[54]

To give the European manufacturers some kind of competitive advantage would not necessarily guarantee that a European-owned HDTV system could block the entrance of the Japanese into the European HDTV market, although some people believe that MAC was an isolationist technology designed mainly to keep the Japanese out of Europe.[55] If one accepts that MAC was an isolationist technology, it must be true that the MAC Directive was also an isolationist legislation. However, one should also bear in mind that once it is fully established, either MAC or HD-MAC would be widely available for open licensing to not only European firms but also foreign companies including the Japanese, as required by European competition policy.[56] Therefore, it would be naive to think that the European HDTV standard will shut the Japanese out.[57]

EU HDTV Strategies

The MAC Directive of 1986 was due to expire by the end of 1991, and the Commission began a process of consultation over the provision of a new directive on satellite broadcasting standards. This triggered further disagreement between affected interests within the Community. Despite the tremendous difficulties imposed by the HDTV affair, the Council of Ministers finally adopted the repeatedly postponed new Directive on 11 May 1992. The adoption of the new Directive set the legal framework intended to secure further development of the European HDTV system. By incorporating the new Directive with a set of other measures, the EU, via the Commission, outlined its strategy for the development of HDTV. This strategy includes several aspects: technological, broadcasting, programmes, and financing.

Firstly, on the technological side, while trying to insert new momentum to promote D2-MAC and HD-MAC the European Commission has left room for R&D through European collaboration dedicated to digital HDTV systems. Differing from the previous MAC Directive, the new Directive emphasises that 'European research efforts must remain in the forefront of all new significant developments, such as a trend towards digital television broadcasting emissions'.[58] Thus, the new Directive describes D2-MAC and HD-MAC as 'not completely digital transmission standards' and, accordingly, HD-MAC was the compulsory official standard applicable to 'not completely digital' HDTV transmission only. In other words, the new Directive does not exclude the possibility of adopting a fully digital HDTV transmission standard in the EU at a certain time in the future. Obviously, this was a retreat or compromise at the Community level as far as the European HDTV strategy was concerned.

Secondly, on the broadcast side, the Commission planned a 'double instrument' strategy to ensure the success of HD-MAC. One instrument was the new Directive, which set a mandate for the whole Community to use HD-MAC as *the* transmission standard for 'not completely digital' HDTV and D2-MAC as *the* intermediate standard for 'not completely digital' transmission in the 16×9 aspect ratio format. Another instrument was the so-called 'Memorandum of Understanding' (MOU), through which the Commission attempted to legally combine satellite operators, equipment manufacturers and broadcasters together to promote the European HDTV system. The originally proposed MOU would create a consortium between the signatories with the participation of the European Commission to coordinate the promotional work and administer the Community incentives, i.e. EU subsidy, related to D2-MAC and HD-MAC over a five year period (1992–96). The Commission, however, failed to get the MOU signed by the actors mentioned above. In light of this, the Commission decided to change direction and asked the industrial groups and broadcasting organisations concerned to agree on a watered-down version of the originally drafted MOU. As a result, on 15 June 1992, the MOU was approved by up to 40 European companies and professional associations, who agreed to join efforts to promote the European HDTV system *on the condition* that EU member states give financial backing.[59] This implies that any failure by the Council of Ministers to approve the subsidy package proposed by the Commission would likely undermine the industry's enthusiasm and effort for HD-MAC.

Table 6.8. Action Plan for the Introduction of Advanced Television Services in Europe (1992–1996), Million ECU.

	1992	1993	1994	1995	1996	**Total**
Broadcasting						
— Transmission	25	90	100	125	120	**460**
Studio upgrades	–	24	30	6	–	**60**
Production	4	32.5	70	59.5	54	**220**
— Production of programmes						
— Conversion of programmes						
Retransmission	4	30	45	19	12	**110**
— Cable networks						
TOTAL	**33**	**176.5**	**245**	**209.5**	**186**	**850**

Source: Adapted from CEC, *Proposal for a Council Decision on An Action Plan for the Introduction of Advanced Television Services in Europe*, COM(92) 154 final, Brussels, 5 May 1992.

Thirdly, in an effort to stimulate concerted action towards development of D2-MAC and HD-MAC services in Europe, the Commission drafted a financial subsidy proposal, called 'Action Plan', for consultation with the European Parliament and to be approved by the Council of Ministers. The 'Action Plan' initially involved ECU 850m to be used to subsidise further development of the HD-MAC system and simulcast broadcasting as well as MAC programme-making over a five-year period from 1992 to 1996 (Table 6.8). In principle, according to the Commission proposal, the fund would only be granted to individual projects on the basis of a bilateral contract between each industrial group and the Commission. It seems that the 'Action Plan' would become the backbone of the final success of the whole European HDTV project, provided the member states could reach an agreement to approve it. This was, however, subject to great uncertainty for the time being because a majority of finance ministers from the member states had repeatedly attacked the new overall EU budget proposal, of which the 'Action Plan' was a part.[60]

It has long been known that the U.K. government had reservations on the idea of using EU funds to promote advanced TV services (including D2-MAC and HD-MAC) in Europe. Most recently, the DTI (Department of Trade and Industry) revealed that the U.K. government was emphasising a technological objection to the Commission's 'Action Plan' because it had doubts about pouring public money into an interim technology, i.e. D2-MAC, which might not be the right format to suggest.[61] Moreover, the British government believed that HD-MAC, a hybrid of analogue and digital techniques, could soon be superseded by fully digital HDTV technologies under development in the U.S.[62] Under pressure from the British government, the Commission agreed to cut down its proposed subsidy package of ECU 850m to ECU 500m by the end of 1992, and promised to consider a report on the prospects for digital HDTV and review its entire HDTV strategy in early 1993.[63]

At the EU telecommunications ministers' meeting on 16 June 1993, the British government finally withdraw its veto of the Commission's 'Action Plan', and agreed to a substantially revised plan with the sum of ECU 228m (£180m) to promote widescreen television services in the Community over a period of four years until 30 June 1997.[64] According to this agreement, EU funds would be available only to broadcasters and programme-makers using the widescreen format (16×9). By agreeing on this subsidy plan, the British government had also won guarantees that Japanese TV manufacturers based in the U.K., such as Sony, would not be excluded from future European HDTV research projects and standardisation discussions.[65] Bearing in mind the fact that Sony has been the leading manufacturer in promoting the Hi-Vision system, this new deal between the British government and other EU countries seemed to be another significant step forward for the Japanese to penetrate the emerging European HDTV industry. Ironically, the EU95 project was initially launched to enable the

European industry to establish a rival HDTV system against Japanese competition represented by the Hi-Vision system.

The above three measures, the new Directive, the MOU and the 'Action Plan' were the three pillars of the Commission's HDTV strategy. However, as discussed above, each pillar has been accompanied by uncertainties. The renewed Directive left room for the development of fully digital HDTV technology; the MOU was downgraded to merely a conditional commitment; while the 'Action Plan' was substantially watered down to a much smaller amount of subsidy by the time of approval.[66]

In addition, having understood the vital importance of software availability to the success of the European HDTV project, the Commission was also promoting indigenous programme-making towards widescreen D2-MAC and HD-MAC broadcasting. More specifically, the Commission decided to encourage programme-making in the 16×9 format through two measures—'Media 16×9' and 'HD Programmes'—under the umbrella of the Community's Media Programme. 'Media 16×9' was launched to concentrate available resources on projects designed to help 16×9 D2-MAC broadcasters to obtain appropriate programmes, whilst 'HD Programmes' were a further extension of 'Media 16×9', and was expected to build up a stock of high definition programmes before the introduction of HDTV services in Europe by 1995.

EU Coordination: *Vision 1250*

Apart from the existence of the Eureka 95 project, there are also other R&D as well as programme-making activities in the EU. To ensure an overall coordination at the Community level among various efforts related to the European HDTV project, the Commission launched the *Vision 1250* HDTV consortium, one of the first European Economic Interest Groupings (EEIGs), in July 1990.

Vision 1250 EEIG represents a major European initiative to assist the transition from the HDTV development stages to the start of HDTV broadcasting services in Europe. It was intended to promote the use of the European HDTV standard throughout the world by coordinating existing European programmes, such as EU95, RACE and ESPRIT, related to the European HDTV technology. Another objective of *Vision 1250 EEIG* was to encourage the European TV and film industry to gain the experience and expertise needed in the production of programmes in accordance with the technical requirement of HD-MAC system. *Vision 1250 EEIG* was initially formed by 16 founding members, of which some are also active EU95 participants such as Philips, Thomson, Nokia, RAI, BTS, BBC, etc. Figure 6.3 shows the organisational structure of *Vision 1250 EEIG*.

According to Fig. 6.3, the most important characteristic of *Vision 1250* is that it has brought together all the related actors for launching HDTV services in Europe, although the consortium is not a coercive or regulatory

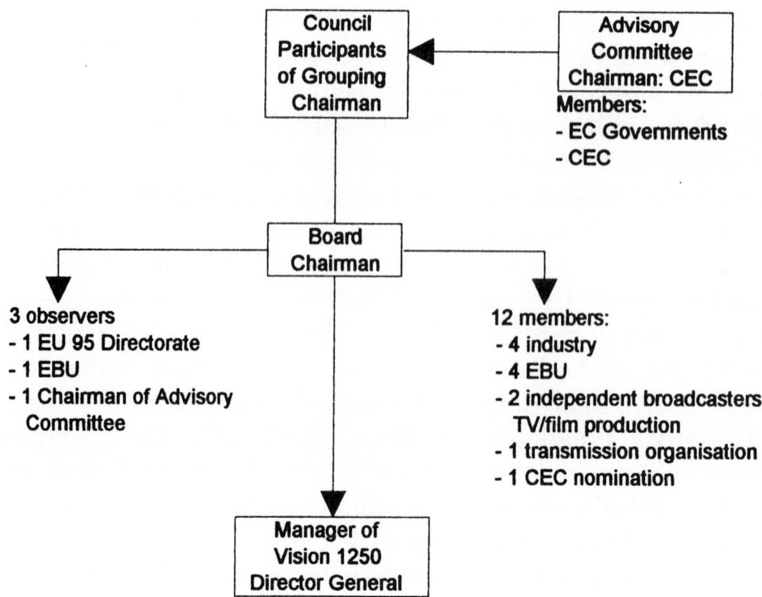

Source: Vision 1250: European Economic Grouping - HDTV (no date).

Fig. 6.3. Organisational Structure of *Vision 1250 EEIG*.

body. An EU official revealed that the Commission participates in the Advisory Committee and gives financial support to contribute towards the running costs of *Vision 1250*.[67]

Because of the vested interests of different industrial groups and broadcast/transmission organisations, it was by no means an easy task for the EU to bring together various efforts and available resources for promoting the D2-MAC widescreen service and developing the HD-MAC transmission system. *Vision 1250* was in part an EU effort to link together the interests of HDTV equipment manufacturers, programme producers, broadcasters, and satellite operators under the same European banner.

CHALLENGES TO EU STRATEGIES AND THE *DÉBÂCLE* OF HD-MAC

In February 1993 a senior official from the European Commission revealed that Europe might have to adopt the forthcoming American digital HDTV system.[68] Although it was subsequently stated that Europe would not necessarily follow the American standard in the future, this statement has effectively marked the failure of the ambitious European HDTV policies since 1986 and, therefore, brought a bitter end to the HD-MAC transmission system.

There were many factors contributory to the *Débâcle* of HD-MAC. Among others, I believe that the impact of EDTV, the implications of the BSkyB affair and, in particular, the impact of digital technology, were the major factors which led to the collapse of HD-MAC.

The Development of EDTV and Its Impact on HDTV

Before HDTV is technologically finalised and commercially accepted in the consumer market, the course of its development has already been made more complicated by new breakthroughs. One of the major factors causing this was the development of EDTV (Extended Definition Television), or sometimes referred to as IDTV (Improved Definition Television). Almost in parallel with HDTV development, various EDTV/IDTV systems have been proposed since the late 1980s, as either interim steps before real HDTV or alternative routes to it, in the U.S., Europe and Japan.

The Proliferation of EDTV

As shown in Table 6.9, many EDTV systems have been developed or proposed on the basis of the configuration of current CTV systems (NTSC and PAL/SECAM). This corresponds to the multiproposal competition for HDTV in the U.S.[69] To summarise the reasons why EDTV systems have been rapidly proposed within the Triad of HDTV development, two aspects are worth mentioning. On the one hand, key technologies such as digital compression (prior to 1990) and flat panel display were not sufficiently developed to allow a genuine high definition system any where in the world. On the other hand, so far true HDTV sets manufactured in Japan have been too expensive to open up a mass consumer market, despite a trend towards price drop.

Amongst all the systems and proposals listed in Table 6.9, only MAC, more precisely D2-MAC, has been *officially* implemented as an interim phase before HD-MAC in Europe, and Clearvision has been launched by some Japanese commercial networks to transmit higher definition information on top of the NTSC signals for five hours per day.[70] However, most

Table 6.9. EDTV Systems/Proposals, 1991.

Japan	W. Europe	U.S.A.
Clearvision	MAC	ACTV
New NTSC	PALplus	Super NTSC
		Spectrum
		MIT
		Narrow MUSE

EDTV systems are either still-born, as happened to those in the U.S., or subject to further development in the laboratories. Europe is the only place where D2-MAC as an EDTV system and interim standard to HD-MAC has been strongly pushed by the manufacturing industry and officially approved and backed by public authorities within the EU.

Main Features of EDTV Systems

Extended or improved definition TV is claimed to be technically better than current CTV systems but less advanced than HDTV. More specifically, EDTV has the following key features.

Firstly, all EDTV systems or proposals have been claimed to be downward *compatible* with either PAL/SECAM in Europe or NTSC in Japan and the U.S. Of course, some kind of encoding/decoding work needs to be done in order to realise this compatibility. In other words, old CTV sets and transmission would not be made obsolete by the introduction of EDTV.

Secondly, EDTV provides higher picture quality. For most EDTV systems, higher resolution or picture quality is achieved by adding extra chrominance and luminance information to conventional CTV transmission. It is believed that EDTV systems have the same number of horizontal scanning lines as those of PAL/SECAM (625) or NTSC (525), although the former have less 'ghosts' and flickers on the screen compared to the latter.

Thirdly, widescreen or bigger picture aspect ratio is another, perhaps the most important, characteristic of most EDTV systems. The screen aspect ratio of 16×9 is not only promised by HDTV but also adopted by most EDTV developers, who believe the concept of widescreen would eventually lead consumers to enter the HDTV world. Reception of 16×9 pictures is possible on 4×3 sets (backward compatibility) but only by leaving two black strips on top and at the bottom of the picture.[71] The concept of widescreen receivers for EDTV is mostly advocated by some leading European consumer electronics manufacturers including Philips, Thomson (*via* Ferguson) and Nokia. These firms claim that widescreen would offer the consumer better viewing experience, and this would make HDTV receivers with the same aspect ratio more appealing in the marketplace in the future; in other words, widescreen receivers is the most important step towards HDTV.[72] Therefore, Philips, Thomson and Nokia have now been firmly committed to promote widescreen TV, and several widescreen models have already been put on sale in Europe.[73]

Fourthly, on most EDTV systems, digital stereo sound is included to replace the previous analogue carrier.

Finally, EDTV systems are believed to be much cheaper to implement compared to genuine HDTV systems.

Despite the above new features offered by EDTV, the reality does not seem to be as encouraging as some manufacturers have anticipated. It might be true that a 16×9 EDTV set is cheaper than an HDTV receiver at current price levels, but it is still more expensive than most of the conventional colour TV sets. Would TV viewers be prepared to invest in a widescreen set just for benefiting from the wider aspect ratio of pictures and slightly improved picture resolution limited by the use of the same amount of scanning lines (625 or 525)? The troubled MAC system in Europe does not suggest a positive answer. As far as digital audio is concerned, EDTV is not the only solution. The introduction of NICAM technology in the U.K. was an example of alternative ways to offer digital stereo audio service.

The Impact of EDTV on HDTV

The impact of EDTV development is not always as positive as conceived by some developers and manufacturers; sometimes it could be challenging to the analogue HDTV systems including HD-MAC in Europe and the more established Hi-Vision in Japan.

In commenting on the relationship between EDTV and HDTV, some commentators believe,

> Extended television systems (EDTV) developed by NHK's competitors, the private commercial broadcasters, have emerged and could well fragment and undermine the Hi-Vision market (the same way extended PAL is expected to slow down the market penetration of HD-MAC in Europe).[74]

In competing with NHK's Hi-Vision strategy, Nippon Television Network, an independent commercial broadcaster, has developed the second generation of EDTV, called New NTSC, in Japan. Unlike Clearvision, New NTSC offers 16×9 widescreen broadcasting.

In Europe, the process of EDTV development in parallel with HDTV seems to be more complicated, because the major industrial participants in EDTV are those who were also the major firms in the EU95 consortium, the European HDTV project. As a completely different approach, the development of PALplus is emerging as a strong challenge to the MAC family, particularly to the collapsed HD-MAC.

Starting from the end of 1988, the PALplus project was aimed to transmit 16×9 pictures through the existing PAL infrastructure. As an enhanced terrestrial transmission system, PALplus was claimed to be commercially available for service in 1995. Currently, the development work of PALplus is being carried out by the PAL Strategy Group, which consists of some public broadcasting authorities including ARD, ZDF, ORF and SRG in collaboration with consumer electronics manufacturers including Philips, Grundig, Thomson, and Nokia from some European countries. In addition, BBC of the U.K. with financial support from the DTI,[75] has also joined force to develop PALplus. So far the PALplus consortium has not given as much publicity to its work as the EU95 consortium. However, the progress of PALplus has shown a fast pace. Although it was claimed that no effort is being made to achieve an 'HD-PAL' to compete with HD-MAC,[76] there exists the technical possibility to upgrade PALplus to 'HD-PAL'.

At the International Broadcasting Convention in Amsterdam in July 1992, the PALplus consortium demonstrated their improved picture quality in comparison with the current PAL standard. It is evident that the progress of PALplus has added more confusion to the European HDTV policy-makers, whose major interest and enthusiasm lay in D2-MAC and HD-MAC.

Implications of the BSkyB Affair

At the beginning of November 1990, BSB (British Satellite Broadcasting) and Rupert Murdoch's Sky Television merged into a single venture BSkyB (British Sky Broadcasting). This event caused wide concern all over Europe and, most particularly, imposed great pressure on the newly defined European HDTV system—HD-MAC.

Having been launched in 1989, Sky Television programmes were transmitted in the conventional PAL format through Astra, a low-powered telecommunication satellite. Sky's DBS transmission in PAL was not prevented by the MAC Directive adopted in 1986, because the latter was limited in scope to high-powered DBS services as defined in the WARC document in 1977. In contrast to Sky, BSB, launched in May 1990 and transmitted *via* the Marco Polo satellite, was the only DBS service using the D-MAC format, a family member of the MAC/packet, in the U.K. It was anticipated that the launch of BSB would increase the popularity of the MAC standards and, therefore, would enhance the European HDTV strategy.

As a result of the merger, the previous five-channel service from BSB was incorporated with the four-channel service from Sky to form a new six-channel DBS service, which are all transmitted in the PAL format from Astra. As part of the deal, the previous small squarials used by BSB were replaced with the bigger and round-shaped Astra dishes. As far as the transmission format was concerned, the BSB-Sky deal was effectively a takeover of BSB by Sky rather than a merger. From a technological point of view, the first, and most interesting, implication of the BSB–Sky affair was that,

> Astra and PAL, the existing TV standard, have won a victory over the more sophisticated D-MAC system used by BSB. The deal marks the virtual end of D-MAC as a European satellite standard.[77]

The victory of Astra and PAL and the collapse of D-MAC, to some extent, resembles the VCR format battle, where the VHS system won the victory over the so-called technologically more advanced Betamax and V2000 systems. In both cases market forces or the choice of the consumers played a significant role.

In terms of financial losses, it was the manufacturers of squarial dishes and D-MAC receivers used for BSB reception who suffered from the merger. Because the BSB equipment was not able to receive signals transmitted from Astra, producers had to stop production and their previous stock could no longer be sold. Philips, who had produced 150,000 receivers for BSB, and Thomson claimed £50m and £20m compensation, respectively, from BSB on the grounds of breach of contract by the latter after the merger.[78] Tatung, a Taiwanese manufacturer, claimed compensation as well.

The failure of BSB has led to criticisms from some industrialists of EU policy-makers for leaving a 'loophole' in the MAC Directive. Timmer, the President of Philips, believes that the Commission's 1986 MAC Directive 'did not prevent the *improper* [emphasised by the author] use of telecommunication satellites for transmission of TV signals in the old PAL standard', and 'it was the cause of a regrettable loss of time and much unproductive debate'.[79] It seems that Philips' official view over DBS regulations was difficult to be agreed by commentators outside the industry. It was argued that broadcasters and satellite operators should have the freedom to choose which system they want to use and this will benefit the consumers,

> Regulate only for safety and antitrust reasons; let markets decide who gets which rights of way and what technology is used. ... there is no reason for regulators to stop broadcasters from choosing among a growing family of technologies instead of insisting on one.[80]

It is believed that the 'loophole' left by the 1986 MAC Directive in wording was only half an opportunity for Rupert Murdoch to steal the lead in launching Sky earlier than BSB. The other half of opportunity was offered by the delayed development of microchips on which MAC receiver production depended.[81] The delay in microchip development was mainly caused by the split into D2-MAC and D-MAC of what was supposedly a unified European standard.[82]

In practice, a coercive Directive may prove to be an unpopular political decision; but a free market approach does not seem likely to become a European policy as far as the choice of DBS including HDTV transmission standards is concerned. Despite this difference between how technology was chosen in Europe, some kind of compromise has already been made within the EU comparing the new Directive for DBS with the 1986 MAC Directive. Firstly, in the new Directive D2-MAC is only compulsory for 'not completely digital' DBS with a 16×9 aspect ratio format. This implies that PAL/SECAM system with a 4×3 aspect ratio may still be used for DBS before 1 January 1995. Secondly, although the new Directive requires that any 'not completely digital' TV transmission of 625 lines must be accompanied by D2-MAC transmission of the same programmes after 1 January 1995, the extra costs of simulcast services will likely be compensated by EU subsidy.

No matter how politically controversial the BSkyB deal was, the new venture has already brought financial benefits to its shareholders. By the time when BSB and Sky Television were merging, both companies were heavily losing money and their independent operations were believed to be at risk. Only 16 months after the merger, BSkyB was reported to expect its first weekly operating profit of £100,000 in March 1992.[83] Moreover, BSkyB has now become Europe's largest DBS television broadcaster.

The Impact of Digital Technology

Doubtless, the BSkyB affair was a big political embarrassment to EU authorities who had strongly pressed ahead the MAC Directive with the so-called regulatory 'loophole'. Another force, which eventually brought the European HDTV system to an end, was the rapid development of digital compression technology.

In the last decade the consumer electronics industry saw rapid technical changes. Since Philips and Sony revolutionised the music industry through their compact disc, DAT (Digital Audio Tape) was launched by Sony. NICAM TV enables the consumer to enjoy digital stereo sound while watching programmes. Again, as announced by Philips and Sony, the recordable DCC (Digital Compact Cassette) and Mini Disc will offer the consumer more choices for digital audio. Based on the well-established computer and compact disc technologies, companies such as Philips and Commodore have released their consumer multimedia products (CD-i and CDTV) which further explore the potential of human–machine interactivity in the consumer digital world.

The progress towards fully digital television, as is happening in the U.S. and Europe, promises more potential than had been anticipated. One of the most evident trends in the development of the consumer electronics industry since the beginning of the last decade has been its shift from the traditional analogue techniques to digital technology. From a technical point of view, both HD-MAC in Europe and Hi-Vision in Japan have been increasingly challenged since 1990, simply because of their analogue nature in the transmission of pictures. Despite the constant efforts of the EU95 consortium to technologically improve and finalise the HD-MAC system, it failed to survive the competition as digital HDTV has become a necessity.

If the trend of technical change has been going towards digital since 1990, why was the European Commission still uncompromisingly committed to promoting HD-MAC, a hybrid of analogue and digital techniques at the beginning of the 1990s? Some Philips managers and EU officials who were in favour of the European system have argued that fully digital HDTV technology may not be ready before the end of this century and this would leave sufficient space for Philips and its collaborators to make money from analogue HDTV in the coming years; they have also criticised those who were in doubt of the EU's HD-MAC strategy for being too optimistic about the progress of digital TV technology.[84] If a fully digital HDTV system works well in the U.S. sooner than expected, would Philips and some EU officials change their views about HD-MAC? It might be easy to change views, but it would not be easy for public authorities to explain why previous EU policies and strategies had led to huge investments into a failed transmission system.

The newly announced HD-Divine project had added more pressure to the European policy-makers and the survival of the more established HD-MAC system prior to its collapse.

HD-Divine (High Definition Digital Video Narrowband Emission) is a fully digital terrestrial HDTV system developed by a Scandinavian consortium, whose members include the Swedish Broadcasting Corporation, Telia Research, Teracom, Telecom Denmark and Telecom Norway.[85] The HD-Divine system was first demonstrated at the IBC conference in Amsterdam in July 1992. Compared to HD-MAC, HD-Divine will enable broadcasters to transmit terrestrially compressed digital HDTV signals through an 8 MHz bandwidth TV channel. It is the HD-Divine group's intention to promote international standardisation through a digital approach. Apparently, HD-Divine became another strong argument for those, such as the British government, who were opposed to subsidising D2-MAC and HD-MAC, to block or scale-down a decision by the Council of Ministers for further backing of the EU95 project.

Compared to the EU95 consortium, HD-Divine is a much younger and smaller project.[86] However, the technological and political influence of the latter seems tremendous in Europe. Technologically, the birth of HD-Divine would undoubtedly speed up the pace of digital HDTV development within Europe. Politically, the EU's HDTV strategy since 1990 was subject to further criticism for its inability to reflect the rapidly changing technical circumstances. In particular, the European Commission's political credibility in identifying and promoting strategic technologies was seriously questioned.

In a broader pan-European scenario, the European technology policy towards HDTV shows even more contradictions than the HD-Divine case has implied. According to my research, new EU strategies for HD-MAC since 1990 were not conceived on the basis that the Commission was unaware of the potential of digital technology. On the contrary, the Commission deliberately took a pro-MAC stance because, as mentioned above, the collapse of HD-MAC would simply be politically disastrous to the Commission itself, in particular those policy domains directly involved.[87]

It is known that the dTTb (digital Terrestrial Television broadcasting) project, proposed for funding under the RACE (Research on Advanced Communications for Europe) programme, has not been financed by the Community, and a feasibility study is still under consideration at the time of writing.[88]

As earlier discussions show, the technological foundation of the proposed European HDTV system was set by the Eureka 95 Project. However, an alternative digital HDTV system was being developed in parallel with

HD-MAC. Under the same Eureka umbrella, two other Eureka Projects, i.e. EU256 and EU625, were also dedicated to defining a fully digital HDTV system including a transmission standard and a production standard. EU625—with a budget of $24m—is also known as VADIS (Video-Audio Digital Interactive System), which has already been joined by 32 European broadcasters, manufacturers, software companies, telecommunications firms and academic institutions with a primary goal to coordinate European work on digital television that could bring out a viable European digital HDTV system.[89] Headed by RAI, the Italian broadcasting organisation, the EU256 Project was intended to develop a digital HDTV transmission system for Europe.

While the current HDTV battle remains messy and far from its final completion, more challenges are coming up from the technological side. In July 1990, at a conference participated by the Japanese manufacturing industry and academic institutions, the second generation of HDTV was planned.[90] The screen display resolution of the so-called second generation of HDTV would be twice as high as that of NHK's Hi-Vision system, and the set could be used as a video telephone display and home computer monitor as well. The newly proposed HDTV system at the conference was promised to be defined within three years, and then presented to the CCIR as another world standard proposal for HDTV. According to this proposal, the new HDTV system was scheduled to be commercialised by 2015. In addition to Japanese participants, MIT of the U.S. also joined force. Due to its technical complexities, the second generation of HDTV may not pose an immediate challenge to the current systems, including Hi-Vision (and HD-MAC before its collapse); however, it would inevitably change people's perception about the previously acclaimed technological and commercial value of the latter.

PHILIPS AND HDTV DEVELOPMENT

As mentioned above, Philips has been the project leader in developing the European HDTV system through the EU95 consortium since 1986. The relationship between Philips and the development of the HD-MAC system was two-fold. On the one hand, the success or failure of Europe's HDTV system would be critical to Philips, a company struggling to bring new gadgets to stimulate the declining consumer electronics industry; on the other hand, Philips' commitment to HD-MAC was one of the decisive factors in determining the fate of the system.

It seems that Philips, as well as Thomson, has adopted a multicommitment strategy—to get itself involved in R&D activities for a fully digital system in the U.S. while actively promoting HD-MAC in Europe. In addition, as discussed earlier in this chapter, Philips has also actively participated in the PALplus consortium to develop an alternative EDTV standard in competition with MAC. This suggests, on the one hand, that Philips and its European collaborators had no intention of missing the digital HDTV bandwagon; on the other hand, they were also very much concerned with the amount of money they had poured into HD-MAC R&D activities during the past few years, and they would explore any opportunity to make the most of HD-MAC by any means, including pressurising the EU to keep the European system going. The multicommitment strategy adopted by Philips suggests that the company was trying to make itself technologically prepared to shift from the analogue HD-MAC to fully digital HDTV manufacturing firstly in the U.S., then in Europe.

It was a widespread Philips argument that a fully digital HDTV system would not be available until the beginning of the next century, and there existed a considerable consumer demand and market for MAC and HD-MAC products. To justify the HD-MAC project, a Philips manager said,

> As we move into the 21st century, undoubtedly, we will move to digital techniques for transmission. But it is clear that the only system capable of showing widescreen and HDTV that can be implemented for the consumer practically this century is the MAC [HD-MAC] system. Of course, digital techniques will follow in the 21st century; they are being developed now and are only on paper; they don't exist as a reality. There is an opportunity to give the consumer 16:9 now and HDTV in 1995, which does not exist with digital.[91]

Apparently, Philips itself also believed that HD-MAC would be a technology of the 20th century, and it could be replaced within about ten years from the time of its launch. However, the timetable for the replacement might be rescheduled. From a technological point of view, digital HDTV development in the U.S. and Europe has already made the EU's HD-MAC standard obsolete.[92] Philips might have been well aware of this prior to the collapse of HD-MAC. It was reported in 1992 that Philips was considering abandoning the HD-MAC system and making a leap-frog to digital television.[93] This speculation was actually fuelled by Mr J. Timmer, the President of Philips, who said that Europe would give up a technological lead in the field of television if the European Council of Ministers decides not to approve the proposed subsidy of ECU 850m by the end of 1992; but Philips would not suddenly be faced with a huge financial loss should this happen.[94]

Doubtless, the collapse of HD-MAC was a great setback to Philips. Prior to the Commission's announcement to give up the HD-MAC transmission standard, Philips had already announced that it would be able to launch its HDTV sets at the price of £3,500 or less in mid-1993.[95] The failure of HD-MAC would inevitably undermine the opportunity for Philips to commercialise all the technologies it had developed for the D2-MAC and HD-MAC systems. Apart from the huge investment by Philips itself, the EU and some member states have also subsidised the European HDTV project since 1986.[96] However, it seems that Philips had passed the political dilemma to the European Council of Ministers as far as HD-MAC was concerned: if the proposed budget could be approved the Council of Ministers would be continuously criticised by the media and private broadcasters for subsidising an out-of-date technology; if they did not approve the subsidy proposal Philips would withdraw from the HD-MAC project and most of the previous HDTV efforts would end up in vain.

HDTV AND INDUSTRIAL COMPETITIVENESS

Despite all the controversies, HDTV appears to be one of the most important (or strategic) technologies in the years to come. It is believed that there is hardly any aspect of life that is not going to be touched by this next generation of television technology.[97] In other words, HDTV will not only reshape the structure of the world consumer electronics industry, but also have significant impact on other related industries.

The Spillover Effect of HDTV

No matter which system is adopted or being developed, industrialists and public authorities all over the world have no disagreement over the point that HDTV is more than consumer electronics and home entertainment. On the contrary, as an industrialist claims, HDTV has become a strategic commercial and technical focus for the fusion of digital media, digital communication and digital computing.[98] Because of the development of HDTV technologies, the conventional boundaries between the consumer electronics industry, computer industry, and telecommunication industry are being blurred or diminished. Nonbroadcast applications of HDTV transmission and display technologies may also have great impact on the future medical equipment, defence industry, etc. In an attempt to lobby the EU and bring European industrial forces together, a senior manager from Philips points out,

> HDTV will have an impact on every facet of economic life. This is why the Japanese have targeted HDTV as a key, strategic technology and have devoted so much energy trying to force their system onto others. They wish to dominate tomorrow's economic world.[99]

Beyond various political controversies in Europe, there seems to be a key issue of who can make money out of the HDTV technologies—either analogue or digital—whichever transmission system and production system is adopted. The history of CTV development has proved that to have production capacity in key components (e.g. CRT) was vital to industrial competitiveness. NTSC was invented in the U.S. and the Japanese use the same transmission standard. However, the American CTV market has now been dominated by Japanese and European multinational companies. It could be the same for the future HDTV industry. The suppliers of key components such as advanced memory chips, high-quality flat panel display (rather than CRT) would possibly dominate the industry. Because of the high costs to develop and manufacture these high-tech components, set assemblers may probably only be able to exploit very limited profit margin, provided they wish to sell their products to the consumer at an affordable price. By the time a fully digital HDTV standard has been accepted all over the world, component supply may be easily organised at a global scale, hence economies of scale could be exploited.

So far, as discussed in Chapter 5, the advanced memory chips industry and LCD (currently the major flat panel display technology) business has been dominated by the Japanese. Europe is very much disadvantaged in both areas. Having withdrawn from the JESSI programme, Philips discontinued its involvement in developing and producing DRAMs (Dynamic Random Access Memories) above 4 Mbytes.[100] This withdrawal might be necessary given the company's current financial situation and its costly restructuring programme; but, as Europe's leading semiconductor manufacturer, Philips has given up a substantial foothold in a futuristic programme.[101]

Substantially differing from its advanced semiconductor policy, Philips has adopted an aggressive strategy in the area of LCD production. Philips is the only significant competitor to the Japanese outside Japan in LCD technology. The company set up its own LCD production centre in Eindhoven in late 1991. However, the gap between Philips and its Japanese counterparts remains wide. The biggest LCD size available at Philips, according to the company's Managing Director of the LCD project Mr J. Stuve, was only 6 inch (15 cm) in 1991, and LCD screens large enough to be used as HDTV displays would not be ready for another 5–10 years.[102] To catch up with its Japanese competitors including Sharp, Toshiba, and Hitachi, which started mass production several years earlier and are more experienced, Philips decided to invest Fl 200m for an active matrix LCD project in November 1991, and hoped to make its first commercial delivery at the beginning of 1993.[103]

While Philips is concentrating on improving its presence in the LCD industry, new technical breakthroughs in other parts of the world may well suggest alternative solutions to designing and manufacturing flat panel display products. Among others, a group of scientists in Cambridge, England recently invented a new technology called Light-Emitting Plastic (LEP). LEP was claimed to have a wide range of implications for flat panel information displays of literally any size.[104] The Cambridge LEP announcement was immediately followed by the electro-luminescent display, an invention of the Stuttgart Institute. It was asserted that manufacturing the electro-luminescent display would require only 15 steps, compared to about 60 steps for manufacturing a large screen LCD panel at present.[105] To make not only CRT but also LCD the technologies of yesterday, Advanced Laser is now exploring the commercial potential of its laser-beam flat TV, a laser-manipulation system. Advanced Laser has already designed a prototype of its first flat screen, called Thinline, with a measurement of 13 inch diagonal and 4.5 inch thick displaying picture quality equivalent to the VGA standard.[106] The company has claimed that the picture quality will be improved to meet the SVGA standards, and no upper limit to screen size has been found. All of these new developments in flat panel displays have happened outside Japan. If any of these technologies proves commercially successful, arguably, Japan's domination in LCD design and production might not add as much competitive advantage to the Japanese manufacturers as it used to do. In a similar way, Philips' ambition of taking a slice of the lucrative flat panel display market may well be endangered if the company puts all of its eggs into one basket, i.e. concentrating on LCD technologies.

Market Potential of HDTV

As far as market potential for HDTV is concerned, there are various estimations. Optimists reckon that HDTV sales in the U.S. alone could total an astonishing $12bn a year in the next century, which would be nearly half as big as the whole of today's American consumer electronics market.[107] Given the estimated size of the HDTV market, the American Electronics Association (AEA) claimed that American firms' share of world chip markets would halve if they do not dominate HDTV sales and, therefore, the U.S. would no longer be able to play a leading role in the future world electronics market without economies of scale and technological synergies from manufacturing HDTV chips.[108]

As shown in Table 6.10, the forecast by MITI of Japan suggests a total HDTV market (including hardware and software) by the year 2000 worth as much as $39.17bn, of which consumer equipment sales alone would account for $22.44bn.

Market development research by BIS Strategic Decisions suggests that 2 million HDTV units and 5.5 million 16×9 (including MAC and PALplus)

Table 6.10. MITI HDTV Market Forecast—2000.

Market	Sales ($bn)
Consumer equipment	22.44
Industrial equipment	6.91
Film/TV exhibition/broadcast	6.66
Film/TV production	3.16
Total	**39.17**

Source: Based on JTEC Report, *High Definition Systems in Japan*, quoted in Komiya (1991), op. cit.

EDTV units would be sold at a retail value of ECU 2.7bn and ECU 5.6bn, respectively, in Europe alone by the year 2000.[109] More optimistic estimation was made by The Economist Intelligence Unit, which forecasts that the HDTV share of the TV receivers market would be 50% for Japan, U.S. and Western Europe by the year 2000, as shown in Table 6.11.

Apparently, some market research results suggest a huge HDTV market in the 21st century; meanwhile, because of the high costs to develop HDTV components and manufacture receivers, genuine HDTV may likely remain a very expensive product in the 1990s, i.e., the size of world market, particularly for consumer equipment, may stay insignificant. In other words, HDTV in the end is a next century consumer product. This obviously contradicts the above mentioned lobbying of Philips group, which repeatedly claimed that manufacturers would have sufficient time to make money out of the HD-MAC system before a fully digital HDTV system is available at the beginning of the next century.

The 'Chicken and Egg' Effect

The above discussion indicates that, on the one hand, HDTV appears to be a strategic technology and, therefore, each of the three regions from the Triad could not afford to ignore it; on the other hand, due to the high costs and technical hurdles to develop an independent HDTV system, the commercial success may not be seen in the short-term although it is extremely lucrative. For the time being, it seems that the creation of an HDTV market is being hindered by a 'chicken and egg' effect.

Firstly, the 'chicken and egg' effect exists between manufacturers and

Table 6.11. Forecast Demand for CTV and HDTV Receivers in 2000 (Million Receivers).

Region	Colour TV Receivers	HDTV Receivers	HDTV % share
Japan	11.0	5.5	50
U.S.A.	24.0	12.0	50
Western Europe	25.0	12.5	50
Eastern Europe & U.S.S.R.	15.0	1.5	10
China	25.0	...	...
Other Asia, Africa & Latin America	40.0	...	
World Total	**140.0**	**...**	**...**

Source: EIU (The Economist Intelligence Unit) (1991), *High Definition Television Progress and Prospects: A Maturing Technology in Search of A Market*, Special Report No. 2189, Business International Ltd., p. 104.

consumers. In Japan, HDTV receivers built for the Hi-Vision system have already been launched on the consumer market. In Europe, prototypes of HD-MAC equipment have also been developed. However, HDTV products from either Japan or Europe appear to be bulky and expensive. For the consumer electronics manufacturers, 'High-definition television exists. The struggle now is to package it in a slim set at an affordable price'.[110] In order to make an HDTV set slim, the traditional CRT display must be replaced with a new flat screen, for which many electronics companies have been making huge R&D investment to find out the right material and right technology. The most widely used flat-screen technology is LCD. For the time being, to build an LCD display with a size large enough to be used as an HDTV screen still remains technically impossible. The slightest manufacturing defect would destroy the whole display panel. Even if the manufacturers can build HDTV screens with LCD, or any other flat-panel technology, the high costs may well keep the consumers away from the market. For instance, Panasonic's 14in 'Flat Vision' (not LCD display) TV set costs about £2,000. Sharp has developed a 14in LCD flat screen and Canon has demonstrated a 24in version built with ferroelectric LCD, but the high costs have prevented both companies from launching their products to the mass consumer market.[111]

Secondly, the launch of HDTV service may only be viable with the existence of a wide range of HDTV receivers—HDTV transmission is hardly meaningful to ordinary CTV receiver owners. But, as discussed above, the HDTV market will be very limited before the end of the century. The EU and some manufacturers recently started a campaign to promote the wide-screen concept; they hope a high level of widescreen (D2-MAC) penetration would make the launch of HDTV services easier in Europe. This campaign has been backed by EU rules (e.g. the new DBS Directive) and financial subsidies, i.e. the revised 'Action Plan'.

Finally, there exists a 'chicken and egg' effect between HDTV broadcasters and programme producers, particularly in Europe. Despite the financial incentives from the EU, the stock of programmes available for HD-MAC transmission is very limited. This situation has been worsened partly due to the late arrival of European production equipment. In May 1990, HDTV Holland, a specialised high-definition programme producer, first approached Philips and asked whether they could supply a completed set of production equipment within one year's time; but Philips could not meet this need. Consequently, HDTV Holland shifted to Sony who promised to provide all the necessary HDVS (High Definition Video System) equipment within three months.[112] Thus, the European system lost a programme producer. It is doubtful whether Europe could be able to build up a considerable stock of HD programmes when full HDTV service is launched, as promised, by the mid-1990s. For programme-makers, on the other hand, it is highly risky to produce software with HD-MAC equipment before there is a normal HDTV service in Europe.

CONCLUSION

To summarise the above discussion, the formation of the European HDTV strategy in its essence was dominated by political consideration, which heavily involves trade protectionism and government–industry coalition. From the inception of the European HDTV proposal until the collapse of HD-MAC as a European transmission system, policy-makers within the EU played a decisive role in picking 'winners'. This has led to distortion of competition between alternative technologies (analogue *versus* digital) and rival companies (local *versus* foreign).

Moreover, to some extent public interest might have been jeopardised, given the sum of public money the EU and some national governments have injected into the industry for developing HD-MAC. The most confusing point was that such an expensive system was abandoned before it was commercially launched. Politically, the whole package of EU HDTV policy and strategy implemented from 1986 to early 1993 ran into opposition from countries (e.g. U.K.), which did not share the view that EU tax-payers should pay for a nonfuturistic technology. It is argued that public subsidy only makes economic sense if ultimately consumers stand to gain from a move to HDTV (Brown *et al.,* 1992, p. 48).

To be sure, the global HDTV arena has been subject to the same uncertainties as discussed above, but the picture of HDTV development in Europe was even worse. In order to provide a legal framework for the establishment of an intermediate step towards HD-MAC, the MAC Directive was adopted in 1986; but within less than three years, in 1989, Rupert Murdoch launched his Sky Television using the PAL transmission system without breaking EU law (the MAC Directive). When Sky Television and BSB merged into BSkyB and, consequently, D-MAC was replaced with PAL in the U.K. in November 1990, the EU had no power to stop it from happening. While equipment manufacturers were crying for a more restrictive legal action to enhance the status of widescreen D2-MAC and HD-MAC for DBS services, the renewed DBS Directive was adopted in May 1992, which left room for developing fully digital TV technologies.[113] MAC's legitimate position as the official standard for 'not completely digital systems' has been further challenged by the development of PALplus.[114]

As far as the European manufacturers are concerned, HD-MAC was not their only stake. On the one hand, European manufacturers such as Philips and Thomson had been promoting HD-MAC and lobbying the EU to do as much as possible to support the system prior to the collapse of HD-MAC. On the other hand, these two firms have also been involved in developing a fully digital TV system in the U.S. and PALplus in Europe. It was evident that, by the beginning of 1993, the issue of whether or not HD-MAC had a future had become a matter for the EU and its member states rather than the manufacturers; leading manufacturers had threatened that the originally proposed subsidy package of ECU 850m over a five-year period from 1992 onwards was a prerequisite for them to continue their commitment to HD-MAC. Failure to agree on such a subsidy package simply led to a decision by Philips in early 1993 to suspend its planned HD-MAC receiver production.

It was argued in this chapter that HDTV is a strategic technology but, with hindsight, the EU has chosen to support an out-of-date system, and the way European public authorities dealt with the matter of HDTV competition was far from convincing. In retrospect, the year 1990 was a critical turning point in terms of international competition and technical change in the area of HDTV. Prior to 1990, the global scenario for HDTV standardisation was dominated by the Japanese Hi-Vision system and the newly established European HD-MAC system, the picture for the U.S. was not clear at all. Under this circumstance, European protectionism against Japanese domination in a strategically important technology area may well justify itself.

However, since GI's revolutionary technical breakthrough in digital compression was achieved in 1990, the FCC managed to channel the HDTV competition towards a fully digital system in the U.S.; this had tremendous impact on HDTV development all over the world. All of a sudden, HD-MAC and Hi-Vision, both technically hybrid systems, began to face the same challenge from digital technology. Instead of reconsidering its position over HDTV, the EU, mainly the Commission, responded to the digital revolution by making its last-ditch effort to rescue HD-MAC until it finally accepted defeat in early 1993.

Contrary to its original intention of picking winners in the HDTV race, EU authorities have fostered losers: the EU had imposed MAC/packet family on the broadcasting industry, but PAL won the victory over D-MAC in Britain; the EU had made HD-MAC the flagship of its technology policy, but digital technology has made it a technology of yesterday.[115] The case of HD-MAC indicates that it is highly dangerous and risky for governments to pick technological winners and foster domestic champions at the expense of competition and further technological advances. Both the relatively pluralistic and flexible policy of the FCC to coordinate the nonsubsidised technological competition in the U.S. and the intrinsically protectionist technology policy adopted by the EU towards HD-MAC in Europe have demonstrated that institutional flexibility is of vital importance to the development and success of new technologies. It is misguided for firms and, especially, government to choose specific technology options which preclude other possible alternatives during the course of rapid technological transition.

Ironically, while the EU has fallen into wide range of criticisms over its technology policy for HD-MAC, Philips and Thomson have managed to become two of the few leading manufacturers in the HDTV battle in the U.S. without any government help.[116] This suggests that European manufacturers, such as Philips and Thomson, may not be as weak as EU policy-makers have conceived; they may well survive the global competition in some key technological areas, such as HDTV.

NOTES

1. *Business Week,* 21 December 1987, p. 48.
2. Note that since 1986 advances in digital compression technology now allow HDTV terrestrial transmission, as with the U.S. and HD-Divine.
3. The European MAC technological family includes several varieties such as C-MAC, D-MAC, D2-MAC, etc. D2-MAC was chosen in the EU as the official European DBS (Direct Broadcast by Satellite) standard for all the member states.
4. *Business Week,* 26 February 1990, p. 56.
5. HDTV is widely talked as 'home cinema'.
6. NTSC was first introduced in the U.S. by RCA in 1953; PAL and SECAM was introduced, respectively, by Telefunken in W. Germany and Thomson in France in the 1960s. See Brown *et al.* (1992), pp. 5,7.
7. *Philips News,* Vol.17, No.3, 29 February 1988.
8. The Independent Broadcasting Authority, and now the Independent Broadcast Commission (ITC), who owns the key patents for the MAC family.
9. This became true in early 1993 when HD-MAC was officially abandoned as the European HDTV system.
10. Indeed, narrow-band MUSE (to be transmitted through a 6MHz channel) was proposed by NHK for terrestrial transmission in the U.S.
11. NHK deliberately chose the date of 25 November, or 11/25 as written in the Japanese order, to symbolise the 1125 lines of the MUSE transmission system.
12. Lang, Y.H. (1990), *Dianshi zai Geming: Mingri de Dianshi Shijie* [TV Revolution: Tomorrow's World of Television], Cheng Chung Book Co., Ltd, Taiwan, p. 104.
13. *The Nikkei Weekly,* 8 February 1992.
14. *The Nikkei Weekly,* 22 February 1992.
15. These will possibly be successful if HDTV programmes prove more popular than those broadcast in NTSC.
16. *The Nikkei Weekly,* 22 February 1992.
17. More recently, Toshiba and Sony announced their full-specification Hi-Vision 32" sets priced at ¥980,000 and ¥900,000, respectively. For more detailed information on HDTV marketing and price trends, see *The Nikkei Weekly,* 22 February 1993 and 19 April 1993.
18. Members of the FCC Advisory Committee include a wide range of interest organisations: selected cable operators, networks, studios, set manufacturers, broadcasters, as well as representatives from the State Department, the Commerce Department, the National Association of Broadcasters, and the National Cable Television Association. See Farrell and Shapiro (1992), Note 24.
19. *Financial Times,* 25 January 1990.
20. For more detailed discussion about Americans' re-entering into the consumer electronics industry, see Dupagne (1990). He argues that the U.S. can re-enter the consumer electronics industry by creating an independent HDTV technology. Tyson (1992) offers a sceptical assessment of this view.

21. This has already happened in some other product areas, such as CD-i, of the consumer electronics industry. As one of the leading American chip makers, Motorola is the major chip supplier to Philips for manufacturing CD-i players. More information about the collaboration between Philips and Motorola is presented in Chapter 8.
22. However, while the Europeans and Japanese are heavily involved in the U.S., American companies have made no presence in either Europe or Japan during the course of HDTV development.
23. In Europe consumers are expected to invest for an interim model, called MAC receiver, if they wish to watch HDTV programmes but can not afford to buy a real HD-MAC set, because HD-MAC signals would not be compatible with PAL/SECAM receivers. In Japan, it is technically possible now that MUSE programmes can be received by NTSC sets if a MUSE-NTSC converter is installed. I will come back to this point later in this chapter.
24. See Farrell and Shapiro (1992) pp. 19–20 for more detailed discussions on this ruling.
25. *Fortune,* 8 April 1991.
26. *Ibid.* But Farrell and Shapiro (1992) argue that, due to the fact that the U.S. has decided to fit HDTV into the already crowded terrestrial television spectrum, the future American HDTV system will be heavily squeezed 'with extra costs, compromised quality, and reduced scope for later enhancements' compared to the DBS transmission methods for HDTV systems in Japan and Europe (p. 20).
27. Except for the $30 million from the Pentagon's Defense Advanced Research Projects Agency (DARPA) for the development of high definition screen technology (Brown *et al.,* 1992, p. 33). The same source has also suggested that the AEA (American Electronics Association) called, in 1989, for $1.35 billion of federal government funding for HDTV. But this was rejected by the Bush administration.
28. *Financial Times,* 1 June 1990.
29. The four European companies' alternative HDTV proposal was submitted to the Eureka Ministers conference in London on 30 June 1986 via their respective governmental representatives.
30. This figure was obtained at the *IBC (International Broadcasting Convention) 1990* in Brighton, the U.K.
31. The EU95 HDTV Directorate was composed by Philips of The Netherlands, Thomson of France, Bosch of Germany, Nokia of Finland, and CISAE (Consorzio Italiano per lo Sviluppo della televisione ad Alta definizione Europea), a consortium of seven Italian companies. See *Philips News*, Vol.19, No.8, 3 July 1990.
32. EU95 Directorate (1991), *Progressing towards HDTV*, published by Eureka 95 HDTV Directorate, June.
33. *Ibid.*
34. Cawson (1992a), 'Sectoral governance and innovation: private interest government and the Eureka HDTV Project', presentation paper for the *ESRC Conference on Government-Industry Relations,* Exeter, 20–22 May.
35. EU95 Directorate (1991), *op. cit.*
36. *The Times,* 10 February 1992. Note that precise figures regarding corporate

spending by big firms, such as Philips, on HDTV development are extremely difficult to obtain, if not impossible. Different sources sometimes suggest substantially different amounts. Reasons for disparity between figures are as follows: (1) Table 6.6 refers only to the estimated costs of Phase I of the EU95 Project; (2) Firms (such as Philips) are reluctant to publicise precise data on actual R&D costs for HDTV; (3) Given the above, *The Times* estimate of £2bn could well be correct.
37. *Financial Times,* 14 April 1992.
38. *Philips News,* Vol.20, No.15, 16 December 1991.
39. Interview with DG XIII official with responsibility for HDTV, Commission of the European Communities, Brussels, 13 November 1991.
40. See Brown *et al.* (1992), p. 30.
41. *Ibid.*
42. Note Taiwan, mainland China, and South Korea have dropped HD-MAC in favour of whichever system wins in the U.S.
43. See Komiya (1991), 'Japan's global HDTV strategy', *Advanced Television Markets,* November.
44. *Screen Digest,* January 1991.
45. *Ibid.*
46. Information in this section comes mainly from Lang (1990). Although some key technologies are supplied by NHK, the future Korean HDTV sets will also be technically adapted to the long-awaited American system.
47. The Industrial Bureau is a branch of the Ministry of Economy.
48. Taiwan adopted the American NTSC system for CTV signals transmission; hence it will be technically easier to work on an HDTV system with 1050 (twice of 525) horizontal scanning lines. Prior to deciding on its own HDTV strategy, Taiwan had approached Piet Bögels, Director of EU95, and asked to license the European HDTV technologies. Having considered the different CTV environment in Taiwan, the EU95 Directorate, instead of supplying technologies to Taiwan, asked Taiwanese firms to invest in Europe for HDTV R&D activities as well as manufacturing if the latter want to buy the former' HD-MAC technologies. Consequently, Taiwan shifted to American HDTV technologies. For more detailed discussions, see Lang (1990), pp. 118–119.
49. The Ministry of Transportation administers the Telephone & Telecommunication Bureau.
50. The MAC Directive refers to the EU Council Directive, *On the Adoption of Common Technical Specifications of the MAC / Packet Family of Standards for Direct Satellite Television Broadcasting,* 3 November 1986, 86/529/EEC.
51. This was put into direct competition with the more established Japanese Hi-Vision proposal.
52. EU Council Directive of 3 November 1986.
53. Philips might still have a fresh memory of its bitter experience with its V2000 format which was defeated by the Japanese VHS in the early 1980s.
54. Interview with DG XIII official with responsibility for HDTV, Commission of the European Communities, Brussels, 13 November 1991.
55. *Financial Times,* 29 April 1991.
56. Note that, as mentioned above, Sony has already built its own HD-MAC receiver to show the manufacturing and broadcasting industry.

57. Interview with DG XIII official with responsibility for HDTV, Commission of the European Communities, Brussels, 13 November 1991.
58. Council Directive, 'On the Adoption of Standards for Satellite Broadcasting of Television Signals', 11 May 1992, *Official Journal of the European Communities,* No. L 137, 20 May 1992.
59. The signatories or parties of the MOU declared that their final commitments will come into effect 'only with the approval by the Community of adequate financial support based on the Action Plan adopted by the Council'. (See *Memorandum of Understanding,* Brussels, 15 June 1992.)
60. *Financial Times,* 16 June 1992.
61. See *Electrical & Radio Trading*, 26 November 1992, p. 3.
62. This does not necessarily imply that the British government has always got it right in terms of HDTV policy-making. On the contrary, until recently, 'the U.K. Department of Trade was both attacking the EU's plans to spend European tax ECUs on subsidising MAC broadcasting, and passing legislation which would make it illegal from the end of this year [1993] to sell a widescreen TV set without a MAC decoder' (*International Broadcasting,* April 1993, p. 19). It seems that, apart from the conflicts between the British government and other national governments at the EU level, contradictions were also created by policy-makers within the British government.
63. See *Financial Times,* 17 and 22 December 1993.
64. *Financial Times,* 17 June 1993.
65. *Ibid.*
66. HDTV, or HD-MAC, was supposed to be the flagship of EU industrial policy. Due to the reduction of the proposed subsidy from ECU 850m to ECU 228m, the 'Action Plan' is now 'less a flagship than a life boat for the original strategy' (*Financial Times,* 17 June 1993).
67. Interview with DG XIII official with responsibility for HDTV, Commission of the European Communities, Brussels, 13 November 1991.
68. *Financial Times,* 9 February 1993.
69. The U.S. situation changed quickly. The FCC has now made it clear that the future HDTV system must be digital, and EDTV systems have been ruled out. Differing from Europe, the U.S. has now firmly committed to a 'single-step' strategy for HDTV.
70. *Screen Digest,* January 1991.
71. In the U.K., Channel 4 pioneered to broadcast feature films in this way and, very recently, the BBC followed suit to transmit some films in the same way.
72. Interview with Vice President of Nokia International, London, 4 April 1992.
73. A 36 inch model with a 16×9 aspect ratio from Nokia was priced around £2000, and a 28 inch model with the same aspect ratio was priced at half of that at the ERT Conference on *Images of the Future* held in London, 4th April 1992. Both models were capable of receiving D2-MAC transmission (provided a separate D2-MAC decoder is installed with an extra investment of about £200) and PAL/SECAM broadcast.
74. Komiya (1991), *op. cit.*
75. DTI has always been opposed to legal compulsion for PAL users such as BSkyB. On the contrary, DTI is in favour of a twin approach: MAC as the preferred European transmission standard for DBS, and an improved PAL

system, i.e., PALplus for terrestrial television. It seems that the DTI approach has more or less been incorporated into the renewed EU Directive on the adoption of standards for satellite television broadcasting.

76. Ziemer, A. (1990), 'PALplus: the downwards compatible enhancement of PAL', *HDTV 91 and Future Television,* Proceedings of the Second Annual Conference, London, December.
77. *Financial Times,* 5 November 1990.
78. *The Guardian,* 20 April 1991.
79. *Philips News,* Vol.21, No.10, 10 August 1992.
80. *The Economist,* 16 March 1991.
81. *International Broadcasting,* April 1993, p. 19.
82. *Ibid.*
83. *Financial Times,* 10 March 1992.
84. Interviews with Philips manager with involvement in the Eureka HDTV Support Group, Philips Consumer Electronics, Eindhoven, 21 October 1991; with Senior Manager of Philips International B.V., Eindhoven, 29 October 1991; with Marketing Manager, Philips Consumer Electronics, Croydon, 23 July 1991; and with DG XIII official with responsibility for HDTV, Commission of the European Communities, Brussels, 13 November 1991.
85. Information in this section is mainly from *International Broadcasting*, August 1992.
86. The HD-Divine group began their research in January 1991 with only five participants from the Scandinavian countries.
87. Note that a pro-MAC stance was not universally adopted by officials from different policy domains within the EU. Apart from the difference between the British government and the other major EU national governments (e.g. the French), disagreement on the EU's HDTV strategy was also existing between different sections of the European Commission. For more detailed discussion see Dai, X., Cawson, A. and Holmes, P. (1994), 'Competition, Collaboration and Public Policy: A Case Study of the European HDTV Strategy', SEI Working Paper No 3, University of Sussex.
88. *Ibid.*
89. See *International Broadcasting,* March 1993, pp. 7, 29; *Screen Digest,* June 1991.
90. Information in this paragraph is mainly from Lang (1990). Participants at the conference include NTT, Nippon Electric, Hitachi, Fujitsu, Mitsubishi, Toshiba, Sony, ITT, Fuji, and Tokyo University, etc.
91. Interview with with Marketing Manager, Philips Consumer Electronics, Croydon, 23 July 1991. This argument was also shared by EU officials in Europe and MITI and MPT officials in Japan, who were responsible for HD-MAC and Hi-Vision promotion, respectively. See Cawson, A. (1995), 'High Definition Television in Europe', *The Political Quarterly*, Vol. 66, No. 2, April–June, Note 21.
92. *Financial Times,* 11 August 1992.
93. *The Independent,* 7 August 1992.
94. *Ibid.*; also *Philips News,* Vol.21, No.10, 10 August 1992.
95. *The Times,* 10 February 1992.
96. For instance, the Dutch government announced to invest Fl 15m ($9m) to Fl

20m to subsidise Philips' R&D activities for HDTV under the auspices of EU95. See *Financial Times*, 7 November 1990.
97. *Screen Digest,* January 1991.
98. Herr, L. (1991), 'Japan sprints ahead in HDTV', *Computer Graphics Times*, March. L. Herr is the President of Pacific Interfaces Inc.
99. Quoted in *Financial Times,* 21 May 1990.
100. According to Lang (1990), to build HDTV sets advanced ICs including DRAMs of 4 Mbytes and above will be needed.
101. In addition, Philips has recently pulled out from MEC (Matsushita Electronics Corporation), a semiconductor joint venture between Philips and Matsushita since 1952, by selling its 35% stake at ¥185bn (£1.08bn) to Matsushita in part to help improve its financial situation. See *Financial Times,* 6 May 1993.
102. *Eindhoven Dagblad* (Eindhoven Daily), 6 November 1991.
103. *Ibid.*
104. *The Sunday Times,* 14 February 1993.
105. See *International Broadcasting,* March 1993, p. 10.
106. *The Sunday Times,* 18 June 1995.
107. *The Economist,* 4 August 1990.
108. *Ibid.*
109. BIS Consumer Electronics Information Service, quoted in Bird (1991), 'The market for advanced television', *HDTV 92 and Future Television,* Proceedings of the Third Annual Conference, London, December.
110. *The Times,* 26 November 1993.
111. *Ibid.*
112. Interview with President of HDTV Holland, London, 10 December 1991.
113. The wording of D2-MAC and HD-MAC systems as 'not completely digital' systems in the EU's new DBS Directive must be purposeful as regards to new development of full digital TV technologies.
114. Note that the proposed service launching timetable for PALplus has been set as the same with that for HD-MAC!
115. It seems likely that MPEG 2, thrashed out over years by many companies, is to underpin digital television worldwide. See *The Independent,* 17 May 1993.
116. Clearly, the reason for the strong position of the European firms is that, unlike the EU, they did not preclude their technology options.

7

THE DIGITAL REVOLUTION AND POLICY CHANGES IN EUROPE IN THE 1990s

The collapse of HD-MAC, as discussed in the last chapter, was not the *coda* to the development of new television technologies within Europe. Contrary to the view of some policy-makers that a proprietary European HDTV system would be vital to improving the international competitiveness of the European industry,[1] the failure of HD-MAC was not only a big financial loss but also a significant political *débâcle* for European policy-makers, in particular some parts of the European Commission such as DGXIII which was responsible for HDTV policy. Apart from institutional failure, the death of HD-MAC was mainly caused by digital technology. The rapid development of digital compression techniques since 1990 has made it possible for high definition TV signals to be transmitted within normal bandwidth (e.g. 6 MHz in the U.S.). To reflect the reality that digital TV technologies promised a new age of broadcasting and communications, EU authorities now made significant changes to their policies and strategies.

Generally speaking, EU policy changes after the failure of its MAC strategy included the following: first, digital TV has been identified as the next generation television technology. Given the fact that the EU did not set a strategy to follow the forthcoming American digital TV standards and its new policies do not exclude any specific new TV technology, it is increasingly obvious that digital TV is the most favoured technology in Europe. Second, digital TV, as the chosen technology, has been viewed by the EU as an integral part of a much broader policy framework, in which the implementation of digital TV will enable homes to be connected for the first time to a large digitised information gateway or the future Integrated Broadband Communications (IBC) network. Third, the European Group on Digital Video Broadcasting (DVB) has been given strong backing from the

European Commission to develop a European digital TV system. Finally, contrary to its regionally orientated HD-MAC strategy, the EU is now prepared to collaborate with its international competitors including the U.S. and Japan in the hope of identifying and agreeing common elements, if not common standards, for the future television systems to be implemented worldwide.

HD-MAC AND EU TECHNOLOGY POLICY

The previous chapter showed that EU authorities had been convinced of the commercial potential offered by establishing a proprietary analogue HDTV system, i.e. HD-MAC for Europe. Leading European consumer electronics manufacturers, such as Philips, used this argument to lobby public policy makers. Shortly after the collapse of HD-MAC, the European Commission admitted in a document, 'it was considered at the time that a window of opportunity of some years existed for this strategy to be successfully implemented'.[2] Unfortunately, this 'window of opportunity' for the Commission's MAC and HD-MAC strategies was quickly overshadowed by the rapid development of digital technologies. In other words, the Commission's strategy was based on a misconception.

From the HDTV format battle the European Commission has learned important lessons. Firstly, it realised that it is the quantity of TV programmes, i.e. software, rather than the quality of image that should be addressed with more emphasis on its previous policies and strategies. With the continued growth of satellite TV services since the late 1980s, 'it transpired that there was a greater immediate need for *quantity* in programmes rather than an emphasis on improving the technical *quality* of the image'.[3]

Secondly, having realised the strategic importance of programme-making, the European Commission now specifically emphasises the adoption of widescreen (16×9) format by broadcasters. This has been reflected in the much revised Action Plan providing ECU 228m of Community funding to be distributed among broadcasters and programme-makers over the period up to 30 July 1997. Compared to the Commission's previous MAC policies, the Action Plan requires only that 'advanced services adopt the 16×9 widescreen format (625 or 1250 lines) irrespective of the European TV standard used and irrespective of the broadcasting mode (cable, satellite, terrestrial)'.[4] In other words, service providers will be able to choose whichever transmission standard they like, whether it is an analogue or fully digital or hybrid (combination of analogue and digital techniques).

Thirdly, as HD-MAC was proposed as a satellite broadcasting format, terrestrial broadcasters were disadvantaged, if not excluded. The lack of enthusiasm of European terrestrial broadcasters to participate in promoting the use of D2-MAC and HD-MAC was an important factor contributing to the failure of the European HDTV technology. In reviewing its HDTV policies the European Commission confirmed this point:

> Terrestrial broadcasters ... felt largely excluded from the European policy since D2-MAC and HD-MAC are only suitable for satellite and cable systems.[5]

Fourthly, and most importantly, the European Commission now admits that the development of digital compression, digital coding and digital multiplexing techniques has made not only multichannel transmission (e.g. 4–5 standard definition digital TV channels to be transmitted in the same bandwidth as a single analogue channel) a reality but also the introduction of fully digital HDTV services highly possible by the end of the 1990s. The Commission noted that,

> Another significant factor has been the growing credibility attaching to the work being done in the U.S. on digital HDTV systems designed for the U.S. terrestrial broadcasting environment. It is now widely believed that the FCC process for selecting a digital HDTV system will lead to the introduction of such services by the end of this decade.[6]

The European Commission now, instead of refusing to believe a fully digital system to be built in a short term, say until the beginning of the next century, acknowledges the potential advantages of digital TV systems. These potential advantages, as identified by the Commission, are: (1) interoperability between services—internetworking between TV terminals, multimedia workstations, personal computers and other information terminals in access to a wide range of information services; (2) clear and stable pictures and sound; (3) more efficient use of the electromagnetic spectrum; (4) flexibility in the provision of TV services allowing a variety of different customer needs to be met; (5) more cost-effective compared with analogue equipment manufacturing in the long run.[7]

Finally, previous EU policies, in particular the 1986 MAC Directive, failed to address the importance of public interest or consumer protection. On the contrary, the whole range of MAC strategies were mainly geared to protect and strengthen the European consumer electronics industry. In proposing new policies after the collapse of HD-MAC, the European Commission has eventually accepted that the "protection of public interest through ensuring free and fair competition and through consumer protection" as one of the most important issues upon which the EU can and must act.[8]

GLOBAL COMPETITION FOR DIGITAL TV

New policy changes for the next generation TV within the EU are being conceived against a global context of competition for digital technologies.

As discussed earlier, global competition for HDTV started in Japan. The European industry took on the Japanese challenge by launching its own programme to develop a proprietary standard under the auspices of the EU. The U.S. decided to join the format battle at a much later stage to develop an independent system compared to Japan and the EU.[9] Ironically, the chronological order for digital TV development followed the reverse direction: the U.S. took the lead to develop a fully digital system from 1990; it was followed by Europe where efforts to develop a digital TV system began to gain official backing in 1993; Japan is now well behind in the race for digital TV—it has not proposed any digital system at the time of writing.[10]

ATSC (Advanced Television System Committee)[11] members agreed on a 'Digital Television Standard for Transmission' on 12 April 1995. However, ATSC only proposes standards to the FCC.[12] Whether or not a standard can be implemented is a matter subject to approval by the latter. In the meantime, the Grand Alliance consortium has delivered the necessary hardware to the Advanced Television Test Centre (ATTC) for extensive testing. While the proposal for a digital HDTV standard is awaiting approval, it is important to note that digital direct broadcast by satellite with standard definition (up to MPEG-1 level) is already spearheaded by companies such as Hughes Aircraft Corporation and Primestar in the U.S.[13]

It is known that the Grand Alliance proposal for HDTV will use MPEG-2 as the video compression standard, which is endorsed by the ISO (International Standard Organisation). As the FCC only regulates terrestrial transmission with respect to TV broadcasting, the adoption of transmission standards by satellite and cable operators are entirely subject to the discretion of these companies themselves. In other words, terrestrially transmitted digital TV services, including the digital HDTV standards put forward by the Grand Alliance, may not be immediately compatible with those offered by DBS and cable networks. It is too simplistic a statement that the Grand Alliance would ensure a single digital TV transmission system.

Although started at a much later stage than in the U.S., digital TV development in Europe appears to be a much more complicated picture. As mentioned in the last chapter, there was no shortage in R&D activities dedicated to digital TV in Europe during the years when the EU was still officially supporting HD-MAC with financial commitment at the Community as well as national government levels. Among others, HD-Divine, the Scandinavian system, has attracted a considerable amount of attention from all parties with an interest in HDTV. HD-Divine, together with the 1990 GI breakthrough with digital compression technology, has played an important role to effectively put a final nail into the coffin for HD-MAC. The VADIS project was also set up to contribute to the ISO's MPEG standard under the administration of the EUREKA programme. In addition to these initiatives, the EU has launched its own dTTb project dealing with terrestrial and cable digital broadcasting under the auspices of the RACE programme. It is pitiful that, as far as European policy-making is concerned, EU authorities including the European Commission did not give enough attention to these projects in relation to its HDTV policies and strategies.

R&D for digital TV has recently gained momentum in the U.K. with NTL (National Transcommunications Limited) leading the venture. NTL, the demerged engineering research arm of the former IBA (Independent Broadcasting Authority),[14] is pushing ahead with its digital terrestrial television (DTT) project. NTL's DTT proposal requires new digital receivers be equipped to receive existing analogue channels, the two BBC channels, ITV and Channel 4. In addition, the NTL system promises to maintain compatibility with existing VHS recorders based on analogue techniques. NTL is now able to export its innovative digital compression technology products, which enables the transmission of new digital satellite services offering 200–500 channels. In parallel with NTL's pilot work, the BBC has also demonstrated its achievement in digital TV transmission technologies. In summer 1995, the BBC succeeded in a test to transmit high quality widescreen pictures digitally from its London-based Television Centre to the Crystal Palace transmitter; the digital signals were received by a TV set with a decoder and a portable set-top aerial.[15]

In the hope to take a world lead in, and reap the widely claimed economic benefits from, digital terrestrial television technologies, the U.K. government set out its proposals for the development of at least 18 digital terrestrial TV channels within two years. The new DTT channels are expected to be made available to between 60–90% of the U.K. population. The main points of the U.K. government's DTT proposals are as follows.

- Six frequency channels, called 'multiplexes', will be available, and each will be able to carry at least three digital TV channels. In addition, seven radio multiplexes will be available.
- Existing terrestrial TV broadcasters (and the forthcoming channel 5) will get guaranteed access to digital frequencies on the condition that at least 80% of their analogue broadcasting programmes must be broadcast on the equivalent new digital services.
- Companies bidding for the digital channels will be assessed according to how fast and how widely they provide digital services.
- Bidders will also be expected to provide subsidy to consumers' purchase of 'set-top' decoders.
- Companies may control up to 25% of available digital terrestrial capacity and 15% of total TV audience with the exception of the BBC. No single company will be allowed to control more than two or three multiplexes.
- The first licence will be free for 12 years.
- Up to 10% of transmission capacity will be allowed for nonbroadcast facilities, such as telecommunications and interactive services.
- The ITC and the Radio Authority have been given the task of allocating channel capacity.

The U.K. government proposals for DTT did not exclude foreign operators. However, operators of the new DTT channels must have majority EU ownership. Despite the government's euphoria at the moment, it is questionable whether or not the newly proposed DTT services will take off in the U.K. There are a number of uncertainties faced by the U.K. authorities.

Firstly, the biggest challenge is the lack of sufficient programmes to fill up the new channels. For the time being, the richest programme source lies in the four analogue TV channels, which will be required to make 80% of their programmes available for digital broadcasting. This leads to a question: if the main contents of the new digital channels are simply repeating the conventional broadcasting programmes, would the consumers be prepared to invest purely for new technology (not to mention to what extent consumers care about the technical difference between analogue broadcasting and digital broadcasting!)? Even if the consumers support DTT services, 80% of the four analogue channels would only account for a small proportion of the available digital channel capacity.

Secondly, it is known that Rupert Murdoch's BSkyB, the largest DBS operator in Europe, is planning to launch its own digital satellite TV services by as early as 1996, which are promised to expand quickly from around 100 to more than 200 channels.[16] Despite that BSkyB has expressed interest in joining a consortium to bid for terrestrial digital channels, its own planned launch of digital services will be an immediate challenge to the government's DTT proposal.

Thirdly, DBS services, digital cable networks and DTT customers will need different set-top decoder boxes. The issue is whether rival service providers and the manufacturing industry can reach a common standard for different services. However, this is an area where state regulators may intervene.

Despite the uncertainties mentioned above, optimistic organisations view the U.K. government's DTT proposals as a window of opportunity. Among others, BT is considering bidding to become a multiplex holder in order to exploit the overlap in technology between digital TV services and its video-on-demand trials in Colchester and Ipswich.[17] Whether or not DTT services will become popular, the U.K. government's new proposals have now positioned the country as one of the few front-runners in the process of global competition for digital TV broadcasting.

Above all, the most significant development in digital TV is perhaps the efforts made by the European Launching Group (ELG) for Digital Video Broadcasting (DVB), which is proposing a European digital TV system conforming to the ISO's MPEG-2 standard for video source coding and multiplexing.

DIGITAL VIDEO BROADCASTING IN EUROPE

The ELG for Digital Video Broadcasting was started at the end of 1991 as an informal grouping with the intention of establishing a fully digital TV system for Europe. The ELG, led by Peter Kahl from the German Ministry of Telecommunications, was joined by broadcasters, equipment manufacturers, signal carriers and radio regulators from several European countries. The ELG later became known as the 'General Assembly', which drafted a memorandum of understanding signed by every participating organisation. In September 1995, the ELG or the General Assembly finally became the DVB, which is now the dominant group in bringing out a set of technical standards for European digital TV. At the time of writing, the number of participating organisations has reached more than 140, compared to 85 at the end of 1993. The DVB Group became particularly popular among European broadcasters. In its recent Green Paper the European Commission suggested that, by April 1994, this Group had been joined by 120 European broadcasters.[18] The ultimate objective of the DVB Group is "to create in Europe a framework for harmonious and market driven development of digital television *via* cable, satellite and terrestrial broadcasting".[19]

The European Commission is represented at meetings of the DVB Group since its early stage but is not a signatory of the Group's Memorandum of Understanding. It is known that the European Commission provides financial support *via* its 'Euro-Image Project' to the daily operation of the DVB Group. The DVB Group's budget is also contributed to by its members.

The organisational structure of the DVB Group, as shown in Fig. 7.1, comprises the following:

- General Assembly of all the members;
- Steering Board of up to 34 elected members;
- A Technical Module;
- A Satellite/Cable Commercial Module;
- A Terrestrial Commercial Module;
- An *ad hoc* Group on Conditional Access.[20]

The DVB Group is now headquartered within the European Broadcasting Union (EBU), in association with the German Ministry of Telecommunications. The close involvement of the EBU, which represents the interest of European public and commercial broadcasters, gives the DVB Group a

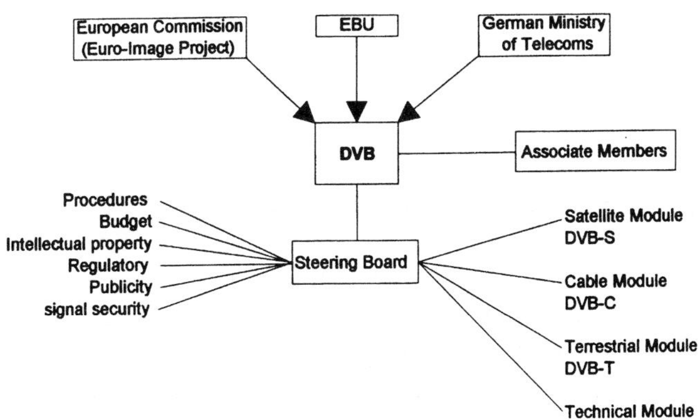

Fig. 7.1. Organisational Structure of the Digital Video Broadcasting (DVB) Group.

unique character distinctive from the Eureka 95 Project. As discussed in Chapter 6, Eureka 95 was dominated by the European consumer electronics manufacturers whilst the interests of the commercial broadcasters and private satellite operators were not given sufficient consideration. This difference between the DVB and Eureka 95 is also indicated by the fact that the majority of participants in the former are broadcasting organisations.

Within the DVB organisation, as indicated in Fig. 7.1, the most important element is the 'Steering Board', whose responsibilities include managing the standardisation procedures and the Group's budget and addressing intellectual property issues, regulatory issues, publicity issues and signal security issues. The four specialised 'Modules' in charge of technical standardisation report directly to the 'Steering Board'.

The Technical Module, initially called the Working Group on Digital Video Broadcasting, closely follows the developments relevant to the ISO's MPEG digital video group. On behalf of the DVB, the Technical Module joins the EBU and the ETSI (European Telecommunications Standard Institute) to coordinate European inputs on the definition of the MPEG-2 generic video standard. It is evident that the DVB's Technical Module plays an important role in promoting the MPEG-2 standard to be adopted in Europe for digital television broadcasting, either terrestrially, via satellite or on cable networks. Progress has already been achieved in that the DVB Group has obtained results on technical specifications for digital satellite and cable broadcasting, although technical specifications relating to digital terrestrial network broadcasting and conditional access, i.e. pay-per-view (PPV) channels, remain to be agreed upon by members of the group.[21] As shown in Fig. 7.1, the Satellite Module, Cable Module and Terrestrial Module are responsible for defining three digital TV transmission standards, called DVB-S, DVB-C and DVB-T, respectively. The DVB Group's approach in establishing a 'three-dimensional' standard family (i.e. satellite, cable network and terrestrial) for digital TV transmission makes a stark contrast with the previous European HDTV strategy, which was set to foster satellite transmission the exclusive way for distributing HDTV signals.

Due to the fact that the DVB Group in Europe, the Grand Alliance and various DBS companies (e.g. Hughes) in the U.S. conform to the same standard package for digital compression or source coding and multiplexing, i.e. the MPEG approach, there is a real opportunity for the whole world to adopt a single standard for digital television. However, as far as digital TV transmission is concerned, some technical differences between Europe and the U.S. should not be underestimated. The FCC has already made it a regulatory prerequisite that any new digital TV system must use the 6MHz bandwidth per channel; while the European channel bandwidth is 7 or 8MHz. It should also be emphasised that digitalisation *via* MPEG-2 does not preclude many technical variations. In other words, MPEG-2 is a generic standard concept and many proposed transmission standards use it as a base, beyond which a great deal of technical parameters can be dealt with different solutions. Therefore, it is possible to have several incompatible transmission standards for digital TV. Further to this, set-top boxes, as required mainly by 'conditional access', are making the standardisation process for digital TV more complicated. It is highly likely that a common scrambling standard for set-top boxes will be agreed in Europe thanks to the EU authorities' willingness to intervene the process with legislation. However, the diversity of scrambling technologies, and for sure the vested commercial interests behind them, will make it an extremely challenging task for any public authority or commercial organisation to unify the set-top box market in the near future.

Another uncertainty in establishing a global standard may come from Japan. Despite the fact that the Japanese Ministry of Posts and Telecommunications (MPT) has attempted to join the race for digital TV, the Japanese consumer electronics industry and the NHK has strongly objected to any government decision on establishing a fully digital standard to replace the hybrid MUSE system. The industry's argument is that the MUSE system is capable of providing adequate HDTV services and daily HDTV broadcasting of eight hours is already available in Japan. Some believe, on the Japanese side, that "digital technology had very great potential for the future and that it was premature to rush into setting standards soon while much of the new potential remained to be explored".[22] Whilst it is true that the Japanese have lagged behind the global race for digital TV standards, one should bear in mind that the Japanese are no less advanced in digital techniques and key components supplying. On the contrary, as argued elsewhere in this book, the Japanese electronics industry, particularly consumer electronics, has been the leading force in providing high-quality products to consumers all over the world. As far as digital HDTV is concerned, the Japanese are already the leading suppliers of advanced memory chips (e.g. DRAMs) and, potentially, the major suppliers of large-size flat panel displays such as LCDs. Even in the U.S. and Europe, Japanese companies would be among, if not the only, major equipment manufacturers once a digital TV system of whatever technical parameters is adopted and launched. Compared to the U.S. and Europe, however, Japan may suffer a competitive disadvantage in audio-visual software provision for future HDTV broadcasting.[23]

Moreover, it is also an uncertainty related to the digital TV standardisation process, including the launch of DVB in Europe, which lies in the complexity of the regulatory environment. Figure 7.2. shows the principal bodies involved in TV broadcasting standardisation at different levels.

The number of organisations involved in the standardisation process for TV broadcasting does suggest that the regulatory environment is extremely complicated. A global standard for digital TV, including digital HDTV, would only be possible if a consensus is reached at and between the three levels: national level, European level and World level. The European Launching Group for Digital Video Broadcasting (ELGDVB, i.e. ELG as in

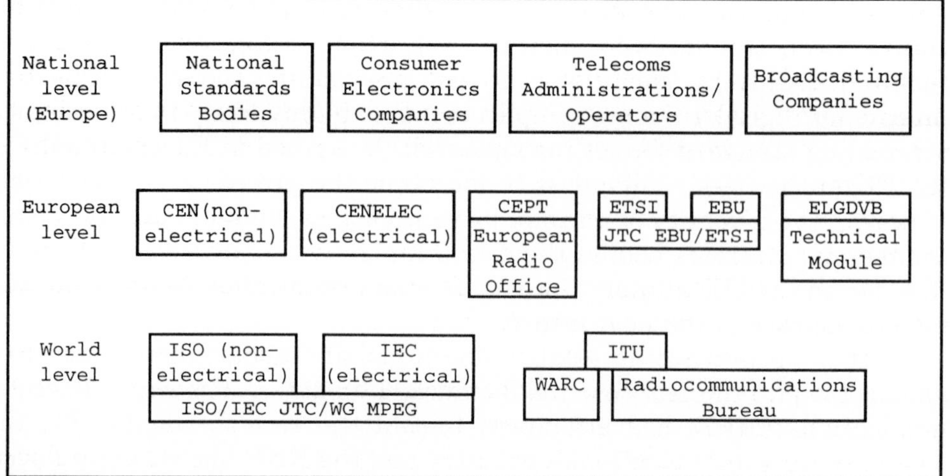

Source: Commission of the European Communities (1993), *Digital Video Broadcasting: A framework for Community Policy*, COM (93), 557 final, Brussels, 17 November, p. 11.

Fig. 7.2. The Principal Bodies Involved in Broadcasting Standardisation.

Fig. 7.1), as seen in Fig. 7.2, is only one of the many players at the European level. Although the European Commission encourages and supports the DVB project, it is not the EU's policy position to back any specific digital TV system by legislation for the time being. This seems to be one of the lessons that the EU authorities have learned from their experience with the failed HD-MAC system.[24] In other words, it could be more difficult to have the European DVB system endorsed across Europe, than the Grand Alliance's digital system in the U.S., where the FCC has the mandate to endorse any terrestrial TV transmission standard.

DIGITAL TV AND TECHNOLOGICAL CONVERGENCE

Related to the process of the digital revolution, the concept of convergence is of particular importance in analysing the process of EU policy changes. Specifically, this convergence 'will involve the film industry, broadcasting services and the television programme industry, cable and telecommunications operators, the publishing industry and manufacturers of information and communications technology equipment'.[25] Firstly, as the next chapter will show, digital techniques transfer various audio-visual materials into the same kind of computer readable data facilitating the emergence of the multimedia industry in the early 1990s. Consequently, conventional distinctions between sectors such as the consumer electronics and the computer industry are blurred due to the emergence of the multimedia industry.

Secondly, digitisation is also transforming the telecommunications industry from an analogue stage to a new digital communications stage. This has made it technically feasible that information, communication and entertainment transmission can be carried out from the same digital network, i.e. convergence of a range of sectors to create the information superhighway.

Thirdly, envisaging this convergence process, the traditional regulatory regimes at either the national or EU level have become removed from operational realities. This perhaps is the most challenging task for the EU authorities to undertake in the years to come. At the present time, the majority of broadcasting, information and communications organisations are treated as separate and distinct sectors and regulated with different terms usually at the national, rather than the EU level.

Challenges to the traditional regulatory environment at the EU level and national government level has already been heralded by the rapid development of the Internet.[26] One of the popular concerns regarding the Internet is the copyright issue arising from the convergence of digitised books, films, music, etc., which can be, at the present time, distributed via the Internet without adequate protection in terms of intellectual property. Recently, the DTI's multimedia task force has organised the media and entertainment industries in the U.K. to form a working party, with representatives from the BBC, IBM, the Pearson Group, PolyGram (the music and film subsidiary of Philips) and Bertelsmann, to present a proposal to the British government introducing a 'transmission right'—a measure to ensure that intellectual copyright owners are paid when their images, words or sounds are distributed on the Internet in a digital form.[27] The same working party is also pressing the British government to lobby the European Commission to make the necessary regulatory changes so that intellectual property rights are addressed adequately at an international level. This does not encapsulate the complexity of the issue. For the time being film and other audiovisual rights are not categorised into intellectual property for protection within the EU. This issue is also complicated by the international nature of the Internet:

> A piece of music downloaded on to a computer in the U.K. can have originated from anywhere in the world, and the country of origin need not necessarily have made provision for a U.S.-style digital 'transmission' right.[28]

After the media, entertainment, data service and telecommunications industries have thoroughly converged into a single digital transmission environment provided by interconnected fibre optical networks, legal challenges facing regulatory authorities at either a national level or an international level would be no less than those caused by the Internet today.

It must be emphasised that, reflecting the process of technological convergence, the EU's new strategy and new policy for new TV technologies are not isolated or stand-alone packages. On the contrary, the EU, *via* the

Commission, seems to be committed to establishing a range of new policies towards the information and communications sector, in which strategies for the next generation of TV are an integral part. In a recent communication to the EU Council and European Parliament, the Commission noted that,

> ... it should be emphasized at the start that television is only one of the 'information services' which will be impacted by new digital technology. It is perhaps not an exaggeration to say that we stand at the threshold of a new revolution in communications in which digital technology will enable the transition to global information networks in which all forms of information—whether in the form of moving or still pictures, sound, text or data—will be made widely available in a flexible manner.[29]

To make and implement new policies in response to the challenge of technological convergence, the EU envisages tremendous difficulties between the three layers. Firstly, given that broadcasting and data and voice communications are undergoing convergence and can be transmitted via the same network thanks to the rapid development of digital compression techniques, how could long-established media and communications laws remain unchanged? Indeed, sectoral convergence needs a unified legislative framework. However, diversified (sometime conflicting) sectoral and corporate interests are difficult to coordinate. In the meantime, relaxation of cross-media ownership rules may lead to violation of competition policies. For example, British Telecom (BT) in the U.K. has been lobbying the British government for having access to the television broadcasting sector so that it could justify a planned investment of up to £15bn to lay down a nationwide fibre optical cable network as the infrastructure for the newly emergent digital information and communications industry. In fear of a possible monopoly of the future digital broadcasting and digital communications sector, the British Government has not yet granted BT the right to enter the TV broadcasting sector.[30]

The second layer of difficulties faced by the EU lies in the uneven development of the telecommunications sector across the member states. The European Commission has already planned to liberalise the telecommunications market by 1 January 1998. However, it has to divide all of the member countries into two groups and let five countries (Spain, Portugal, Ireland, Greece and Luxembourg) finish their liberalisation process at a later stage. Noticeably, the major telecommunications operators are preparing themselves so that they will get the full benefit, or at least not lose out, from the liberalised communications market in Europe after 1998. The most favoured new strategy of the big telecoms operators is strategic alliance with local communications companies. This has already caused concern at both the national level and EU level as far as competition policy is concerned.

Following the successful formation of the Concert joint venture between BT and MCI (the second largest long-distance telecoms operator in the U.S.) across the Atlantic, Deutsche Telekom and France Telecom have proposed another joint venture under the name Atlas to offer a global data transmission service. The two leading European operators have also planned to take a 20% stake worth $4.2bn in Sprint, the third largest U.S. long-distance carrier, to form a powerful global alliance known as Phoenix.[31] Despite its own cross-border alliance with MCI, BT has filed a formal complaint to the European Commission claiming that the proposed Deutsche Telekom and France Telecom deal would enable the two companies to gain a dominant position in the future communications market. Mr Van Miert, the current EU competition commissioner, told a recent G7 meeting that he was concerned about the moves to close markets and 'there was evidence that most of the EU's telecommunications companies were resistant to opening up their networks'.[32] Whilst the EU's policy position remains undecided, the American government has already approved the Phoenix alliance between Sprint and the two European partners.

It appears that the big national telecoms operators will likely edge the relatively smaller operators out of the market in the long run. Most recently, in cooperation with Bell Atlantic, BT has made another bid for a stake in Belgacom, the Belgian state-owned public telephone company. This simply adds another item to BT's European alliance list: it has formed alliances with a range of companies such as Viag of Germany, Albacom of Italy, Banco Santander of Spain, as well as big operators in Scandinavia.[33] These big telecoms operators will, either by themselves or via joint ventures and strategic alliances, play a leading role in providing the main infrastructure for delivering digital information and communications services. If cross-media rules are going to be relaxed without any condition in the near future, they may also insert substantial control of digital TV broadcasting as well.

The third layer of difficulties faced by the EU authorities is a dilemma: liberalisation (or globalisation) *versus* protectionism. Apart from its telecom liberalisation policy, the development of digital technology has provided with a possibility to create an integrated broadband communications network across the traditional sectoral boundaries and national borders. The 1995 G7 Conference in Brussels also expressed the urgency to build a borderless information and communications network, i.e. the Global Information Infrastructure (GII). As mentioned before, the EU's new strategy for the next generation TV technology recognises the importance of international cooperation or coordination between Europe, the U.S. and Japan in order to establish a worldwide system. However, these developments, in terms of technology and policy, are overshadowed by an opposite stream of policy-making. The EU Directive (1989), entitled *Television without Frontier,* specifically requires that all TV broadcasting channels of EU countries should have a minimum of 50% of their broadcasting programmes produced within the EU. This Directive is unlikely to be abolished in the near future. On the contrary, the EU, lobbied by the French government, is considering extending the 1989 Directive to including the newly emergent Video-on-Demand service.[34] The likelihood is that any future digital TV services may be ruled by similar, if not the same, policy. European protectionist technology policy, as shown elsewhere in this book, failed to rescue the European VCR technology in the early 1980s, and it was not successful in promoting HD-MAC in competition against the Japanese MUSE standard and the more futuristic digital alternatives pioneered by the U.S. These are the lessons and bitter experience to be remembered by European policy-makers in a new age of integrated digital broadcasting and digital communications.

CONCLUSION

The digital revolution, and consequently the convergence of different domains of the information and communications sector potentially provides a lucrative opportunity for hardware manufacturers and software providers. As far as the former are concerned, Philips is one of the front-runners in some aspects of the new digital information age.

Compared to its bitter experience with the V2000 system in the early 1980s, as discussed in Chapter 4, Philips did not suffer a heavy loss from the collapse of HD-MAC. On the contrary, Philips may have accumulated considerable experience from its heavy involvement in R&D activities relating to HD-MAC, and this experience could be used for its exploration of digital television elsewhere (firstly the U.S. then Europe).[35]

As a matter of fact, both Philips and Thomson, as European firms, play an indispensable role in the American Grand Alliance project for fully digital TV. Philips has also been promoting its interactive TV technology based on the CD-i system in the U.S. since the early 1990s. Undoubtedly, the involvement of Philips and Thomson in digital TV development in the U.S. would enable these two companies to become leading manufacturers of digital TV equipment for Europe in the future.

Philips may not lose out from the ongoing digital TV race in that it is determined to become a leading European supplier of flat panel displays, a widely believed critical component for constructing the future 'hang on the wall' type of HDTV nicknamed 'home cinema' or 'infotainment wall'. In addition, Philips' advanced semiconductor business has been back in black since 1994 and the company has decided to expand this business sector. Any digital TV or HDTV set cannot be built without sufficient supply of advanced chips.

In its 1994 Green Paper on the audio-visual sector, the European Commission has elaborated that Europe may not be able to catch up with the U.S. in the newly emergent information society should it fail to build up a rich stock of audio-visual materials, mainly film and TV programmes. PolyGram, the music and film branch of Philips, is one of the top players in the world as far as entertainment software is concerned.

It is not the author's intention to advocate Philips' new competitive position in bringing about the next generation TV, which is an important part of the emerging information society. However, the interface between corporate strategies of Philips and public policy of the EU, as indicated by the development of HDTV and digital TV in Europe, does show much more flexibility and, to a certain extent, a cunning characteristic on the part of the former.

Technological and sectoral convergence is also being materialised by the various optical disc-based multimedia products. At the vanguard of the newly emergent multimedia industry, Philips has developed its CD-i system, which aims to bring interactive entertainment into the home. In sharp contrast with the EU's controversial technology policy for HD-MAC, the next chapter will examine the issue of how Philips managed to improve its competitive strength through a better designed corporate strategy, rather than lobbying for EU policy support, in the area of multimedia based on optical disc technologies since the 1980s.

NOTES

1. See Bangemann (1992).
2. Commission of the European Communities (1993), *Proposal for a Directive of the European Parliament and of the Council on the Use of Standards for the Transmission of Television Signals,* COM(93) 556 final, Brussels, 15 November, p. 2.
3. *Ibid.*, p. 3.
4. *Ibid.*, p. 4.
5. *Ibid.*, p. 3.
6. *Ibid.*, p. 4.
7. See Commission of the European Communities (1993), *Digital Video Broadcasting: A Framework for Community Policy,* COM (93) 557 final, Brussels, 17 November 1993, p. 4.
8. *Ibid.*, p. 2.
9. Chapter 6 showed that the U.S. had supported the Japanese HDTV system before the CCIR meeting held in May 1986. After the Japanese proposal was blocked by the European governments a group of European companies led by Philips and Thomson proposed an alternative HDTV standard to compete against the Japanese system. Under this circumstance the FCC decided to develop an American HDTV standard different from both the Japanese system and the European proposal.
10. A recent report by the Japanese Ministry of Post and Telecommunications (MPT) suggests digital terrestrial television broadcasting will be possible in Japan between 2000 and 2005. The same report also recommends that digital HDTV broadcasting by satellite should start after 2007 so that a window of opportunity is maintained for the analogue Hi-Vision technology, the first HDTV package in the world. See *Screen Digest,* May 1995, p. 97.
11. ATSC was formed by the Joint Committee on Inter-Society Coordination (JCIC) to establish voluntary technical standards for advanced television systems such as HDTV. JCIC members include the Electronic Industries Association (EIA), the Institute of Electrical and Electronics Engineers (IEEE), the National Association of Broadcasters (NAB), the National Cable Television Association

(NCTA) and the Society of Motion Picture and Television Engineers (SMPTE). Note that Philips and Thomson are members of ATSC.

12. Any TV standard or system would have to be recommended by the Advisory Committee on Advanced Television Service (ACATS) to the FCC. ACATS was appointed by the FCC and the FCC reserves the right to reject all or part of any ACASTS recommendation.
13. *Screen Digest* reported that, as of April 1995, Primestar had 385,000 subscribers for its digital medium-powered service. June 1995, p. 123.
14. The British government's *Broadcasting Act 1990* separated the transmission and engineering research responsibilities of the IBA. The IBA's regulatory responsibility for independent TV transmission was transferred to the ITC (Independent Television Committee).
15. *Financial Times,* 11 August 1995.
16. BSkyB has now reached 4 million homes, equivalent to about 10% of U.K. households.
17. See *The Sunday Times,* 13 August 1995.
18. Commission of the European Communities (1994), *Strategy Options to Strengthen the European Programme Industry in the Context of the Audiovisual Policy of the European Union,* Green Paper, COM(94) 96 final, Brussels, 6 April, p. 24.
19. Commission of the European Communities (1993*), Digital Video Broadcasting: A Framework for Community Policy,* COM (93) 557 final, Brussels, 17 November 1993, p. 15.
20. Commission of the European Communities (1993*), Digital Video Broadcasting: A Framework for Community Policy,* COM (93) 557 final, Brussels, 17 November 1993, p. 16.
21. Commission of the European Communities (1994), *Strategy Options to Strengthen the European Programme Industry in the Context of the Audiovisual Policy of the European Union,* Green Paper, COM(94) 96 final, Brussels, 6 April, p. 24.
22. *Ibid.,* p. 20.
23. The Japanese ambition to increase their share in audio-visual programme stocks, particularly films, has suffered a drawback due to Matsushita's recent sale of MCA to a Canadian soft drinks company. Matsushita's takeover of MCA, together with Sony's purchase of Columbia pictures, was seen as a strategic move of the Japanese consumer electronics industry towards getting control over consumer software supply in response to new innovations in the next generation consumer electronics products.
24. As discussed in the last chapter, the EU passed its MAC Directive and issued a Council Decision in 1986, immediately after the launching of the Eureka 95 Project, to provide legal backing for the newly proposed European HD-MAC system. This measure was seen by many commentators as an immature policy intervention which hindered innovation in alternative technologies.
25. See Commission of the European Communities (1994), *Strategy Options to Strengthen the European Programme Industry in the Context of the Audiovisual Policy of the European Union,* Green Paper, COM(94) 96 final, Brussels, 6 April, p. 1e.

26. The world's largest network of computer networks interlinked via the national and international telephone networks. The Internet was started in the U.S. in the late 1960s as a scientific research network and has now grown into the most popular digital highway network with a great commercial potential. The argument that whether or not the Internet should be commercialised seems to be won by the 'yes' side as the number of virtual business malls opened on the nonproprietary network is steadily getting larger and larger.
27. *Financial Times,* 5 July 1995.
28. *Financial Times,* 10 July 1995.
29. Commission of the European Communities (1993*), Digital Video Broadcasting: A Framework for Community Policy,* COM (93) 557 final, Brussels, 17 November 1993.
30. Note that the Labour Party in the U.K. has voiced that it would let BT have access to the broadcasting sector if it gets elected during the general election due in 1996.
31. *Financial Times,* 28 February 1995.
32. *Ibid.*
33. *The Independent,* 1 July 1995.
34. Video-on-Demand (VoD) is a new way of home entertainment: the viewer dials up a video distribution company's phone number asking for a specific film/programme to be sent via the telephone line (either the conventional copper wire network or a new cable network subject to availability in a particular area) into his/her home television screen with a charge. VoD saves the viewer's travel from the home to the video rental shop. In the U.K., BT is experimenting VoD services with 2500 residential homes in Winchester and Ipswich. BT is currently banned by British lay from entering the television broadcasting industry but is allowed to provide VoD services. Obviously, the British interpretation of the nature of VoD is substantially different from the French interpretation.
35. Note that HD-MAC was a hybrid system incorporating both analogue and digital techniques. For instance, all audio channels of HD-MAC were digital whilst pictures would be transmitted in analogue form. In addition, all prototypes of HD-MAC sets were built with a large number of advanced computer chips.

8

MULTIMEDIA DEVELOPMENT AND PHILIPS' GLOBAL STRATEGY

Almost in parallel with the global race for HDTV and digital TV, major technical changes have brought about another format battle in the consumer electronics industry since the second half of the 1980s: the introduction and rapid development of CD-based multimedia systems. Compared to the HDTV and digital TV sector, as discussed in Chapters 6 and 7, competition and collaboration for multimedia has been less subject to government intervention, either in Europe or anywhere else. In other words, the success or failure of a particular multimedia system will be determined mainly by market forces. It is argued in this chapter that the wide range of interfirm collaboration for R&D and strategic partnerships for production and marketing is the major characteristic of the emerging multimedia industry. This chapter provides a new perspective to look at the issue of how new technologies were globally manoeuvred by allied firms from the consumer electronics industry or, more broadly, the whole information technology sector.

Even before some key multimedia technologies, such as full screen, full motion (FSFM) video, were fully developed,[1] some multimedia products had already been launched on the consumer market. After Commodore of the U.S. launched the CDTV (Commodore Dynamic Total Vision) system in April 1991, Philips followed suit with its CD-i (Compact Disc-Interactive) format in October 1991. Among other latercomers, Matsushita's 3DO has increased more competitive pressure on Philips' CD-i format since the early 1990s. Despite the failure of CDTV, as well as the collapse of the Commodore company itself, it is anticipated that it will be a long time before the format battle for interactive multimedia comes to an end.

This chapter is centred on the discussion of certain issues concerning the development of the CD-i system against the background of the ongoing multimedia format battle. In particular, Philips' new corporate strategies, internal structure as well as marketing performance related to the CD-i project, will be analysed. As one of the few priority areas within Philips,[2] the CD-i project is also understood in the context of the whole group's struggle to survive the global competition at a time of increased Japanese domination in the consumer electronics industry and international economic recession.

Although, in contrast with HDTV, the EU has hardly made any commitment to protecting CD-i as a European technology, some aspects of the European audio-visual policies and related EU media programmes will be briefly reviewed. It is also a task of this chapter to trace the history of optical disc[3] technologies, in which videodisc, CD (Compact Disc), and CD-ROM (Compact Disc-Read Only Memory) were the major breakthroughs which laid down the technical foundation for CD-i and other CD-based multimedia products. Meanwhile, Philips, the pioneer and industry leader of optical disc technologies, has learned lessons from its own experience, including failures and successes, in the historical development of optical disc technologies.

TECHNICAL CHANGE AND COMPETITION IN THE OPTICAL DISC INDUSTRY

In the IT (Information Technology) sector,[4] two essential technologies for information storage have been developed: magnetic discs/tapes and optical discs. Magnetic discs are mainly used for the computer industry, and magnetic tapes for the audio-visual industry. Although optical disc technology was introduced to the consumer market in the 1970s,[5] a fully digital consumer technology did not become a reality until CD was launched in the early 1980s. Prior to the emergence of the Compact Disc technology a brief format battle competing for the analogue videodisc storage standard was concluded with LaserVision accepted in some parts of the world. Most importantly, videodisc is widely believed to be the 'seed' of the more popular CD technology.

Due to the close technological linkage between videodisc and Compact Disc, it is worthwhile to make a brief historical review of the former.

The Rise and Fall of Laser Disc

The history of optical technology can be traced back to Thomas Edison's voice recordings on a wax cylinder in 1877; Reginald Friebus was the first person who actually brought optical technology into reality by inventing the optical videodisc in 1929.[6] However, real industrial applications were not realised until the 1960s.

In the 1960s R&D had enabled a laser to be used instead of a stylus to read recorded signals. The emergence of laser technology opened an entirely new era for the home entertainment industry. In the early days of videodisc technology, many big electronics companies were independently involved with R&D. There were four major videodisc systems in two different modes in the 1970s and early 1980s before the introduction of CD technology. They were Philips' LaserVision, RCA's (Radio Corporation of America) CED (Capacitance Electronic Disc, also known as SelectaVision), JVC's VHD (Video High Density) and another system proposed by Matsushita. However, the Matsushita system never reached the market-place and production of RCA's CED system was halted in 1984, three years after its introduction, due to the huge losses incurred in keeping prices low and organising software production.[7] Both CED and VHD used a stylus to read signals recorded on the disc. Therefore, any scratch on the surface of the discs could damage the quality of sound and pictures.

Philips made its first commercial launch of the LaserVision system under the Magnavox brand name in the U.S. in 1978, and Pioneer followed suit to manufacture and market the system in the same year. In 1982, Philips launched LaserVision as a medium to play prerecorded video programmes in Europe with the intention of reinforcing its unpopular V2000 VCR system. As discussed in Chapter 4, neither the V2000 system nor the LaserVision videodisc player became a successful consumer product in Europe. It is believed that, in comparison with CED and VHD, the LaserVision system used a laser beam instead of a stylus to read the recorded information on the disc, and achieved higher picture resolution and better sound quality.

The main reason why videodisc did not succeed on the consumer market in most parts of the world was the challenge from VCR technology which had been well established by the late 1970s and early 1980s. More specifically, the following were the major factors which led to the defeat of laser videodisc players, particularly in Europe.

Firstly, short playback time and the lack of prerecorded software compared with VCR machines made videodisc players a less attractive alternative.

Secondly, closely related to the above point, most film companies had a fear that the introduction of videodisc technologies might eventually undermine their well-established cinema business. Consequently, no substantial commitment was made by film studios to put their films on videodiscs, and the availability of prerecorded software on VCR tapes, in particular on VHS tapes, were never matched by that on videodiscs.

Thirdly, the inability to record off-air programmes from television broadcasts severely undermined the market opportunity for videodiscs and players.

Fourthly, most of the functions offered by a videodisc player could be substituted by VCR; it was again the consumer who made the choice between competing technologies.

Finally, in part the failure of videodisc systems was caused by the lack of standardisation among discs and players.[8]

In 1982, the introduction of the Compact Disc technology, in particular the completion of the specifications for CD-Audio standard by Philips and Sony, made the marketing of laser videodisc more complicated. In an effort to repair the poor image of the LaserVision system,

> Philips' response was to confuse matters once more by renaming the product and gearing up for a relaunch under the CD Video banner. This in the hope that the consumer would associate CD-V with the now hugely successful CD.[9]

Shortly after the launch of CD-V,[10] Philips adopted the Pioneer term, Laser Disc, to replace CD-V and categorise various videodisc products. It might be still too early to say whether the introduction of CD-V or Laser Disc was a success or another failure; the essence of the problem, however, was not simply a matter of renaming the product but what kind of new experience and advantage a new product could offer the consumer at an acceptable price level.

Due to considerable investment commitment by Pioneer, Laser Disc as a consumer system has been widely accepted in Japan and some adjacent east Asian areas (e.g. Hong Kong and Taiwan). After the launch of CD, Pioneer introduced a new machine which was able to play both video discs and CDs in 1983; and four years later, in 1987, the company launched another improved system, the so-called 'Combi-Player' or 'Multi-Disc Player',[11] which offered the capability of playing-back 12" video discs and 5" CDs. As a result, videodisc players have reached a significant level of household penetration in Japan and the U.S. sales of videodisc players have been growing in some other areas such as Hong Kong and Taiwan in recent years.

Because of the fact that both CD and CD-V discs can only be read by laser, it was technically difficult for JVC to make its VHD compatible with the new technologies and, therefore, the company was greatly disadvantaged in competition with laser discs. Consequently, VHD failed to become a consumer product, and Laser Disc has gained substantial industry support as major consumer electronics firms such as Sony, Mitsubishi, Toshiba, and Hitachi are now producing both Laser Disc players and videodiscs (Lang, 1990, pp. 58–59).[12] This makes a big contrast to the situation of the early 1980s when Philips failed to open the American and European markets and Pioneer was almost alone in marketing the LaserVision system. Table 8.1 presents a list of production growth and penetration rates of Laser Disc players in Japan.

The LVAP study, as shown in Table 8.1, suggests that production and home penetration of optical video players has gained a steady growth in Japan since the beginning of the last decade. Other sources suggest that around 250,000 units of laser disc players had been sold in the U.S. market by 1988.[13] In parallel with the growth of hardware production, the software business increased rapidly as well. For instance, in 1990, up to 4,060,000 video discs were produced in Japan every month, of which 66% were from Pioneer; and in Taiwan, 20,000 discs were sold every month, not to mention the growing rental business (Lang, 1990, p. 51). It looks as if the LaserVision system has been established as an important consumer electronics product with a considerable market potential in the South East Asian region, despite the previous failures experienced by Philips in launching the system in the U.S. and Europe.

The technology to make recordable or rewritable laserdiscs was subsequently developed, and Laser Disc recorders have been introduced. For instance, Sony developed its erasable videodisc recorders which are being used by NHK; Matsushita has sold a number of write-once videodisc recorders in recent years; in collaboration with ITT, Pioneer introduced its videodisc recorders in 1990, which can be used for HDTV recording; in summer 1989, Philips announced a new technology to manufacture videodiscs with a durability of recording up to 1000 times.[14]

The 'Red Book' and the Birth of Compact Disc

While Philips' R&D team were working on the LaserVision system in the 1970s, another product concept was developed within the optical disc project—to shrink the size of disc from 12 inch for mixed audio and video recording to less than 5 inch for audio only. This new version of optical disc technology, firstly called ALP (Audio Long Play) and then renamed Compact Disc in 1974, offers consumers digital sound and gained astonishing success

Table 8.1. Accumulated Production and Home Penetration of LaserVision Players in Japan, 1981–1991.

Year	Accumulated production	Home penetration rate
1981	10,000	–
1982	40,000	–
1983	100,000	0.3%
1984	230,000	0.6%
1985	460,000	1.2%
1986	760,000	1.9%
1987	1,090,000	2.7%
1988	1,580,000	4.0%
1989	2,280,000	5.7%
1990*	3,280,000	8.2%
1991	4,780,000	12.0%
1992	7,080,000	17.7%

Note: * Figures from 1990 onwards are estimation.
Source: LVAP (LaserVision Association Pacific), October 1989, quoted in Lang (1990).

in the consumer electronics industry all over the world during the last decade.

During the period 1974–1979, Philips defined most of the important technical parameters and demonstrated its first working model of the Compact Disc system to the international press. However, the risks involved in attempting to set a new standard against Japanese competition forced the Philips management to make a critical decision to join forces with the Japanese industry, which had already gained about 90% market share of the world Hi-Fi industry by the late 1970s. Having lost ground to the Japanese VHS system during the home VCR format battle, in which Philips was initially a technological pioneer, the Philips management decided to achieve a global standard for CD even before the technology itself was perfected and any marketing plan was made. In searching for a new strategy to commercialise the CD system, some senior Philips managers realised that,

> If ten companies invent a new technology and introduce working prototypes that are, to the naked eye, virtually identical, all ten versions can still be totally incompatible with each other in a host of technical details. If the ten inventors do not agree to resolve those technical differences, then the marketplace—first in the form of manufacturers, then in the form of consumers—makes the choice. When the market chooses among ten, there is one winner, and there nine losers. And sometimes, there are no winners at all.[15]

After careful consideration, Philips finally entered into collaboration with Sony in August 1979 for standardisation of the CD system. The reason why Philips chose Sony rather than any other company, such as Matsushita, may be multilayered; but Sony's digital decoding technology, more specifically digital correction, was badly needed to further purify the sound produced by the CD system. Under the new agreement signed by Philips and Sony, the two rival firms agreed access to each other's secrets and to share all their technical know-how in the area of optical disc research. In doing so, Philips and Sony set up a mixed R&D team joined by managers and technical staff from both sides to work on the same objectives.

The Philips–Sony alliance for CD was seriously challenged by two important forces.

Firstly, the existence of other digital audio systems, in particular JVC's AHD (Audio High Density) with the full support of Matsushita, the former's parent company, could challenge the ambition of the Philips–Sony venture. In 1979, the Digital Audio Disc (DAD) Committee, a subcommittee of the MITI, came into existence[16] in Japan with the task of choosing a standard for the digital audio industry amongst competing systems. Philips and Sony agreed the definitive CD system and submitted it to the DAD Committee in June 1980, when there were no less than three different digital systems under consideration by the Committee. Surprisingly, a turning point was reached in January 1981 when JVC's parent company Matsushita announced that it would adopt the Philips–Sony CD system.[17] This announcement meant that the Compact Disc system would be supported by the three largest manufacturers from the consumer electronics industry, and other systems would immediately lose their ability to compete for the world standard of digital audio. Following the action of Matsushita, the total number of CD player manufacturers increased to 39 at the beginning of 1983 (see Table 8.2).

Secondly, the prospect for CD was threatened by the growing resistance from the music industry. While Compact Disc was given increased publicity, the music industry, particularly that of the U.S., became increasingly concerned about the new media for music, which was inevitably to challenge their well-established business. The music industry showed hardly any initial interest to shift to CD for their records mainly due to a royalty of three cents per disc demanded by Philips and Sony:

> The almost unanimous resistance of the record industry threatened to become organised several months before Philips and Sony's 1982 product launch.[18]

In fear of being rejected by the music industry, Sony decided to give up claiming royalties for Compact Disc and left Philips as the sole recipient. Because of that any organised industrial conspiracy to boycott a new technology would be a criminal offence under the American antitrust law, a collective action by the music industry against the CD system did not actually happen in the U.S.[19] Consequently, many record companies became licensees of the Compact Disc format one after another.[20] In the process of promoting the CD technology Herbert von Karajan's enthusiasm should not be underestimated. As a famous classical music conductor of the Berlin Philharmonic, Karajan's attitude played an important role in convincing many other musicians and critics that CD was a better technology than the conventional LP and compact cassette.

Having got the substantial support of major consumer electronics manufacturers and the music industry, Philips and Sony launched the Compact Disc system on the Japanese market in October 1982, when Marantz, a Philips branch in Japan, and seven other companies were able to supply CD players and CBS/Sony and PolyGram presented about 150 disc titles.[21] By Autumn 1983, Compact Disc was already on sale in the U.S., Western Europe, and most countries of the world, with no competition for becoming a global standard. One of the most important achievements of Philips and Sony during the CD campaign was that a substantial number of hardware manufacturers and music companies had already become licensees prior to the official launch. After the Matsushita announcement to support CD in January 1981, the number of licensees soared to 40 companies, of which 30 were player manufacturers and 10 disc producers, by March 1982. By the time of launch the number of player licensees increased to 37.

Table 8.2. Compact Disc System Partners, 1 August 1983.

Company	Country	Hardware	Software
Akai	Japan	x	
Alpine	Japan	x	
Asahi Electronics	Japan	x	
Bang & Olufsen	Denmark	x	
CBS	USA		x
CBS/Sony	Japan		x
Central Tape	Japan		x
Clarion	Japan	x	
Crown Radio	Japan	x	
Digital Images	USA		x
Dong Won (Silver)	Korea	x	
Forward Technology Industries	UK	x	
Foster Electric	Japan	x	
Fujitsu-Ten	Japan	x	
Fulet	Taiwan	x	
Garrard	UK	x	
General Corp.	Japan	x	
Goldstar	Korea	x	
Grundig	W. Germany	x	
Hitachi (Lo-D)	Japan	x	
Kisho Corp.	Japan	x	
Kyocera Corp.	Japan	x	
Marantz Japan Inc.	Japan		
Matsushita (Technics) incl.	Japan	x	
JVC, Viktor, Teichiku	Japan	x	x
Minnesota Mining & Manufacturing Co. (3M)	USA		x
Mitsubishi (Diatone)	Japan	x	
Nakamichi Corp.	Japan	x	
New Nippon Electric (NEC)	Japan	x	
Nimbus	UK		x
Nippon Columbia (Denon)	Japan	x	x
Nippon Gakki (Yamaha)	Japan	x	
Onkyo	Japan	x	
Philips	Netherlands	x	
Pioneer	Japan	x	x
PolyGram	Netherlands/ W. Germany		x
PR Records	UK		x
RCA	USA		x
Samsung	Korea	x	
Sansui	Japan	x	
Sanyo (Fisher, Otto)	Japan	x	x
Sharp (Optonics)	Japan	x	
Shin-Shirasuna	Japan	x	

Table 8.2. continued.

Company	Country	Hardware	Software
Sonopress	W. Germany		x
Sony incl. Aiwa (Wega)	Japan	x	
Studer (Revox)	Switzerland	x	
TEAC	Japan	x	
Thomson (Dual, Nordmende, Saba, Telefunken, Brandt, Continental-Edison, Pathe-Marconi)	France	x	
Toolex Alpha	Sweden		x
Toshiba (Aurex)	Japan	x	
Toshiba-EMI	Japan		x
Trio (Kenwood)	Japan	x	
Total	**55**	**39**	**16**

Note: In brackets: alternative brand names.
Source: Based on Press Information on Compact Disc from Philips.

Table 8.2 shows the number of total licensees as 55, of which 39 were player manufacturers and 16 disc producers at the beginning of August 1983. In August 1985, the total number of licensees increased to 89 (53 for CD players and 36 for discs). Because of the wide acceptance of CD by the music industry, the number of disc titles reached 1,000 in October 1983, exactly one year after the first launch in Japan with only 150 titles to offer.

The amazing success of the CD digital audio technology may also be illustrated by the following sales figures: until June 1990 about 74 million CD players had been sold worldwide since the introduction and in 1990 alone 33 million players were expected to be sold.[22] In the case of PolyGram, the specialised record subsidiary of Philips, CD sales increased from 10% in 1985 to 37% in 1989 of the company's total sales.[23]

CD-DA (CD-Digital Audio) was established as a world standard, and that means any disc can be played back on any player produced by any company all over the world. From a technical point of view, this world standard was achieved by Philips and Sony before the commercial launch. The whole standards package agreed by the two companies in 1982 was called the 'Red Book'. Specifically, the 'Red Book' defines CD audio standard as follows:

- disc capacity and data transfer rate;
- single side;
- data format;
- modulation and error correction;
- physical dimension of disc;
- physical dimension of pits, CLV (Constant Linear Velocity) organisation and density.

The 'Red Book' ensures that consumers all over the world would be using the same standard without being worried about problems caused by incompatibility as happened to CTVs and VCRs in their early days.

The 'Yellow Book' and the CD-ROM Industry

As an extension of the Compact Disc technology, CD-ROM (Compact Disc-Read Only Memory) disc was designed mainly for large-volume information storage; the CD-ROM player is essentially a PC peripheral device. In December 1983, Philips and Sony agreed on the standard package for CD-ROM, which was called the 'Yellow Book'.

The amount of information that a single CD-ROM disc can store is 650 Mbytes, the equivalent of approximately 1,000 double density floppy disks. As a PC peripheral, the CD-ROM player requires a software retrieval package running on the PC, and the data stored on the CD-ROM disc can be read, down-loaded onto hard or floppy disks, or printed out.

CD-ROM stores digital computer data instead of music signals on the same compact disc and hence established itself as the second generation of CD technology. There is no doubt that CD-ROM was an important development which offered a new way of large quantity information storage for computing. However, the establishment of CD-ROM as another industrial standard of CD technology appeared to be much more difficult compared to CD-DA. The major reason was that, on the one hand, both Philips and Sony are mainly consumer electronics firms, and they had no industrial power to simply impose their CD-ROM standard onto the computer industry without the consent of the specialised computer giants. On the other hand, the computer industry itself has been plagued by incompatible standards from its beginning. For instance, the world personal computer market has been mainly categorised into two completely incompatible camps—the MS-DOS operating system backed by IBM and the Macintosh system designed by Apple. Mainframe computers are also diversified. Despite the failure of Philips and Sony to standardise the CD-ROM industry, CD-ROM technology was introduced in the early 1980s.[24]

In 1985, the leading consumer electronics groups (Philips and Sony) and the computer hardware manufacturing and software leaders including IBM, Apple, Digital Equipment and Microsoft gathered together at Lake Tahoe in the High Sierra mountains of the U.S. to discuss issues about setting up the CD-ROM standard. As a result, the two sides reached an agreement on encoding information on the CD-ROM discs and the High Sierra CD-ROM agreement has been widely accepted as a world standard ever since. The physical and recording characteristics and specifications of the CD-ROM standard were included in the 'Yellow Book', which was jointly defined by Philips and Sony. The 'Yellow Book' was specified in accordance with the 'Red Book' specifications but at a higher level of error correction techniques. More specifically, the physical format of CD-ROM was agreed to as ISO

(International Standards Organisation)-10149 and the logical file structure was agreed to as ISO-9660.

CD-ROM products are now widely used for professional purposes. In particular, they have become very popular in libraries. For instance, by the end of 1986 there were about 100 CD-ROM products on the market, of which 40% were aimed at the library market (Davis and Daum, 1987). However, fragmentation in the CD-ROM industry follows the pattern of incompatible standards in the PC industry. To fit into various computing environments, the CD-ROM systems have been configured by different file-handling methods. Consequently, CD-ROM drives are not always compatible with each other.

As a matter of fact, some technical aspects like the method of information storage, retrieval and the make of disc drive were not included in the CD-ROM specifications contained in the 'Yellow Book'. In an attempt to ensure compatibility of their future interactive multimedia system, i.e. CD-i (Compact Disc-Interactive), with the CD-ROM discs, Philips and Sony, joined by Microsoft, launched an intermediate CD format, called CD-ROM XA in 1988. CD-ROM XA was defined as a method of storing information on future CD-ROM discs so that they will play on CD-i players.[25] In other words, CD-ROM XA was proposed to add audio, graphic, and future extensions of full-screen, full-motion video to the current CD-ROM format at the technological level of CD-i. It was the designers' intention that CD-ROM XA would be used as a PC peripheral product and, therefore, mainly targeted at the professional market. However, CD-ROM XA as a half-way technology between CD-ROM and CD-i has been slow in gaining acceptance by manufacturers and software designers.

THE PROLIFERATION OF MULTIMEDIA

As shown in Fig. 8.1, multimedia is a further development on a combined base of CD-DA and CD-ROM, and it represents an entirely new era of Compact Disc technology. Meanwhile, the development of multimedia systems also heralds the coming of the next generation consumer electronics and entertainment industry.

For the time being, the major systems available in the marketplace include CDTV (Commodore Dynamic Total Vision) launched by Commodore,[26] CD-i (Compact Disc-interactive) brought out by Philips together with Sony, Photo CD developed by Kodak in collaboration with Philips, and 3DO of Matsushita, AT&T and Time Warner. Consumer products from the first two systems became available in 1991, and Photo CD players went on sale in September 1992. 3DO was put into the consumer market in the U.S. in October 1993. DVI (Digital Video Interactive) was heralded in 1987 as an independent multimedia system, but it has turned out to be no more than a pure digital video compression technology. Many other computer firms also have the intention of jumping on the multimedia bandwagon by developing

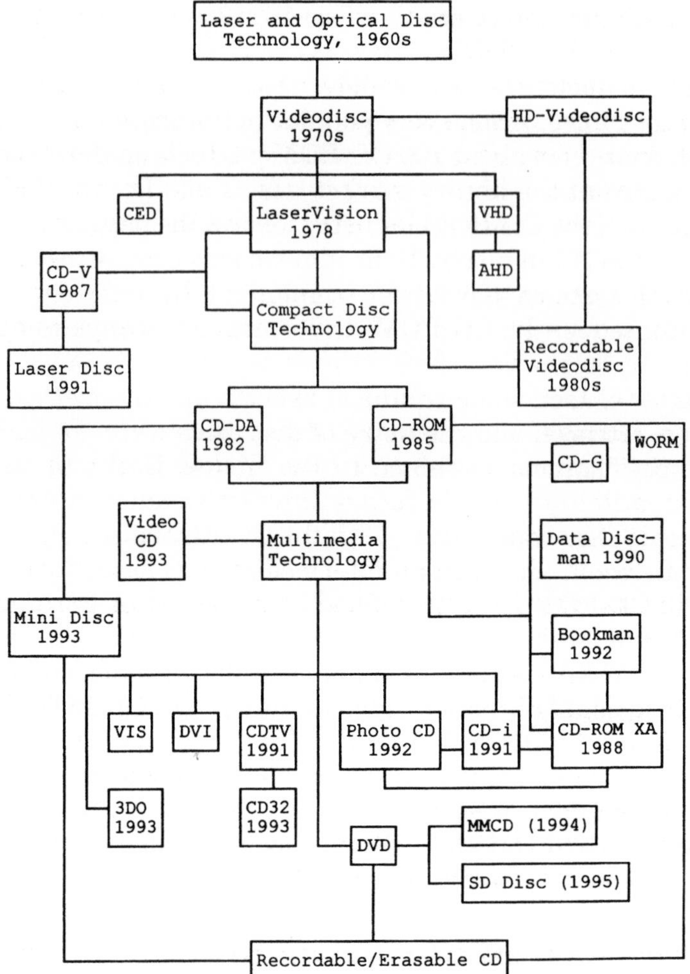

Fig. 8.1. Evolution of Optical Disc and CD Technologies.

their own independent systems. For instance, IBM and Apple, two rival firms on the personal computer market since the last decade, have signed an agreement to set up a joint venture, called KALEIDA, to develop multimedia products. Additionally, Apple itself has announced a series of plans to develop multimedia products for the consumer market mainly by collaborating with Japanese firms including Sony, Sharp and Toshiba. In addition to its joint venture with Apple, IBM has been active in seeking other partners (e.g. Toshiba and Time Warner) to jointly develop multimedia systems.

On the side of the corporate market, the concept of MPC (Multimedia Personal Computer) has become widely known from the mid-1980s. An MPC is an IBM or IBM-compatible personal computer running Microsoft Windows

3 (or higher versions) with multimedia extensions incorporating a CD-ROM drive with an extra sound card. The MPC standard as such has been agreed by the MPC Council and supported by a group of hardware manufacturers (including Philips) and software houses (in particular Microsoft).

In short, in the same way as the effect of HDTV on blurring the sectoral boundaries between the consumer electronics and computer industries, multimedia has provided another opportunity for integration across these two industrial sectors. Without extensive government intervention and premarketing standardisation at the industry level, independent or incompatible multimedia systems have been launched to compete with each other in the marketplace until a *de facto* industry standard has been chosen by the consumer.

Despite a concerted attempt by Philips to follow the CD-DA strategy to define the CD-i standard before commercial launch, from a technical point of view, the birth of a multistandard multimedia industry has in practice corresponded to the chaotic CD-ROM industry. Rather than going into specific technical details of the multimedia systems, I only touch on some general characteristics of the new interactive technologies and the emerging infant industry in this chapter.

Defining Multimedia

Multimedia, by its name, means a complicated media form which embraces elements of different conventional media forms. As massive R&D efforts have been committed to bring about a new industry, various descriptions have been suggested to identify the nature of multimedia. However, so far there is no universally accepted definition for multimedia. In his recent study, Feldman (1991) presents a definition after having examined many other definitions for multimedia,

> We now have an identity for multimedia, a working definition against which we can measure developments and potential impacts. We can describe it as the seamless integration of data, text, images and sound within a single digital information environment (p. 6).

What is missing in this definition is 'interactivity', a crucial feature which a genuine multimedia system must have.[27] To some extent, the level of interactivity between the machine and the user tells how sophisticated the multimedia system is. Another point about the above definition is the term 'images'. In referring to 'image', one must clarify whether it is still picture or animation or full-screen full-motion video used in a digital information environment. Sometimes, electronic firms use still or animated pictures or partial screens of moving images to conceal their inability to incorporate FSFM digital video, or FMV, into their so-called multimedia systems.

In this chapter, I use the term 'multimedia' to describe a fully digital information system which combines sound, text, graphics, and pictures (still, animation and full motion pictures as required by the software programme) on a single Compact Disc, and allows the user to operate it in an interactive way. When a multimedia player is connected to a TV set or Hi-Fi system or home computer monitor and used mainly for entertainment purposes, it is a consumer product; and it will be a professional product if it is used for corporate purposes such as training, POI (Point of Information), POS (Point of Sale), electronic book, etc. The major focus of this chapter is the former category, i.e. multimedia systems as new consumer electronics products. Most consumer multimedia systems as consumer products embrace a common set of major characteristics including:

- interactivity
- edutainment (short for education and entertainment)
- full digital
- a built-in computer operating system
- technologically based on CD-ROM
- backward compatible with CD-DA discs.

Moreover, every consumer multimedia system can be connected to a colour TV set or Hi-Fi system in the home.

The Major Systems

Since the mid-1980s, various multimedia systems have either been launched or planned. As far as the consumer market is concerned, CDTV of Commodore, CD-i of Philips, and Photo CD of Eastman Kodak were the first to be launched. In addition, because of the leading position of Intel in digital video compression technology and its potential influence upon the future consumer market, the company's DVI system is also worth mentioning.

The first consumer multimedia CDTV (Commodore Dynamic Total Vision) was launched by Commodore during the occasion of *CD-ROM Europe 91 Conference & Exhibition* in London in May 1991. Being the first to bring its multimedia to the consumer market does not necessarily mean Commodore had completely overcome all the major technical difficulties facing the multimedia industry. On the contrary, the launch of CDTV was not accompanied by FSFM video, which many believed to be one of the key elements in a consumer multimedia system. Having been aware of the potential competition on the multimedia market, Commodore initially hoped to make major inroads into the market with a head start of two years over Philips.[28] By the time of its European launch, Commodore had only 25 software titles available for sale, and this figure was promised to be increased to 50 at the beginning of summer 1991 and 100 before the end of the same year.[29]

The second consumer multimedia system to follow CDTV in the marketplace was the CD-i (Compact Disc-interactive) system. Apart from the work related to the specifications of CD-ROM, both Philips and Sony were independently involved with investigation and development for an extended system of the CD-DA format with the purpose of opening the interactive multimedia market. The two companies joined forces and combined their research efforts and technical expertise at the beginning of 1985. This cooperation provided the technological basis for the future CD-i specifications—the 'Green Book'. Another important event was Matsushita's participation in the research work on the integrated circuits to be used for manufacturing CD-i machines in the same year.[30]

Philips and Sony introduced their draft specifications for the CD-i standard, known as the 'Green Book', in March 1986. In July 1987, these two companies finally completed the specifications for the CD-i system. However, the launching time for the CD-i consumer model was repeatedly postponed since autumn 1987, and the first consumer launch was not made until October 1991 in the U.S. CD-i was not launched in Japan and Europe (U.K. first) until the end of April 1992. The 'Green Book', unlike the 'Yellow Book', was fully specified and included aspects such as definition of file format, directory structure and location, record access, etc. The CD-i specifications contained in the 'Green Book' have laid down the guidelines which must be followed by any software house in mastering or pressing CD-i programme discs. It was promised by Philips that the 'Green Book' ensures that any CD-i disc could be played back on any CD-i machine, although compatibility with rival systems is completely out of the picture.

After the official launch in the U.S. in October 1991, the CD-i machines were launched by Philips in Japan and the U.K. at the end of April 1992.

In collaboration with Philips, Eastman Kodak developed its Photo CD system, which is compatible with the CD-i format.[31] Photo CD was launched in summer 1992. This system allows up to 100 35mm photo snaps to be electronically stored on a normal Compact Disc, and be displayed on any colour TV screen. With the Photo CD system the user can also enlarge different parts of a photo image on the screen. Therefore, as a consumer product, the Photo CD system was mainly intended as a family electronic album. Like CD-i discs, Photo CD discs can also be operated on a CD-ROM XA drive.

Compared to Canon's still video camera, or otherwise known as Ion (Image Online Network),[32] the Photo CD system offers much higher screen resolution. Consequently, Photo CD has opened a new way for DTP (Desk Top Publishing) as well. The system has already attracted attention from the major computer software houses and publishers.

As one of the earliest multimedia technologies, DVI (Digital Video Interactive) was originally developed by RCA's David Sarnoff Research Centre located in Princeton, New Jersey and the first owners of this technology were RCA and General Electric (GE). As an independent multimedia system, DVI is a combination of CD-ROM with video compression. Running the standard 5-inch Compact Disc, the DVI system was said to be able to offer 72 minutes of FSFM video in digital code. The first announcement of the DVI system was made by RCA and GE in 1987.

In October 1988, all the rights of the DVI technology were bought over by Intel, the leading American chip maker.[33] For the time being, Intel is strongly supported by IBM in promoting the DVI system. Recently, Olivetti, the Italian computer group, decided to produce PCs for the DVI system in Europe. DVI players produced by Japanese firms including Intel (Japan) and Applix and by Olivetti have already been shown to the industry; and players specially designed for the Apple Macintosh computer environment were planned to be marketed in the first quarter of 1991.[34] Thorn EMI has also shown interest to support the DVI system by installing DVI chips in its future hardware and marketing DVI products in the U.K. Differing from the Philips/Sony CD-i system, DVI was initially targeted at the professional and business market with a possible extension to the consumer market.

The DVI player to be launched on the professional market would be mainly built as a peripheral product to a personal computer. DVI offers another solution to store full-motion video, 3D graphics, and multitrack audio on a single CD-ROM disc. For operating the DVI system, the user will have to buy four pieces of hardware components:

- a personal computer
- an add-on video board
- an add-on audio board
- a CD-ROM drive or another data storage device.[35]

As a retailer of the DVI technology, Thorn EMI demonstrated the functionality of the DVI system at the *CD-ROM Europe 91 Conference & Exhibition* in London. This system consisted of a complete 386 PC, a built-in CD-ROM drive, an Intel DVI card, and two external speakers. The cost of the DVI boards that plug into a PC was about $2,000,[36] and the total outlay of a complete DVI system package would be around at least $5,000 in comparison with the originally promised $1,000 launching price of Philips' CD-i system.[37]

It is reported that Intel's DVI machines have already been installed in Terminal 2 at the De Gaulle Airport in Paris.[38] The DVI machines were used in connection with Compaq 386 PCs as an interactive information service system at the airport. Travellers can easily find out their way and various services at the airport by operating the DVI information system, which provides audio, text and high-quality pictures in an interactive way. Despite various professional applications, there has been no sign of an immediate consumer launch for the DVI system.

Apart from CDTV, CD-i, Photo CD, and DVI, a number of other multimedia systems have also been announced. For instance, IBM and Apple have signed a joint venture agreement to develop an independent multimedia system for the personal computer market. With its newly issued

QuickTime as a powerful authoring tool with its own digital compression system, Apple revealed its ambition to challenge the future consumer electronics market by launching a series of personal interactive electronics gadgets in cooperation with the Japanese consumer electronics industry. One of the announced Apple products has been known as PDA (Personal Digital Assistant), a hybrid of consumer electronics, personal computer, and personal telecommunication techniques. Apple recently formed a new product division, called Apple PIE (Personal Interactive Electronics), to promote its new personal information systems, particularly the PDA series. This hand-held interactive PDA system was intended to be launched to the market by the middle of 1993 with an expected annual sales of 5 million units, and Apple wished to establish its system as the *de facto* industrial standard for the next generation interactive multimedia products.[39]

As a previous supporter of Philips' CD-i system, Tandy of the U.S. recently stopped making CD-i players and shifted to its own technology. Together with Zenith and Microsoft, the biggest computer software house in the world, Tandy announced its VIS (Video Information System) in August 1992. Inevitably, this format will be another competing technology to the CD-i system. As a CD-based multimedia system, VIS will run under Microsoft's DOS/Windows environment. Sources from Zenith, the VIS player maker, suggests that up to 50 software companies had already agreed to support the VIS format and about 100 software titles had been put into development.[40] It is promised by VIS developers that players would be available for sale in the U.S. during the 1992 Christmas sale season, and the European launch would be due in 1993.[41] For the time being, it seems that the Video Information System offers everything a multimedia system is supposed to do except for full-screen full-motion *video* itself!

To add further confusion to the newly born multimedia market, Sony launched its Data Discman in 1990 and Bookman in 1992. Data Discman is a hand-held player running smaller sized compact discs of 8 cm in diameter (compared to 12 cm for ordinary CDs) with a built-in LCD display. A new model of Data Discman introduced by Sony in November 1991 was capable of playing back sound in addition to texts and Graphics. Until the beginning of 1992, sales of Data Discman reached 150,000 units in Japan, and another 180,000 units in the U.S. and Germany since the launch. In the meantime, some 60 electronic book titles were made available, and 400,000–500,000 copies of these titles were sold.[42]

While Data Discman was reaching the market, Sony announced another new system called Bookman in 1991, and promised to launch the new product in the U.S. by autumn 1992. Bookman was technically a CD-ROM XA drive with an LCD display panel, speaker, keyboard, and a cursor pad. Differing from Data Discman, the Bookman machine uses standard 12 cm compact discs and is compatible with CD-Audio.[43]

Another Format Battle?

The above discussion indicates that the proliferation of multimedia technologies has not been followed by standardisation. As far as the consumer market is concerned, no consensus has been reached at the industry level. After the launch of CDTV, CD-i and Photo CD at the beginning of the 1990s, another format battle has become inevitable. Consequently, rather than being excited, the consumers who have the interest to explore the new interactive 'edutainment' world of multimedia may be confused.

As Fig. 8.2 shows, there are three categories of multimedia: multimedia systems in connection with computers, with TVs, and those with LCD displays. The first category are mainly targeted at professional applications; while systems using TV displays are mainly consumer products. Portable multimedia products with LCD displays are primarily designed for the corporate market but could be extended to the consumer market if prices drop quickly.[44] In most cases these multimedia systems are incompatible with each other.

In the computer world, there are two major streams of multimedia development based on IBM compatible and Macintosh computers, respectively. So far numerous multimedia authoring systems based on the DOS/Windows and OS/2 operating systems have been demonstrated for the PC industry. Aimed at creating an industry standard including hardware and software, the MPC (Multimedia Personal Computer) Marketing Council was set up in 1991 with the support of a dozen computer and consumer electronics firms such as Microsoft and Philips. In parallel with the MPC council, another organisation called IMA (Interactive Media Association) was recently formed to promote the standardisation for PC multimedia formats. The latter was joined by big computer firms such as IBM. Corresponding to the recent IBM/Apple alliance to create an open system for the computer industry, these two firms also signed an agreement to design a

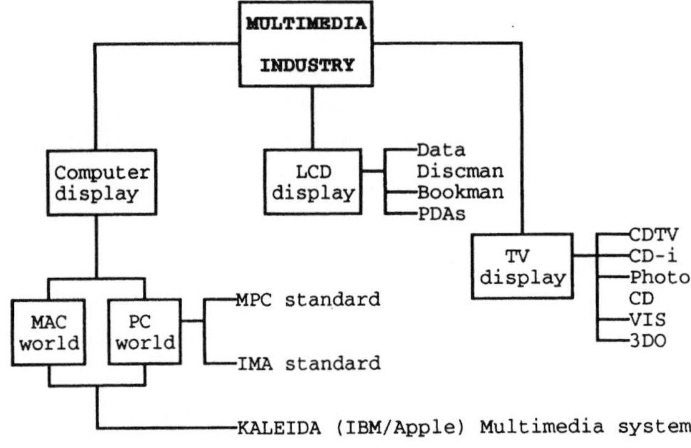

Fig. 8.2. Proliferation of Multimedia Standards.

multimedia system by using expertise from both sides through their joint venture, called KALEIDA, as mentioned above. According to an IBM manager, the KALEIDA multimedia joint venture is a long-term partnership, and an independent multimedia system may not be developed in the near future.[45]

So far no organised effort has been made at an industry level to unify or bridge various incompatible multimedia formats developed for the consumer electronics market. As mentioned above, the three commercially available multimedia systems (CDTV, CD-i, and Photo CD) are all backward-compatible with CD-Audio, and used in connection with colour TV sets in the home, but there is no cross-format compatibility amongst them except that Photo CD discs can be played back on CD-i machines.

There were speculations suggesting that Apple would support the CD-i format by making its future PDA products compatible with CD-i. Mr Jan Timmer, President of Philips, was believed to be personally dealing with this matter.[46] Should this become a reality CD-i would gain a substantial competitive advantage against others in the format battle.

The history of the consumer electronics industry, particularly that of audio and video, strongly suggests that the market favours only one system amongst several contenders in most cases except for colour television.[47] In the audio area, Philips championed the compact magnetic cassette and Compact Disc (together with Sony) almost without any competition. Whereas, in the case of home video recording, JVC swept the whole world with its VHS standard which killed off the technologically superior formats including Betamax and V2000, as discussed in Chapter 4. If the same logic is applicable to the consumer multimedia industry, many companies will, perhaps, have to bear heavy losses after the format battle, although at present it is hard to tell which system is likely to emerge as the winner.

From a technical point of view, most multimedia systems (at least for consumer applications) have been developed on the basis of the CD-ROM specifications, and every system is built with a computer 'brain' inside it. These two factors are vital to explain why multimedia systems are not compatible with each other. Neither the CD-ROM industry nor the computer industry is completely standardised. Because most multimedia systems were either developed by large computer companies alone or by consumer electronics firms (sometimes in collaboration with the former), compatibility between independent systems could hardly be achieved unless special bridging efforts are made.

In the consumer market, competition started with available products designed and produced for CDTV, CD-i, and Photo CD. Although DVI units have also been available, according to a Philips manager, they are exclusively shipped to the professional market hence they do not compete directly against CD-i for the time being.[48] Because of the fact that Photo CD was developed by Kodak in collaboration with Philips, and its discs are compatible with CD-i players, Photo CD does not challenge CD-i. On the contrary, as assumed by Philips, Photo CD will reinforce CD-i's acceptance by the market. Therefore, CDTV appeared to be the only real competitor against CD-i at the early stage of home multimedia development. As a proprietary multimedia system of Commodore, CDTV physically resembled the CD-i system, and it offered similar software features. Due to the fact that Commodore initially identified the Amiga computer users as the major target to promote CDTV sales, the consumer multimedia market became further complicated, and this might eventually affect Philips' marketing campaign for CD-i. Philips still maintains its intention to define CD-i as a consumer electronics product like television and Hi-Fi, distanced from the concept of home computer. This point will be further discussed later in this chapter.

In Philips' view, the development of DVI is a potential threat to CD-i in the professional market if Intel drops its prices and popularises its system.[49] In the future market, the IBM/Apple joint venture to develop an independent multimedia system and the Apple/Sony/Sharp alliance for PDAs imposes a more serious threat to the CD-i system as seen by Philips:

> In our strategic planning, we regard DVI as a possible threat. Another threat is the recently announced IBM/Apple [cooperation to develop a] multimedia machine. I don't know how it looks like. It might be a big threat, although they will not launch their system until 1993. It is not available today, but you have to look into the perspective of a couple of years. ... We also know that Apple is working with Sony on something which imposes more than enough threats.[50]

On the other hand, potential threat could be turned into an advantage for CD-i if cross-format compatibility is achieved with Apple's future consumer multimedia products.

It seems that the format battle for a universally accepted multimedia system cannot be resolved in the near future. In other words, CD-i is not only facing the competition from CDTV (now CD32) for the time being but also substantial threats from other systems including Matsushita's 3DO and the newly announced Tandy/Microsoft/Zenith VIS consumer multimedia system. Optimistic industrialists hold the position that the current format battle between different systems will be resolved within a couple of years, and market force will eventually favour one format.[51]

Public Policy and Multimedia

The current multimedia format battle has so far not been evidently subject to government intervention. In Europe, the EU has not imposed any legislative or protective procedures upon the industry as it has done for the HDTV sector since the mid-1980s. Consequently, multimedia system makers have been competing within a more or less *laissez-faire* market environment and CD-i, as a European technology, has not been the subject of many promotional efforts by either the EU or any member state. However, it is still too early to say whether EU authorities will not shift their policy from the currently 'hands-off' approach to some sort of protectionist approach. Some evidence already suggests the latter.

First of all, the EU has a notorious history of intervention or protectionism in the consumer electronics industry. Apart from its protectionist procedures applied to the European VCR industry since the early 1980s and its HDTV policies implemented since the second half of the 1980s as discussed, respectively, in Chapters 4 and 6, the EU also applied antidumping policies for imported Compact Disc players. Aiming at protecting the indigenous industry, the European Commission started an antidumping investigation in 1987 concerning imported Compact Disc players made by foreign companies, in particular those from Japan and South Korea. As a result, provisional and definitive duties were imposed in 1989 and 1990, respectively.

Will the EU apply the same or similar antidumping policy to the future multimedia market should a similar trading situation come up again? It is still too early to make a final judgement.

Secondly, in 1986, the European Commission supported the European industry initiative to set up an Optical Disc Forum to consult and inform European organisations on technical developments and the progress in achieving standardisation of data handling on CD-ROM.[52] DG X and DG XIII of the EU have maintained a close watch over developments of the European audio-visual and other information technology sectors since the late 1980s. Regular consultations with organisations from the concerned industries have been made by the EU, although direct intervention has not been seen in the multimedia hardware industry so far.

Thirdly, certain cooperative activities for interactive media development have taken place with EU support. During the occasion of *The First European CD-I Conference* held in London in June 1990, the Media Investment Club of the EU announced that a European CD-i development project which would have the support and funding from the Media Club was to be set up in September 1990; and this European programme was to promote the creation and production of new media and audio-visual products with the investment of ECU 700,000 for 1990–1991 in the preproduction and development phases of new media interactive products.[53] Apart from the European Commission itself, this Media Investment Club was also joined by Philips, Thomson, RAI, the French Antenne 2, and half a dozen of other organisations.[54] As an example of the EU financial sponsorship, Line TV, a European production company, has been producing a CD-i disc on *Viking History* with financial help from the Commission's Media Investment Club.[55] Under the umbrella of the IMPACT (Information Market Policy Actions) programme the EU launched a call for proposals for share-cost projects to produce interactive multimedia information services in June 1991. Information from DG XIII of the European Commission indicates that

Table 8.3. Provisional and Definitive Rates of Duty on Imports of Compact Disc Players from Japan and Korea.

Exporting Company	Provisional Rate of Duty %	Definitive Rate of Duty %
Japanese Exporters:		
Nippon Columbia Co Ltd	17.5	17.0
Funai Electric Trading	8.4	8.9
Kenwood Corporation	19.3	23.3
Matsushita Electric Industrial Co Ltd	33.9	26.3
Onkyo Corporation	8.5	8.3
Pioneer Electronic Corporation	28.8	26.3
Sanyo Electric Co Ltd	27.8	26.5
Sony Corporation	15.9	10.1
TEAC Corporation	6.4	12.7
JVC	20.9	17.9
Yamaha Corporation	23.7	27.5
Sharp Corporation	33.9	32.0
Toshiba Corporation	33.9	31.0
Chou Denki	33.9	17.8
LUX Corporation/Alpine Electronics Inc	0	0
Marantz Japan Inc	0	0
Other Japanese exporters	33.9	32.0
Average rate of duty	**20.39**	**18.68**
Korea Exporters:		
Inkel Corporation Ltd	20.1	14.4
Goldstar Co Ltd	32.5	26.1
Samsung Co Ltd	23.0	10.7
Haitai Co Ltd	21.3	19.4
Other Korean exporters	32.5	26.1
Average rate of duty	**28.88**	**19.34**

Source: BIS Mackintosh (1990).

a total of 317 proposals involving 1,158 participant organisations from all 12 EU countries submitted to the Commission for EU funding.[56] The finally selected projects would possibly receive up to a maximum of ECU 250,000 each from the EU during their definition phase (1 January to 30 June 1993) in comparison with a total budget of ECU 6.5m available.[57]

It is also known that Esprit, the EU information technology programme, was involved to set up a special project called Multiworks. Multiworks was aimed at setting standards to ensure manufacturers' products could function together and hence a European multimedia terminal was to be created.[58] This project was supported by Philips, Bull, Acorn, and Northern Telecom.

Philips' CD-i technology has particularly been favoured by the Dutch government. Under the Ministry of Education and Science's 'New Media' policy, Philips entered a contract with the government in 1990 to supply a total of 30,000 computers to primary schools all over the country. Another aspect of the 'New Media' policy was the setting up of a CD-i project between the government and Philips in 1990. As a pilot scheme of this CD-i project, a CD-i programme on English language teaching for primary education, called 'No Problem', was carried out during the period April 1991 to January 1992. As revealed by Philips, a total investment of 600,000 guilders was set aside for producing this English language disc, of which 250,000 guilders was provided by the Ministry of Education and Science.[59] Further decision on public procurement of CD-i equipment to be installed in Dutch schools has not been announced by the time of writing. It is evident that any kind of government backing may be very important to help Philips open the education market for its CD-i system.

CD-I AS A KEY PROJECT AT PHILIPS

Since the announcement of its full functional specifications in June 1987, CD-i has been identified by Philips as one of its key consumer electronics projects.[60] It is believed that the success or failure of the CD-i project will heavily contribute to whether Philips could retain long-term survival in the increasingly competitive global consumer electronics industry.

Implications of CD-i

The implications of CD-i are many-fold. First of all, as an independent consumer multimedia product, CD-i appears to be one of the most important new gadgets to stimulate the growth of the saturated consumer electronics market. By the late 1980s, most conventional consumer electronics durables such as Hi-Fi systems, CTVs, VCRs and CD players have reached a considerable penetration level in major consumer markets including Japan, the U.S. and Western Europe. Market potential and profit margins for these mature consumer electronics technologies have already become very limited. Therefore, CD-i, if it wins the current format battle, would be one of the most important elements to create new consumer demands hence to offer room for the consumer electronics industry to grow. It is anticipated by the industry that multimedia will eventually turn consumer electronics technology from a linear stage into a multidimensional or interactive stage. However, during a period of worldwide economic stagnation, the speed of acceptance for any new gadget on the consumer market might well be slowed down, and inevitably, CD-i would be affected as well.

Secondly, and most importantly, CD-i is vital to the struggling Philips group. As discussed in Chapter 3, Philips has been regarded by some commentators as being good at bring about technological breakthroughs but very poor at commercialising its new technologies. Both the LaserVision system in the optical disc area and the V2000 format in the magnetic video recording sector are often taken as two classic examples. The successful introduction of the CD-DA format re-established Philips' leading position in optical disc technologies. It has been repeatedly claimed by the Philips management that the company wishes to extend its CD success into CD-i in the 1990s.[61] One aspect of the current Philips restructuring programme, namely 'Operation Centurion' as discussed in Chapter 5, was to give much more emphasis on market-orientation or commercialising the company's new technologies including CD-i. It is expected that after Philips suspended its planned production of HD-MAC receivers, the controversial European HDTV technology, other companies such as Thomson and Nokia would share the loss of investments; but the development costs of CD-i has not been shared by other companies, apart from a limited technological contribution from Sony. Although there are no published figures concerning the costs of CD-i R&D, as well as financial and manpower help offered by Philips to third-party software developers, the sum of investment by the company is undoubtedly substantial.[62] Taking into account the financial difficulties experienced by Philips in the first few years of the 1990s, one may easily come to the conclusion that the company cannot afford a failure for CD-i. Maybe this partly explains why Philips has committed an unprecedented managerial, as well as financial, effort to ensure success for CD-i.

Operational Structure of the CD-i Business

The huge investment may not necessarily be sufficient to make a strategic product a success. On the contrary, proper managerial commitment is indispensable. In the case of CD-i, Philips has created a globally stretched organisational structure specifically accountable for the development, production and marketing of the new system.

The first layer of the CD-i organisational structure is the setting up of Philips Interactive Media Systems (IMS). Philips IMS is an important Business Unit (BU) under the administration of the Consumer Electronics Division. Major R&D work related to the CD-i system has been done by IMS. Meanwhile, IMS is also responsible for CD-i equipment production, sales and marketing. With its headquarters in Eindhoven, Philips IMS has many national branches which are respectively concerned with the special needs of sales/marketing on local markets.

In parallel with IMS, Philips has also set up a holding company, i.e. Philips Interactive Media International (PIMI), in charge of the company's global interactive media publishing activities. Under the banner of PIMI,

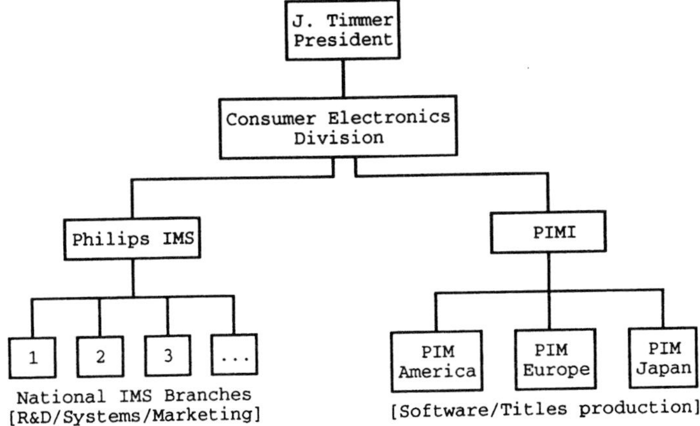

Source: Based on interviews with Philips managers.

Fig. 8.3. Philips' Managerial Structure for CD-i.

three regional branches including Philips Interactive Media America (PIMA), Philips Interactive Media Europe (PIME) and Philips Interactive Media Japan (PIMJ). PIMI's major task is to promote software development for the CD-i system worldwide.

Traditionally, Philips' corporate organisations were created to meet the requirement of hardware development and production (PolyGram was an exception). The paralleled organisational structure (IMS and PIMI) for CD-i indicates that the Philips management have now realised the indispensable role of software to make an innovative consumer product successful.

PHILIPS' GLOBAL STRATEGY FOR CD-I

As discussed above, CD-i was intended as another global standard following the success of CD-DA. In Philips' words, CD-i is a world product.[63] Accordingly, the scope of Philips' strategic planning for CD-i is being carried out at a global level. Philips' global strategy for CD-i includes the following major aspects: collaborative R&D with electronics companies (notably Motorola) to build key components, a wide range of strategic licensing agreements with independent software developers and publishers to make CD-i titles, joint venture agreements, and strategic partnerships with major consumer electronics firms to get industrial support for marketing the CD-i system.

Industrial Collaboration

It would be wrong to imply that Philips had never considered industrial or commercial cooperations with other companies before the 1980s. On the contrary, in its history, as shown in Chapter 3, the company was involved in many cooperative agreements, including a number of joint ventures located in different parts of the world. As far as Philips' strategic planning is concerned, it is important to point out that only after the early 1980s has industrial collaboration become the core of the company's corporate strategies. This is the lesson Philips learned from the bitter experience with its LaserVision and V2000 systems. Indeed, the sweeping victory of the CD-DA format in the music industry led to the belief among Philips management that cooperation with major Japanese firms is now crucial to make a consumer electronics product successful. This strategic change within Philips has been noticed by some commentators:

> Philips, a Dutch electronics company, is discovering that cooperation with its Japanese rivals over new-fangled consumer products is a better strategy than head-on competition. After its V2000 video recorder and LaserVision video discs went on the blink in the face of rival systems from Japan, Philips negotiated with its Japanese competitors and agreed on common standards for compact discs—launched by Philips and Sony in 1983.[64]

As far as CD-DA and CD-i is concerned, it might not be Philips' major purpose to enter collaborative agreement with Japanese firms from a technological point of view. In fact, Sony came on board when Philips had almost completed the development work for the CD-Audio system. It was estimated that Sony contributed about 15% of technical know-how to finalising the CD-DA format.[65] However, joining forces with Sony was crucial for Philips to make CD-DA a global standard for the music industry in the early 1980s. In recalling the process of the above strategic change, Philips managers put special emphasis on the company's 'old' relationship with its Japanese rival Sony:

> We've learned from the VCR experience that if you cannot go with your own system, you have to find out joint ventures or agreements with other companies for alliances so that we can work together to set a standard. That's exactly what we have done with CD-Audio. Without Sony, maybe we would have never made CD-Audio such a success. It was very important from a strategic point of view to make a clear statement to the outside world that Philips and Sony were setting a standard. Consequently, we got a lot of supporters, and we had the standard. It is exactly the same that we are doing today with CD-i and Photo CD.[66]

Starting with CD-Audio, Philips has built a very special relationship with Sony. The two companies have been sharing all the technical expertise and patents related to laser optics with each other.[67]

In Table 8.4, Philips' cooperation with Sony was mainly involved with precompetitive R&D activities, and that with Matsushita was strategic partnership. It is evident that CD-i has already been supported by the top three giants as far as the consumer electronics industry is concerned. The cooperation between Philips and Motorola was aimed at VLSI (Very Large Scale Integrated Circuits), and the central task was to develop a new chip

Table 8.4. Philips' Industrial Collaboration for CD-i Hardware.

Name of venture	Partner(s)	Year of set up	Nature of venture
Philips/Sony	Sony	1985	Cooperation
JNMS	Kyocera	1985	Joint venture
Philips/Matsushita	Matsushita	1989	Cooperation
Philips/Motorola	Motorola	1989	Cooperation
Sun-Philips New Media Group	Sun Microsystem	–	Joint venture
Kodak/Philips*	Kodak Eastman	1990	Cooperation
Philips/C-Cube	C-Cube	1991	Cooperation
Philips/GTE	GTE	1992	Cooperation
Philips/JVC	JVC	1992	Cooperation

Note: *The cooperation between Philips and Kodak was to develop Photo CD, another independent multimedia standard. Because of the compatibility between CD-i and Photo CD discs, this cooperation was regarded as part of Philips' CD-i strategy in competing against rival systems.

in order to provide full-screen full-motion video for realising the full capabilities of the CD-i system. Very recently, Philips and Motorola announced that they were to set up a joint chip development centre, known as Motorola Philips Design Centre (MPDC), in Eindhoven to accelerate the development of future integrated chips for CD-i and related multimedia products.[68] As one of the leading firms in digital compression technology, C-Cube Microsystems would develop real-time video compression chips for Philips as well. It is believed that, in the field of advanced semiconductors, Philips' cooperation with Motorola and C-Cube Microsystems was critical to bringing the industry standard video (i.e. MPEG) onto the CD-i machine.

As a consumer product, CD-i's further integration in the television broadcasting world may create more interactivity and excitement. GTE ImagiTrek, a Division of GTE Vantage Inc., and Philips recently announced their joint development relationship on technology and software to simultaneously combine information provided by a CD-i system and off-air TV programmes. This was viewed as an important step towards the future of interactive TV.

In an effort to diversify the CD-i related product range, Philips recently announced its cooperation with JVC to jointly develop a CD-Karaoke system based on the MPEG-1 full-motion video standard. Discs designed for this Karaoke system will be playable with CD-i machines with an FMV capability. This agreement seems to be a sort of compensation to Philips' loss of face in the early 1980s when its V2000 system was killed off by JVC's VHS, provided CD-i wins the multimedia format battle in the future.

Apart from the aforementioned industrial cooperation ventures, Philips has also licensed its CD-i technology to a large number of hardware manufacturers. A source from Philips (Taiwan) indicates that more than 100 companies worldwide are licensed to manufacture CD-i hardware equipment.[69]

The driving force for Philips to adopt a global collaborative strategy could be understood from the perspectives as follows.

Firstly, as revealed in the above quotation from Philips, the bitter experience of failure in launching the V2000 format and the LaserVision system forced Philips to rethink and question its traditional corporate strategies.

Secondly, since the beginning of the 1980s not only the consumer electronics but also other industrial sectors (e.g. the computer industry) have been seeing increased global and regional industrial groupings or alliances led by big multinational companies from each sector. As the biggest European consumer electronics company and the third largest one (after Matsushita and Sony) in the world, Philips would not like to be excluded from the mainstream of global competition.

Thirdly, technical change and new product innovations since the 1970s has been characterised by strong forces towards global standardisation. To achieve a global standard for a particular product, technological superiority is of great importance but might not be sufficient.

Fourthly, it is becoming increasingly rare for a single company to introduce and maintain a product as a world standard.[70] This new characteristic of technology development has made it a necessity for companies, including Philips, to join forces with each other for launching new products or systems. In this aspect the consumer electronics industry has already experienced a number of successful and unsuccessful examples.

Finally, the increased cost for R&D is another important factor to force rival firms to collaborate and share risks.

The question whether or not Philips and its partners continue to benefit from the CD-i system, the third generation of compact disc technology, in the future still remains to be answered, despite Philips' efforts to promote industrial cooperation related to the CD-i project. As a multimedia system, CD-i's success may depend on a number of factors of which software development is widely regarded as the most important one.

Strategic Licensing and Partnership

The development of consumer multimedia, however slow, is further confirmation of the widely believed notion that the consumer electronics industry is increasingly characterised by the synergetic effect between hardware and software—to a certain extent the importance of software surpasses that of hardware. Philips paid a high price to learn this lesson (e.g. the failure of LaserVision and V2000). In the case of CD-i, Philips has been facing severe competition from rival system builders not only in technologies (e.g. FMV) but also in software availability. Philips has claimed from the very beginning that the success of CD-i depends on the availability of sufficient quantities of good software.[71] To meet this requirement, the Philips management dedicated substantial efforts to the software side of its CD-i project. As shown in Fig. 8.3, half of the company's managerial structure, i.e. PIMI (Philips Interactive Media International), is committed to software development and titles production. Strategically, Philips has been licensing its CD-i technology to independent software developers, and entered a number of joint venture/partnership agreements with publishing and entertainment companies worldwide.

Table 8.5 shows Philips' major strategic partnership arrangements for CD-i software development. It is worth mentioning that about half of these partnerships were joint venture agreements. Apart from the 80%-owned PolyGram music company, these joint ventures for CD-i titles development represent Philips' most important strategic move into the software business.

Table 8.5. Philips' Industrial Collaboration for CD-i Software.

Name of venture	Partner(s)	Time of set up	Nature of venture
JIM	Pony Canyon/	1988	Joint venture
OptImage	Sun Microsystems/ Microware Systems Yamaha	1989	Joint venture
SPIN UK	Shell UK	1989	Joint venture
Denshi Media Service	Toppan Printing	–	Joint venture
Philips/CRC (CenturyResearch Centre)	CRC	1989	Cooperation
Maxwell Multi Media	Maxwell	1990	Joint venture
Philips/Bertelsmann	Bertelsmann	1990	Cooperation
Capitol Disc Interactive	Capitol Video Communications	1990	Joint venture
Telecity CD-i N.V.	LIM/NOB/Telecity N.V.	1990	Joint venture
Philips/RCS Milano	RCS Milano	1990	Cooperation
Philips/ISG (Interactive-Systems Group)	ISG	1990	Cooperation
Philips/Nintendo	Nintendo	1991	Cooperation
Philips/Sino United PublishingHoldings Ltd. (SUPH)	SUPH	1992	Cooperation
Fathom Pictures	Fathom Pictures	1992	Joint venture
Philips/Eaglevision	Eaglevision	1992	Cooperation
Philips/Echo Publishing Company (EPC)	EPC	1992	Cooperation

To enlarge the allied camp for CD-i, Philips has also made all of its software authoring tools available for any independent developer to license.

With a strong push by Philips, three independent groupings joined by CD-i hardware manufacturers and software developers have now been set up in the three major markets including Japan, North America and Europe. Firstly, CD-i Consortium Japan (CDICJ) was formed in March 1991, and has now been joined by up to 200 companies, of which 50 are hardware manufacturers. In the wake of the Japanese initiative, the CD-i Association of North America (CDIA) was announced in October 1991. Membership of the CDIA has now reached 100. As a sister organisation of CDICJ and CDIA, the European CD-i Association (ECDIA) was launched in April 1992. ECDIA has already been joined by over 70 European companies with an interest to explore the potential of the CD-i business.

Despite the fact that all three CD-i associations have declared their independent status, a strong influence from hardware manufacturers, in particular Philips, has been exerted on each of them. Both the CDIA and the ECDIA are headed by Philips managerial staff.[72] So far, CD-i is the largest camp in the consumer multimedia industry as far as the number of industrial supporters and licensees is concerned. Other systems, such as failed Commodore CDTV format and Matsushita's 3DO as well as the newly launched VIS system, have not been supported by a comparable business camp.

TECHNOLOGICAL IDENTIFICATION OF CD-I

To create a market for a new product requires the industry or company concerned to create a proper image for the new product in advance. In other words, if the concept or image of a new product is not clearly distinguished from other products, the new product may not be easily accepted by the consumers. In the consumer electronics industry, multimedia is an entirely new concept which cannot be easily compared with any conventional product. Accordingly, it is extremely difficult, as claimed by some industrialists, to explain the idea of multimedia to the consumer. Sometimes companies adopted very different approaches to present their systems to the market by attempting to give distinctive identities. The inabilities of Philips and Commodore to bring out full-motion video for their CD-i and CDTV products during the early stage and their differentiated views over the relationship between multimedia and computer products might well explain the complexity of the multimedia industry.

The Controversies over FMV

The first and, perhaps, most important technological aspect to identify a consumer multimedia system is full-motion video (FMV) since consumers are now accustomed to full-motion images from cinema, television and home video.

FMV is now regarded as one of the essential elements of any interactive CD system labelled as multimedia, it has been the major technical bottleneck facing every system since the beginning of multimedia development.

In the current multimedia industry, the dominant digital video compression technologies have been standardised through the MPEG (Motion Picture Experts Group of the International Standards Organisation) and International Electro-technical Commission.[73] MPEG-1 allows for up to one hour's full-screen full-motion video on a standard Compact Disc with a quality comparable to VHS.[74] DVI was a precursor of MPEG-1 and was favoured by computer and professional multimedia system developers; MPEG-1 has been widely adopted as the FMV standard for consumer multimedia applications. Both Philips and Commodore have announced that they will upgrade their CD-i and CDTV systems, respectively, in accordance with the MPEG-1 standard, although this still did not necessarily mean that the two formats would become compatible with each other. Philips has taken the lead in promoting the international standardisation of digital video compression technology through MPEG.

Before MPEG-1 was made available for licensing, CD-i had been under constant challenge from DVI because of the latter's capability to cope with FSFM video. By the time when DVI was announced with more than an hour's FSFM digital video compression, Philips had only achieved six minutes of FSFM video on its CD-i system by the end of the 1980s. Although afterwards Philips repeatedly told the industry that they had overcome the video compression difficulty and pushed CD-i FSFM video far beyond DVI,[75] the first consumer launch of CD-i in the U.S. failed to offer the consumer CD-i players with FMV. Instead, the first consumer CD-i players were built with an empty slot at the back of each machine, and buyers were expected to upgrade their CD-i machines after FMV has been made ready in the future. While the industry and first buyers were eagerly anticipating the upgrading of the CD-i system with FMV as the company had promised, Philips launched the same machines without FMV on the Japanese and European markets in April 1992. When CD-i was being launched in London, Philips, together with the FMV chip maker Motorola, made another firm promise to add FMV to its CD-i players by the end of 1992.[76] This indicates that Philips always believed FMV was a necessity but the company could not offer this crucial technology because the required digital compression technology was not yet ready when it was launching the CD-i system on a global base.

Without FSFM video, could CDTV or CD-i satisfy the consumer's expectation for multimedia? To justify the first CDTV launch without FSFM video, a senior Director from Commodore explained that

> Only those who are in the multimedia profession understand about motion video, partial screen and full screen. If you go out of here in Kensington high street and ask about multimedia, most people won't even know what multimedia is, not to say whether it is motion or not motion. So we must look at this from the consumer's standard point.[77]

This view, however, neglects the point, as mentioned above, that consumers are accustomed to seeing full-motion video on their televisions, even if they do not worry about the technology that produces it. Commodore also argued that if people want to sit down and watch movies they can have a video recorder and get all kinds of movies they want, because VCR works very well.[78] This official explanation about the lack of FSFM video on the CDTV system contradicts the fact that Commodore had been trying by all means to find a proper FSFM video solution for its multimedia system. As a matter of fact, Commodore was quick to licence C-Cube's MPEG technology when the latter finished its development work for the MPEG-1 chips.

After the CDTV launch Philips had no hesitation in pointing out that the Commodore machine was indeed inferior, partly because of the lack of and inability in handling FSFM video.[79] However, in fear of being left behind Commodore in the marketplace, Philips launched its CD-i system without FSFM video.

Intel's DVI was believed to have a consumer implication as far as FSFM was concerned. However, in contrast to Commodore and Philips, so far Intel has not built a consumer model for the marketplace. During various international multimedia events Intel has been keeping a relatively low profile and has made less publicity about its development work for DVI.

As far as software development is concerned, FMV might not be necessary for some applications, but this cannot deny the fact that FMV is vital to realisation of the full potential of interactive multimedia. At the first European CD-i Association General Meeting in London, a delegate from a German film company expressed criticism at the lack of FSFM video on the CD-i system for the time being. This film maker claimed that without FMV, video CD-i means nothing to the film industry in particular.

CD-i and the Computer

The second important aspect in identifying consumer multimedia is whether the system should be conceptually different from the computer, particularly games computers used in the home. Philips adopted a very different approach from that of Commodore at this point.

At the time of its commercial launch, Commodore made it very clear that, in connection with a television screen, CDTV was a new generation of consumer electronics product. Despite the fact that inside the box of a CDTV player there was a built-in computer operating system, Commodore managers insisted that CDTV should never be confused with a home computer. Commodore required electrical retailers to put CDTV players at a certain distance from the shelves where computers were displayed.

However, within less than a year after the first launch, Commodore admitted that they 'have made a big mistake'.[80] Consequently, they changed the name of CDTV into 'Amiga CDTV'. The reason why Commodore made this strategic change was that the company wanted to make CDTV a success on the basis of current Amiga computer users. To accompany this strategic shift, Commodore launched its A570 CD-ROM drive which was capable of running CDTV discs. This radical change has actually made CDTV become a peripheral of Amiga computers. Commodore regards this as 'double market opportunities'—Amiga computer users plus TV owners.[81]

Philips sees the future of multimedia in a different light. Philips anticipates the future success of CD-i to be realised on top of the huge colour television market. Therefore, the Philips commercial campaign slogan is '*Without CD-I you will lose half of your TV*'. For the time being Philips remains firmly committed to this 'TV approach' as far as the consumer market is concerned.

As mentioned above, multimedia is a mixture of audio, video, and computer. The built-in driving engine is a computer operating system or the 'brain' of a multimedia machine. However, physically, a consumer multimedia machine, such as the CD-i player or CDTV player, looks like a home VCR or CD deck in its design. It is so simple to operate that even a child can enjoy the system with a remote control, joy-stick or mouse. To some extent, consumer multimedia is a great breakthrough in the application of traditional computing technology to consumer electronics product design. It is a common objective for most consumer multimedia system builders to give the consumer the combined benefits of audio, video, and computer power in a multimedia format. However, Philips management has no intention of having their CD-i machine confused with a computer, and they believe that CD-i does not feel like a computer. Because of the simple design and easy user interface on the CD-i system, a Philips IMS manager claimed that,

> You don't even have to tell the consumer [about the operating system]. You just tell them to listen, and not to be worried about it. There is a big engine [inside the box] which is called computer. Forget about the word 'computer'. The system works by just pressing a single button. Simple and self-explanatory demo discs or software discs will explain themselves. The great advantage is that the consumers don't need to read a big booklet like the menus of MS-DOS and Lotus 1-2-3. For the latter you need to study the menus before you can enjoy the software. If you have a CD-i software, it will explain itself when it is being used.[82]

In accordance with the above Philips vision over the image of consumer multimedia products, CD-i machines are either displayed separately or in the CD players section in retail stores.

FROM LABORATORIES TO THE MARKETPLACE

It does not seem to be very difficult for innovative firms such as Philips and Sony to jump from one technology to another in their R&D laboratories. In order to bridge the CD-i world with the CD-ROM world, these two firms designed the CD-ROM XA format. However, it is a completely different matter to build an easy path for a new technology to move from its in-house birthplace into the marketplace, where technical superiority does not necessarily meet consumer preferences. First of all, which market—consumer or professional or educational market—should CD-i manufacturers start with? If it is the consumer market, is there an easy way for CD-i to reach the consumer market? What is the Philips management's perception about the future of CD-i?

Which Market?

Philips declared that CD-i is now technologically ready for large-volume production. However, hardware manufacturers have not reached a consensus on which specific market they should give the first priority or, in other words, where the products could be more likely accepted. It is no secret that Philips and Sony, the two leading firms which have been involved with the CD-i business for several years, hold substantially different views in understanding and choosing the priority marketing areas.

Philips sees that the future of CD-i mainly lies in the consumer market. In other words, Philips anticipates the future success of CD-i to be realised on the basis of the huge colour television market. For the time being Philips has been firmly committed to its 'TV approach', as discussed above, in conquering the consumer market. When Commodore was changing its CDTV marketing strategy and directing commentators' attention to its acclaimed large size of Amiga computer users' market in spring 1992, Philips was quick to point out the fact that the size of CTV users' market all over the world is much larger than that of not only the Amiga computers but also the whole computer users' market.

To Sony, the above controversies might not be immediately relevant because the company has been convinced of a completely different strategy for CD-i marketing. Despite the fact that Sony has collaborated with Philips in developing and finalising the CD-i system, Sony's vision over CD-i marketing is substantially different from Philips' marketing strategy. Sony believes that the first market opportunity of CD-i lies in the corporate market rather than the consumer market. During the occasion of *The 3rd Multimedia Conference on CD-I*, Sony only presented a portable CD-i model with a 4-inch colour LCD display at the exhibition in London at the end of April 1992. Sanyo and Kyocera of Japan did the same. These portable models, according to a senior Sony manager, were not aimed at the consumer market but at the corporate market simply because these machines would cost much more (up to £800–900 per unit) than an ordinary Philips CD-i player (£599 as a launching price).[83] It seems very unlikely that Sony would come into the consumer market in the near future. Obviously, Sony and Philips were collaborators at the R&D stage before CD-i was technically finalised; while in the marketplace the two companies would meet again as competitors, and Sony is not prepared to spend money on public campaigns for CD-i in a way that Philips has been doing. In other words, the two companies can join forces in their laboratories when it is necessary but they make money out of the research results in totally different ways.[84]

In autumn 1992, Philips also built and demonstrated its own portable CD-i model with a 6-inch colour LCD display. This Philips model was priced at £1,299, and was planned to satisfy the demand from the corporate market.

The difference between the above marketing strategies towards the launch of multimedia systems imposed great uncertainty on the emerging industry and might also cause confusions among the consumers, of whom the majority were not at all familiar with the term 'multimedia'. Philips' experience in launching its LaserVision system might be worth mentioning here. In recalling the reason why the LaserVision launch was not successful, a Philips spokesman from the company's Press Office said,

> I was there when the LaserVision system was being launched. The difficulty with LaserVision, as far as I see it, was that nobody at that time could decide whether it should be a consumer or professional system, and what field it should be used. It was a new thing. We could use it for education, for instruction of banking, and so on. That was the professional side of the system. Or, should it be used just for movies at home? The idea switched constantly between choosing the professional market and the consumer market. When we started launching, it was on the consumer side in the U.S. Then it turned out to be a wrong choice because it did not work. If we had chosen another way, things might have been different.[85]

It seems that the choice of the priority market area for launching a new product is not irrelevant to the final success of this product. However, at this stage it is still too early to say which company's decision was on the right track as far as Philips' and Sony' CD-i marketing strategies are concerned.

The 'Marketing Hurdle'

To identify the right market for a new product is one thing, but to persuade the consumer to buy the product is another. Due to the complexity of the CD-i system, which involves audio, video, graphics and text, it is alleged by manufacturers that to properly explain the functionality of the system seems extremely difficult. In particular, interactivity and *edutainment*, the most important features of a multimedia system, could hardly be advertised in a conventional way. Philips regarded this difficulty as a 'marketing hurdle' facing the current global CD-i campaign.[86] With regard to the introduction of CD-i, Timmer, the Philips President, acknowledged that,

> The introductory phase for such a product is long. Especially when, as with CD-i, it is a product that is difficult from a marketing point of view. CD-i cannot be explained on paper; you have to demonstrate it. This kind of introduction takes time.[87]

The introduction of CD-i in the marketplace was not only time-consuming but also costly. It was acknowledged by Philips that a great deal of money has been put into creating opportunities to attract the customers in front of the player and let them experience the technology themselves, because

> You can't do the public campaign [for CD-i] by way of conventional advertising. you need to put people in front of the technology, and show them what it does. That is where a lot of the marketing money in America will go. I think you will see the same in Europe and anywhere else. ... It does need a lot of work to put into the creation of the awareness of what the technology does; you need to educate the consumer. That is the problem we have.[88]

Apart from educating the consumer to create a public awareness of the CD-i technology, Philips also gave special training courses to the shop assistants working in the CD-i retail stores all over the world. According to Philips, the number of CD-i outlets was expected to reach 2,000 in the U.S. by the end of October 1992, and 5,000 in Europe by the end of the same year, not to mention those in Japan and the rest of the world.[89] As far as manpower and financial investment is concerned, it does not appear to be an easy task for Philips to offer training courses for so many outlets.

As the launch of CD-i involved spending a great deal of money,[90] according to a Philips manager from the U.S., Philips also tries to utilise the campaign as an opportunity to improve consumers' awareness of the Philips name.[91] Initially, all of Philips' CD-i equipment have been branded as 'PHILIPS'. Prior to the CD-i launch, Philips' consumer electronics products were also marketed under other brand names such as Magnavox in the U.S.[92] This new move could be of long-term importance for Philips to improve the popularity of its brand name in competition against the leading Japanese brand names such as Sony. This could be more or less a kind of compensation to the very costly worldwide CD-i campaign should the alleged spin-off effect be achieved as far as the brand name issue is concerned.

The 'Software Hurdle'

In parallel with the above discussed 'marketing hurdle', Philips has also been faced with a 'software hurdle'—the greatest challenge is, perhaps, the availability of software in making CD-i a successful consumer product.[93] To overcome this hurdle, as mentioned above, Philips has established a powerful managerial structure in which software development was given equal weight from inside; and, outside the company, Philips has formed a substantial number of strategic alliances and partnerships with publishing and entertainment companies to create software titles (as partly shown in Table 8.5) during the past few years. However, the level of complexity and difficulty involved in creating a single CD-i title is far greater than any other conventional electronic entertainment programme. Would Philips and the whole CD-i industry be able to jump over this 'software hurdle'?

To express his optimistic view about the future of CD-i, Timmer told a Press Conference,

> What is involved is the creation of a completely new medium [CD-i], in which we are building on the success of the Compact Disc. Only in a few years' time will we be able to judge the success of such a product.[94]

The above quotation represents a popular Philips perception that the success of CD-i could be built on that of CD-DA. How convincing is this claim? As discussed earlier in this chapter, one of the most important factors contributing to the overwhelming success of CD-DA was the support of the music industry or, in other words, the rapid increase of software availability. When Compact Disc was first launched in Japan in October 1982, some 150 disc titles were made available by PolyGram and CBS/Sony.[95] One year later, in October 1983, the number of CD titles reached 1,000. Within a few years' time the number of CD disc titles had boomed and the music industry was dominated by the Philips/Sony CD system.

When CD-i was launched in the U.K. and Japan at the end of April 1992, there were 34 titles on sale, of which the majority were chosen from the American catalogue, and only eight titles were in Japanese. The promised growth rate of software titles by Philips was an average five per month. A few months after the U.K. commercial launch, a mere 57 available titles were listed in the CD-i catalogue for summer/autumn 1992 sales (see Table 8.6).[96]

GLOBAL COMPETITION AND COLLABORATION FOR DIGITAL VIDEO DISC

Conflicts over format are almost inevitable when a new technology is about to become a consumer product nowadays. It is not an exaggeration to say that format battles are already a matter of life as far as the consumer electronics industry is concerned. Whilst competition between rival multimedia formats (e.g. CD-i and 3DO) in the consumer electronics market

Table 8.6. CD-i Titles (UK Market), Summer/Autumn 1992.

Category	Ready	Coming	Total
Games	12	6	18
Music	7		7
Kids	25	1	26
Special Interests	13	2	15
Total	57	9	66

Source: Philips IMS, Dorking, UK.

remains unresolved, another optical disc-based hi-tech format battle for Digital Video Disc (DVD) is already on the way. DVD is an optical disc of the standard compact disc size storing digitised film(s) or video programme(s), currently being promoted by leading consumer electronics manufacturers, the film makers and the computer industry. The competition in the newly emergent DVD industry is between two main camps: one side is led by Philips and Sony, and the other is led by Toshiba and Time Warner. This new format battle is reminiscent of the VCR format battle of the late 1970s and early 1980s, as discussed in Chapter 4. Despite the alternative effort made by JVC to bring out a digital version of its VHS, called DVHS,[97] DVD and its recordable successors have been either developed or proposed by electronics and entertainment companies in the hope of capturing a potentially lucrative market for digital video in the home.

In collaboration with Sony, Philips experienced huge success in launching the Compact Disc-Digital Audio (CD-DA) standard with hardly any competition due to the nonexistence of alternative standards for the same technology. But subsequent developments of CD-based products/systems have been subject to format battles or low level of compatibility with each other. CD-ROM, the 'Yellow Book' standard, has many varieties to meet the computer industry's need for information/data storage. As a further development of CD-DA and CD-ROM technologies, CD-i, the third generation of CD-based technology jointly defined by Philips and Sony in their 'Green Book', is only one of the competing formats for the multimedia industry. As discussed earlier in this chapter, most multimedia systems such as CD-i, CDTV (later CD-32) and 3DO are all capable of offering full-screen full-motion digital video mixed with other types of information stored on the same compact disc. The current competition for DVD is a new stage of development in the area of digital video technology, which is an important part of the emergent multimedia industry.

Announced in 1994, the Philips/Sony standard for DVD is called Multimedia CD (MMCD), which was initially proposed by Philips, Sony and 3M as a new generation of computer data storage medium to replace CD-ROM but also has an application for the consumer electronics industry. There are two types of MMCD: a single-layered disc with the capacity of storing up to 3.7 gigbytes of data or 135 minutes of digital video of broadcasting quality (equivalent to five times the capacity of today's standard compact disc for digital audio and CD-ROM). The second type of MMCD is a dual-layer disc capable of storing 7.4 gigbytes (or 270 minutes of digital video with broadcasting quality) which doubles the data capacity of the single-layered disc.

The Super Density DVD system was announced in early 1995 by the SD Alliance. From a technical point of view, the SD (Super Density) disc system offers more data capacity than the MMCD disc. An SD disc has two sides, which are created by bonding together two super-thin discs with half the thickness of a normal CD disc. Two sides of the SD disc are capable of storing up to 10 gigbytes data—5 gigbytes on each side. It is believed that the ability of the SD Alliance to fit a feature film on one side of a disc was the main feature which has attracted the support of the Hollywood film studios.[98]

Due to differences in many aspects (e.g. the material and structure of the disc, the way the disc is read by the laser beam etc.), MMCD and SD are not compatible with each other. In order to avoid a format battle, attempts have been made by the two camps to explore a common standard for DVD.[99] It is also the hope of the computer industry that a single DVD standard should be adopted. A group of leading American hardware and software companies, including Apple, IBM, Compaq, Hewlett-Packard and Microsoft, have claimed that both the MMCD and SD formats should meet nine requirements for DVD, these are:

- single standard for TV- and PC-based applications;
- backward compatibility with current CDs;
- forward compatibility with future read/write and write-once discs;
- a single file system for both entertainment and PC-based content and for read-only, read/write and write-once discs;
- cost comparable to current CD-ROM drives and discs at equivalent volumes;
- no mandatory container, i.e. caddy or cartridge;
- reliable data storage and retrieval for read-only, read/write and write-once media;
- high capacity extendible with future enhancement;
- high performance for both sequential (e.g. movies) and nonsequential (e.g. random-access computer data) files.[100]

A format battle could have been avoided had the computer industry's 'wish list' indicated above been met. However, the conflicting interests of the two sides have so far excluded any immediate hope of standardisation. It appears that another bloody format battle is gathering momentum. The ongoing digital revolution suggests that digital video, rather than analogue video represented by the VHS standard, will be the future. It is also a convincing trend that digital video will be distributed not only in the consumer electronics market but also in the computer and communications sectors. In other words, digital video in the future will be an integral part of an interconnected open 'infotainment' network, and a leading supplier of digital video equipment or technologies would have a strong foothold in the future information and communications industry. This is perhaps the primary reason why a compromise between the Philips/Sony camp and that of Toshiba/Time Warner has not been reached—every company prefers being the technology/market leader to becoming a technology follower. However, as happened to the VCR industry, the market may choose only one format and the short- and long-term consequences on the loser can be extremely devastating. Both Philips and Sony have already had bitter experiences with their V2000 and Betamax systems, respectively. More importantly, the rivalry between competing standards for the same market could be a double-edged sword. In the early 1990s, Philips developed its DCC (Digital Compact Cassette) system and Sony launched the Mini Disc format. Both systems were intended by the two companies to be recordable/erasable digital audio media. So far the two systems have failed to make substantial market penetration, given the fact that both DCC and Mini Disc are technically advanced. In a similar way, the Philips/Sony camp and the SD Alliance are at the moment taking great risks to engage themselves in a format battle for DVD. The chance for both the MMCD format and the SD disc to succeed seems to be extremely slim.

The industry response to the forthcoming DVD format battle at the moment is mixed. For example, Gateway 2000, a leading American PC manufacturer, announced in June 1995 that it will commit itself to the Philips/Sony standard by integrating the latter's next generation Multimedia CD drive into its computers. Gateway 2000 managers believed that 'the Philips/Sony MMCD format offers the best cost- and time-to-market and superior backward compatibility'.[101] In addition to the support from the CD-ROM drive industry, support for the Philips/Sony camp has been voiced by the Lion Group, a giant Malaysian conglomerate. The reason why the computer peripheral equipment manufacturers support Philips and Sony lies mainly in that the MMCD format is backward compatible with conventional CD-ROM drives. In other words, the proposed MMCD drives would not make the widely used CD-ROM products obsolete. The number of supporters on the SD Alliance side is also growing. Among others, Warner Music Group and Warner Home Video, both Time Warner subsidiaries, have formed their 'Advanced Media Operations' to make SD discs. In addition, two Korean companies, Samsung and SKC, have recently announced that they intend to back the SD format. But the decisive force behind the SD Alliance at the moment is the support from the film industry.

If there is anything in common between the strategies of the Philips/Sony consortium and the SD Alliance, it would be strategic alliance. The history of the consumer electronics industry since the 1970s shows that any format battle for new technology may not be won by a single firm without the help of strategic alliance or industrial collaboration. Both sides of the DVD format battle may have learned the lesson from history and they are trying to extend the range of strategic alliances to as many companies as possible. There are also differences in terms of competitive strategies between the two rival consortia for DVD.

Judging from the sectoral specialities of the main members of each side, the SD Alliance has gained more support from a host of Hollywood studios and, therefore, stress the strategic importance of marrying hardware and software in order to benefit from the synergy between the two aspects. This is confirmed by a senior vice-president of Thomson Multimedia:

> This [DVD] is the biggest leap forward in video technology since the introduction of the VCR. It is a symbol of an era of new cooperation between software and hardware companies.[102]

Obviously the SD Alliance is similar to JVC's strategies during the VCR format battle in the late 1970s and early 1980s. As argued in Chapter 4, one of the major elements of JVC's strategy in winning the VCR format battle was the company's success in gaining the support of the film companies, who agreed to put their titles onto VHS tapes for rental. If a range of Hollywood studios put a large number of films onto the forthcoming digital video discs of the SD Alliance format, this may well become an important competitive advantage against the Philips/Sony camp. Compared to the SD Alliance, the Philips/Sony side have their strategic focus on technological convergence, the most fashionable concept of the new information and communications sector. Until now Philips and Sony, two leading consumer electronics manufacturers pioneers in optical disc technologies, have already secured support from several leading computer and CD-ROM drive manufacturers. This suggests that Philips and Sony have proposed their DVD standard to benefit from not only the consumer electronics industry but also the computer sector, although mainly the computer peripheral equipment suppliers. These two industries, as discussed earlier, have already penetrated each other via a number of multimedia products.[103]

Despite the antagonism between the Philips/Sony consortium and the SD Alliance for high density CDs, a compatible erasable CD (CD-E) format for conventional density disc has been agreed by a group of ten companies and, among others, Philips, Sony and Matsushita from the two competing camps for high density CD, are leading members. It is expected that CD-E drives will be made available from early 1996, and they will be able to play back all current CD-ROMs.[104] There is hardly any evidence suggesting that this single standard for conventional density CD-E can be extended to a single format for high density digital video disc. On the contrary, immediately after the conventional density CD-E standard, the SD Alliance recently announced that they had developed an erasable/rewritable high density CD system, which is cable of recording 2.6 gigbytes of data on each side of the double-sided disc.[105] This move may lead to an extension of the current format battle for DVD to erasable high density CD between the two rival camps and, therefore, the optical disc industry will be further divided in the years to come.

Overall, the ongoing global competition and industrial collaboration for digital video disc is opening a new area for the development of the consumer multimedia and the computer industries. However, before the consumers see the benefits from new innovations in optical disc technologies, DVD also adds more confusion to the fledgling multimedia industry. Firstly, disagreement between the two competing camps over DVD standards is a negative factor for building up the consumer's confidence in the new technology. This will undoubtedly prolong the taking off period for DVD products, if there were a market for them. Secondly, competition from alternative technologies should not be underestimated. As mentioned earlier, given the involvement in DVD by its parent company, i.e. Matsushita, JVC has announced its DVHS format—a digital version of its VHS home video system. JVC's proposed retail price of £250 for the DVHS machine, if proved realistic, will be a strong challenge to either the MMCD system or the SD format or both. Any future recordable version of DVD will unlikely be priced at lower than £250. Third, at the moment neither MMCD nor SD discs capable of recording programmes whilst the DVHS system offers playback and recording features. The ability to record off-air programmes would be an important function for any new generation of video machine, given the rapid development of multichannel digital satellite broadcasting in many parts of the world at the moment. Nevertheless, the future for the two rival DVD systems remains uncertain.

CONCLUSION

To conclude this chapter, I would like to emphasise that, firstly, there seems to be a long way to go before the consumer multimedia industry is completely standardised, not to mention the whole multimedia sector. In other words, several incompatible formats would be competing against each other in the marketplace with hardly any strong government intervention. A single global multimedia standard for the home market has yet to be seen. Being the proprietary patent owner of the first available consumer multimedia system, the CDTV format, Commodore initially intended to conquer the consumer market and establish CDTV as the world standard. As discussed above, Commodore changed its strategy and diverted CDTV to the Amiga computer users market. Although Commodore managers, such as those interviewed, were enthusiastic to mention the number of Amiga computers in use, Amiga computers have only achieved considerable market penetration in the U.K. and some other European countries. In other crucial markets including the U.S. and Japan, the Amiga computer is not popular at all. Maybe because of this reason, nobody from Commodore talked about CDTV as a world standard any longer, and CDTV was dropped in mid-1993. In commenting on the consumer multimedia standardisation issue, a senior manager from Commodore expressed that a single world standard seems to be impossible; on the contrary, many incompatible systems would coexist and compete with each other in the marketplace.[106]

Secondly, from a purely technical point of view, the process of standardisation for multimedia products will be deeply affected by digital video compression technologies. On the one hand, there does seem to be a *de facto* standardisation process for digital compression technologies through the ISO's MPEG group. On the other hand, although some multimedia systems adopted the same video compression standard, they still remain incompatible with each other. Both Philips and Commodore have licensed and supported the MPEG standards, as mentioned earlier in this chapter, but this did not change the fact that CD-i and CDTV were two independent rival formats. In the meantime, the development of digital video disc by the Philips/Sony consortium and the SD Alliance since 1994 has added further confusion to the standardisation process for optical disc-based digital video and multimedia products. Similar to the competition process involving CD-i, CDTV (now CD32) and 3DO, a long-term contest between the MMCD and SD disc formats (if not further contenders come up) in the marketplace is anticipated. This would inevitably prolong the taking off period for the DVD industry.

Thirdly, as far as the CD-i campaign is concerned, Philips has created a huge network of international collaboration and partnerships with other firms in order to get a wide range of industrial support for hardware and software development. This was in part Philips' new corporate strategies established by the company's management during the 1980s as discussed in Chapter 5. The very costly CD-i project might not be able to bring immediate financial benefits to Philips, but its influence on further changing the company's corporate culture and strategic manoeuvring appears to be profound. In promoting CD-i, Philips was not fighting alone but together with a large team of hardware manufacturers and software developers all over the world. This is in stark contrast to those years when Philips was suffering from a lack of industrial support for its V2000 technology.

Fourthly, the global competition process for multimedia technologies is an ongoing one. It has become increasingly complicated as new contenders are coming up to join the battle. Apart from the aforementioned 3DO system established in part by Matsushita and the Tandy/Microsoft/Zenith alliance for VIS, IBM and Sony recently unveiled their cooperation to promote Sony's standard for CD-based multimedia technology. It was speculated that the IBM/Sony deal could possibly create the basis for a powerful electronic industry alliance to support Sony's new consumer electronics products.[107] It seems that this new move has imposed another serious challenge on the fledgling CD-i system.[108] Hypothetically, the IBM/Sony alliance sheds more light on demystifying Sony's attitude towards CD-i: Sony might be interested in collaborating with Philips for CD-i, but it was more keen on setting up its own multimedia standard for not only the consumer electronics industry but also the computer industry.[109] Compared to Philips, Sony has much less stake in the CD-i business anyway. While Philips, as far as multimedia is concerned, has been solely committed to the CD-i system, Sony has already expanded its multimedia ambition into several camps—with Philips for CD-i, with IBM for its own multimedia standard, and with Apple for the PDA series. The previously mentioned cooperation between Philips and Nintendo appeared to be a good signal for CD-i.

Finally, no matter how confusing the multimedia industry is, pioneering firms such as Philips, Commodore (disappeared now from the industry), Matsushita, Apple, Sony, IBM and many others have been trying to revolutionise the way of home entertainment. The idea of combining digital audio, video, graphics and text on a single compact disc and offering a high level of human–machine interface is pushing the consumer electronics industry into a new era. Multimedia, in a narrow sense, puts together the previously separate gadgets (e.g. Hi-Fi system, VCR, TV, and home computer) on a single system and hence integrates the consumer electronics technologies. From a much wider perspective, multimedia has also been blurring the demarcation between the consumer electronics industry, semiconductor industry, computer industry, and telecommunication industry and this potentially integrates the whole IT sector. This process is the so-called 'technological convergence'. Together with the ongoing digital TV and digital HDTV revolution, as suggested in Chapters 6 and 7, multimedia development may eventually create a borderless IT world, in which 'edutainment', 'infotainment', 'teleworking', 'personal comms', 'home cinema', 'interactive TV', 'Video-on-Demand' 'global village' linked by an information superhighway, etc., will become some of the major elements of learning, working and entertainment in a new information society.

NOTES

1. Note that, by the time of writing, chips based on MPEG-1 and MPEG-2, two international standards for full-motion digital video, have already been made by electronics companies such as C-Cube Microsystems. MPEG-1 offers motion video with a quality level equivalent to that of VHS video; MPEG-2 is video of broadcast quality.
2. As part of its long-term strategy, Philips wants to achieve, by the year 2000, the goal of generating 25% of its turnover in media, of which the CD-i electronic multimedia project is a key area. For other key technology areas, see Chapter 5.
3. On the one hand, as a convention, most people use 'disk' to refer to magnetic devices and 'disc' to refer to optical devices. On the other hand, some people suggest using 'disc' to name consumer products, and 'disk' for nonconsumer products. Here, in this chapter (as well as other chapters of this book), 'disc' is used throughout to denote various products related to optical technologies.
4. The IT sector as a whole includes four major branches: the semiconductor industry, computer industry, consumer electronics industry, and telecommunications industry.
5. Philips started R&D work on optical disc technology in its laboratories in 1969, and first demonstrated the LaserVision system to the consumer electronics industry in autumn 1972. This event firmly established Philips as the industry leader in optical disc technologies ever since.
6. For more details, see Helgerson, L.W. (1987), *Introduction to Optical Technology*, Image Technology Consumer Handbook, The Association for Information and Image Management, Printed in the U.S.

7. Par, P. (1991), 'The Long and Winning Road', *Complete CD & Hi-Fi Buyer*, March.
8. See Feldman, T. (1991), *Multimedia in the 1990s*, BNB Research Fund Report, No. 54, the British Library, p. 42.
9. Par (1991), *op. cit.*
10. CD-V is a hybrid product comprising part CD and part LaserVision and, therefore, offers digital sound and analogue video clips on a standard 5 inch compact disc. Philips used CD-V, later Laser Disc, to name the full range of optical audio and videodisc. In other words, CD-V is a compact disc with a video segment. See BIS Mackintosh (1988) for more detailed introduction about CD-V.
11. See BIS Mackintosh (1988).
12. The same source also indicates that, until November 1989, 36 hardware manufacturers and 53 software producers had been registered in the LaserVision Association Pacific, and 18 companies were selling LaserVision players.
13. BIS Mackintosh (1988).
14. Lang (1990), p. 53, and BIS Mackintosh (1988).
15. Nayak, P.R. and Ketteringham, J.M., 'Philips' Compact Disc: The Dutch Giant Thinks Small', a research paper supplied by Philips Interactive Media Systems (No date), Eindhoven, p. 13.
16. *Ibid.*, p. 21.
17. Press Information on Compact Disc from Philips.
18. Nayak and Ketteringham, *op. cit.*, p. 25.
19. *Ibid.*
20. Sony's stake in CBS records and Philips' in PolyGram helped to ensure that at least two leading music companies were prepared to back the new format.
21. Information in this paragraph is mainly from documents supplied by Philips Press Office.
22. Presentation by J.D. Timmer, President of Philips, at the First European CD-i Conference held in London during 18–19 June 1990. BIS Mackintosh research result indicates that 75 million CD players were in use in 1989 and the world sales of the same year was 28 million units.—See *Mackintosh Report* (1990).
23. See *Financial Times*, 6 March 1991.
24. It is a widely known weakness of the 'Yellow Book' that neither the Philips–Sony alliance nor the High Sierra Group could manage to fully define the CD-ROM standard, i.e. to cover file handling and data retrieval.
25. Fox, B. (1991), 'Opening the Door to Interactive CD', *Complete CD & Hi-Fi Buyer*, March.
26. Note that CDTV is now renamed CD-32.
27. That is, in contrast to a linear mode of presentation, interactive multimedia puts the user in control of the manner in which the information is accessed.
28. *Electrical and Radio Trading*, 28 June 1990.
29. Interview with Director of Special Projects, Commodore International, London, 21 May 1991. Actually, the promised number of CDTV software titles was not realised. By June 1992, there were only 81 titles available on sale, according to Commodore's press information.
30. See Bruno, R. (1987), 'Making Compact Disks Interactive', *IEEE Spectrum*, November.

31. Photo CD discs can be played back on a CD-i machine, but CD-i discs cannot be operated on a Photo CD player. Because the Photo CD system conforms to the 'Red Book' specifications, it is capable of playing CD-DA discs.
32. Instead using Compact Disc to store images, the Canon Ion still video camera records up to 50 still video images on each floppy disk. Through Canon's Ion-PC or Ion-Mac kits, the Ion still video images can be stored on a computer for desk top publishing applications. See *The Guardian*, 7 May 1992.
33. In 1987, General Electric and Thomson signed an agreement, under which the former would purchase the latter's medical equipment business and, in return, the latter would buy the former's consumer electronics business. The interesting point during this transaction was that GE retained its DVI business rather than selling it to Thomson. To explain this GE gave its reason by saying that it was 'making a major commitment to commercialize and promote the technology' (*CD-I News*, September 1987, p. 2.). However, one year later, in 1988, GE sold off its DVI technology to Intel. In the meantime, Thomson announced that it would manufacture and introduce CD-i hardware in 1988.
34. *Screen Digest*, January 1991.
35. *CD-I News*, April 1987.
36. Intel has declared that the cost of adding DVI to a PC would fall to $500 by 1992.—See *Financial Times*, 11 October 1990.
37. *New Scientist*, 31 March 1990.
38. See *Complete CD & Hi-Fi Buyer*, April 1991.
39. *Financial Times*, 24 June 1992. Note that, until the mid-1990s, Apple's PDA products have failed to take off in the market-place.
40. See *Screen Digest*, October 1992, p. 220.
41. *Ibid.*, November 1992, pp. 254–255.
42. *The Nikkei Weekly*, 22 February 1992.
43. See *Information World Review*, No. 71, June 1992, p. 15.
44. There are some overlaps between these three categories. For instance, there are several portable CD-i models (Sony, Kyocera, Sanyo, and Philips) designed with LCD displays, and they are mainly sold to the corporate market for the time being because of their high prices. Consumer multimedia products also have professional models.
45. Interview with Manager of Multimedia Solution Centre, IBM (U.K.) Ltd., London, 14 October 1992.
46. Interview with Vice President (Marketing), C-Cube Microsystems, London, 14 October 1992.
47. Note that the camcorder industry is currently dominated by two competing formats: JVC's VHSC system and Sony's 8mm system.
48. Interview with Senior BU Manager, Philips IMS, Eindhoven, 17 October 1991.
49. *Ibid.*
50. *Ibid.*
51. Interview with Vice President (Marketing), C-Cube Microsystems, London, 14 October 1992.
52. Stanford-Smith, B. *et al.* eds (1987), *Report on The Intensive Workshop on CD-ROM*, A Joint Project by the CEC (DG XIIIB) and EURIPA, July.
53. IMS *Newsletter*, No.3, 20 July 1990, Philips Consumer Electronics B.V., Eindhoven.

54. See Presentation by I. Maxwell at *The European CD-I Conference* held in London during 18–19 June 1990.
55. See *CD-I Bulletin*, the Publication of the European CD-i Association, Vol.1, No.1, May 1990.
56. See *X III Magazine: News Review*, Issue Number 4/92, p. 9.
57. *Ibid.*
58. See *The Times*, April 25, 1991.
59. Information in this section comes mainly from *Philips News*, Vol.20, No.8, 17 June 1991, and Vol.21, No.1, 20 January 1992.
60. CD-i, DCC (Digital Compact Cassette) and HDTV (High Definition Television) are regarded as the three key consumer electronics projects within the Philips group. These three new technological areas are also expected to play a substantial role in further development of the consumer electronics industry as a whole.
61. When asked about the future market potential of the CD-i system, a number of Philips managers from the U.K. and Eindhoven of The Netherlands used the same argument during interviews that CD-i would be successful because CD-DA was a great success.
62. It was estimated that so far Philips invested $200m in the production of CD-i software alone.—*Screen Digest*, November 1992, p. 251.
63. In the same way, other companies have also claimed their systems as global products.
64. *The Economist*, 2 May 1987, p. 72.
65. Interview with Senior BU Manager, Philips IMS, Eindhoven, 17 October 1991.
66. *Ibid.*
67. *Ibid.*
68. *Philips News*, Vol.21, No.11, 7 September 1992.
69. *Zhongyang Ribao* [Central Daily News, Taiwan], 14 August 1992.
70. Philips' new digital tape technology DCC was launched in collaboration with Matsushita. Most interestingly, Philips and Sony reached an agreement in October 1991 to support each other's new digital audio system, DCC and Mini Disc which are believed to be two rival systems competing for the next generation of world standard for the digital music industry.
71. *Philips News*, Vol.21, No.8, 22 June 1992.
72. The Chairman of the Board of CDIA is B. Luskin, who is also the President of Philips Interactive Media of America; while J. Lynn-Evans, the Acting Chairman of ECDIA, was a manager from Philips (U.K.) based in London. Note that Evans left Philips in 1995.
73. Intel recently coined another name, called 'Indeo', for its digital video technology to be used by Microsoft for video application programmes running on Windows. For more detailed report see *Financial Times*, 12 November 1992.
74. A senior manager from C-Cube Microsystems, leading chip-maker for digital video, mentioned that the current MPEG-1 standard offers a picture resolution equivalent to one-fourth (or 352×240) of that a normal PAL TV screen can offer, and the next generation of this standard, called MPEG-2, which was promised to be available by June 1993, could offer the resolution of PAL TV pictures. Interview with Vice President (Marketing), C-Cube Microsystems, London, 14 October 1992.

75. Fox (1991), *op. cit.*
76. FMV finally became available as a £150 add-on in late 1993, and among the first software titles available were conventional (linear) movies including *Top Gun*.
77. Interview with Director of Special Projects, Commodore International Ltd., London, 21 May 1991.
78. *Ibid.*
79. Interview with Senior Manager for Corporate Strategic Planning, Philips International B.V., Eindhoven, 29 October 1991.
80. Interview with General Manager of CDTV Development Europe, Commodore International, London, 9 June 1992.
81. CDTV was dropped in 1993 when Commodore decided to launch the CD-32, a new games console based on a new version of the Amiga.
82. Interview with Senior BU Manager, Philips IMS, Eindhoven, 17 October 1991.
83. Interview with General Manager of the Video Disc Player Group at Sony Corp. of Japan, London, 29 April 1992.
84. *Ibid.*
85. Interview with Manager of Philips Press Office, Eindhoven, 30 October 1991.
86. *Philips News*, Vol.21, No.6, 11 May 1992.
87. Cited in *Philips News*, Vol.21, No.10, 10 August 1992.
88. Interview with Manager of CD-i Consumer, Philips IMS, London, 25 June 1991.
89. *Philips News*, Vol.21, No.13, 19 October 1992.
90. It was estimated that Philips' marketing budget spent on CD-i in the U.S. alone was over $20m in one year's time from October 1991. See *Screen Digest*, October 1992.
91. Cited in *Philips News*, Vol.20, No.13, 4 November 1991.
92. But Philips reverted to the Magnavox brand for low cost CD-i players in the U.S. in November 1993.
93. See *Philips News*, Vol.21, No.6, 11 May 1992.
94. Cited in *Philips News*, Vol.21, No.10, 10 August 1992.
95. Figures in this section are mainly from Philips Press Information.
96. It is important to note that music software could be readily and cheaply transferred to CD-DA format. In the case of CD-i, however, *new* titles have to be created and new skills have to be developed.
97. In the beginning of April 1995 JVC announced its new generation of VCR, the DVHS system, which was promised to record simultaneously up to six different digital TV channels; types of the same length can be used to record digital programmes of between three and a half and 49 hours depending on the recording qualities required. The DVHS machine takes digital TV signals from satellite, cable or terrestrial antennae and plays back from the tape into a set-top box, which will decompress and convert the information from digital format into analogue TV pictures. The cost for such a machine, as promised, will be about 250 British pounds. This is much cheater than most top range VHS machines in the shops today. For more details see *The Sunday Times,* 9 April 1995.
98. *The Observer,* 4 June 1995.

99. In early summer 1995 Philips had negotiations with Toshiba to explore the possibility of merging MMCD and SD to made a single standard. Talks between the two sides ended without any agreement.
100. Reported in *Screen Digest,* June 1995, p. 139.
101. Quoted in *Screen Digest,* July 1995, p. 149.
102. Quoted in *The Observer,* 4 June 1995.
103. For instance, as a consumer electronics product the CD-i machine actually has a computer CPU (Central Processing Unit) in it. In the meantime, personal computer is getting increasingly popular in the home to play games as well as other functionalities. Some computer companies have developed new products, such as ICL's PCTV, to combine personal computers with TV screens for home use.
104. *Screen Digest,* June 1995, p. 140.
105. *Ibid.,* p. 139.
106. Interview with General Manager of CDTV Development Europe, Commodore, London, 9 June 1992.
107. See *Financial Times,* 10 November 1992.
108. If Sony extends its strategic alliance strategy for its CD-ROM-based multimedia technologies to cover the newly launched Mini Disc technology, the survival of Philips' DCC system might be affected as well.
109. Sony recently announced its intention to enter the computer games market in competition with Sega and Nintendo.

9

CONCLUSIONS

The major concern of this book was to show how new technologies and the competitiveness of firms were affected by various forces from the corporate level, industry level and government or supragovernment level within the context of the European consumer electronics industry since the 1970s.

Chapter 2 provided a literature review covering a wide range of theoretical perspectives showing differentiated understandings about the effectiveness of corporate strategies, industrial collaboration and government policy in promoting new technologies and improving industrial competitiveness. The role played by different forces at the corporate level, industry level and national or supranational government level, as well as the interplay between these forces, in the process of managing new technologies were summarised by two 'conceptual frameworks' in Chapter 2. The issues raised in these theoretical discussions have been further investigated in the subsequent chapters, where a great deal of empirical evidence related to the circumstances inside and outside Philips was analysed.

Drawing upon the experience of Philips, this book has reported the results of research on a number of consumer electronics technologies including the V2000 VCR format, the proposed European HDTV standard, i.e. HD-MAC, the CD-i multimedia system and the competition for digital video disc standardisation. It was shown in this book that the global competition and collaboration between firms for these new technologies was closely associated with government–business relations, which were embodied in the interplay between corporate strategies and public policy. In an attempt to assess to what extent government can intervene in the process of technical change, Chapters 5 and 7 have shown that government technology policy should not preclude competition in choosing new technologies in the marketplace. In order to provide a background knowledge for better understanding the complicated strategic manoeuvring over these technologies at the corporate level, I have also examined the historical growth of Philips company and its organisational restructuring process.

This concluding chapter summarizes the major findings of the study and indicates some suggestions for future research.

CORPORATE STRATEGY AND TECHNICAL CHANGE

It has been shown that the corporate strategies adopted by firms are of great significance to the success or failure of their new technologies and new products. Different ways of understanding the nature of corporate strategy in the literature since the 1960s, as discussed in Chapter 2, partly reflected the historical changes in firms' strategies and the global competitive environment.

Having presented a historical account of Philips' corporate development since the last few years of the 19th century to the 1960s, Chapter 3 indicated that the company had followed an expansionist strategy characterised by its rapid technical/product diversification and organisational stretch on a global scale. In accordance with this strategy, the Philips management strongly emphasised the importance of technological superiority over competitors based on its own in-house R&D activities. Philips had successfully gained international competitiveness *vis-à-vis* European and American firms and grown from a small Dutch family business into a large multinational corporation in the electric lighting industry by the late 1960s and early 1970s. However, Philips failed to translate its success in the lighting industry into every major area of its consumer electronics operations.

By the 1970s, the Japanese consumer electronics industry had become a fully-fledged competitive force, and the major firms began to move from their home market to other highly competitive markets, particularly the U.S. and Western Europe. This newly emerged Japanese competition was underestimated by the Philips management until the early 1980s, when the company's technologically superior V2000 home video system lost the format battle to JVC's VHS system, as discussed in Chapter 4.

Philips' V2000 failure provoked a wide range of criticisms about the company's long-standing strategy which features technical diversification and organisational expansion. Some commentators believed that Philips, as well as Thomson Consumer Electronics, have been good at mastering new technologies but poor at marketing the products that contain them, and this is regarded as one of the 'European flaws', so long as the consumer electronics industry is concerned. This research has demonstrated that, as presented in Chapter 3, such a criticism may be applicable to some later circumstances but could not be justified by Philips' own history prior to the 1970s. To be good at mastering new technologies, I believe, was a major impetus for Philips' growth and success in a number of areas, rather than the reason why the company has failed in marketing some of its innovative products. The setbacks Philips suffered have a much more complicated structural origin.

RATIONALISATION AND ORGANISATIONAL CHANGES OF THE FIRM

Philips has inherited a long-standing structural problem from its own history. It has been found during this study that Philips' organisational structure was one of the major factors affecting the company's efficiency and competitiveness in commercialising its new technologies. This problem arose, sometimes, because of various changes in the external competitive environment; but it stemmed mainly from the company's own strategies.

On the one hand, it was found that Philips was established with a dual managerial structure, in which commercial (marketing) management and engineering (technical) management developed along two parallel hierarchies initially headed by the two Philips brothers, Gerard and Anton. This unique managerial structure was followed by the coexistence of the NOs (National Organisations) and PDs (Product Divisions) from the Second World War and throughout the following decades. Accordingly, the Philips organisation has grown into a 'matrix' structure—with the NOs as the rows and PDs the columns—which lasted for several decades. The major features of the Philips matrix in the early days was discussed in Chapter 3.

Although matrix structure became a fashionable management style for big multinational firms in the 1970s and early 1980s, conflicts always existed between the commercial side and technical side within the Philips matrix structure. As the empirical studies have shown in Chapter 5, these conflicts greatly reduced the accountabilities of each side and necessary communication between the technical people and the marketing people within Philips was sometimes blocked. This partly explained the inefficiency of the Philips management in launching new products. As an important part of Philips' restructuring programme, 'tilting the matrix' has been one of the top priorities which preoccupied the top management since the 1980s. It was found (in Chapter 5) that the decision-making power of the NOs has been substantially transferred to the PDs and, consequently, the latter are responsible for not only production but also marketing all over the world. In contrast, the previously powerful NOs have now been left with the sole responsibility for administering public relations and legal affairs in each country where Philips has business operations.

On the other hand, Philips' expansionist strategy had enabled the company to operate in over 60 countries all over the world by as early as the 1960s. One of the most obvious drawbacks of this strategy was the increased autonomy of the National Organisations, which were located far away from their parent headquarters. This autonomy held by the NOs reached such a point that some of them became *de facto* independent companies during and after the Second World War in response to the Nazi occupation of The Netherlands. Some commentators regarded Philips as a 'loosely organised federation', in which the Philips headquarters in Eindhoven, to some extent, lost control over some national subsidiaries. Chapter 4 has suggested that, among many other factors, the rejection of the V2000 system by Philips North America significantly contributed to the eventual failure of the Philips technology during the VCR format battle. Many readers would find it surprising that a national branch had the power to refuse to follow its parent company's strategy for an important new technology against fierce competition from the alternative Japanese systems.

As discussed in Chapter 2, some recent managerial theories reject the 'headquarters mentality' and have proposed a decentralised and highly flexible management structure model; others advocate a 'new federalist' theory suggesting a 'dispersed centre' through 'distributing power' to each branch of a multinational corporation. These approaches do not easily explain the Philips case. Departing from a decentralist path, Philips has taken another path to reform its own corporate structure: the current President assumes direct responsibilities for some core technologies such as CD-i and DCC; the corporate headquarters in Eindhoven has reasserted firm control over all of the subsidiaries (including Philips North America) by significantly stripping off the decision-making power of each National Organisation.

By the time of writing, the mid-1990s, Philips appeared to be a much healthier company compared to the situation of the early 1990s. Having seen a profitable year in 1994, the company reported a significant improvement of net profit in the first quarter of 1995.[1] From a geographical point of view, further financial improvement, as indicated by the President of Philips, would be largely contingent upon the company's performance in the Asia-Pacific region. Research reported in this book shows that it is Philips' long-term strategic target set in the early days of its restructuring process to have one-third of turnover generated in Asia. This strategic target may well be realised in the 1990s. Recent Philips forecast suggest that about 25% of the company's sales would be achieved by its operations in the Asia-Pacific region by the year 1998, a sharp rise from the 15% of 1994 and a three-fold increase on the 8% of 1990.[2]

THE PRINCIPLE OF HOSTILE BROTHERHOOD[3]

While many big firms, such as Philips, are struggling to put their own organisational house in order, they have also realised that 'do-it-yourself' in developing or marketing new technologies is a highly risky business as sectoral boundaries are eroding due to increased economic internationalisation and technological uncertainties. In response to a rapidly changing international economic and technological climate, strategic alliances or interfirm collaboration has become a fashionable way for firms to survive the competition since the 1980s. This book has, on the one hand, discussed some major issues about strategic alliances (e.g. the motives for strategic alliances, the role of state in promoting and regulating collaborative activities) in general terms by presenting a variety of views by many authors in Chapter 2; on the other hand, a great deal of the empirical studies in this book have been devoted to investigating Philips' collaboration with other competitors all over the world.

First of all, the historical study of Philips, in Chapter 3, showed that the company had been engaged in collaborative arrangements with competitors from the U.S. and Europe in the early days of the electric lighting and consumer electronics industries up to the 1940s. Philips' alliances with other companies were heightened by its active participation in the 'cartel movement', which dominated the international lighting industry in the 1930s. This suggests that interfirm collaboration[4] was not a new phenomenon as far as Philips was concerned. However, strategic alliances had not been adopted by Philips as a strategic means to develop and commercialise its *new* technologies until the early 1980s.

Each of the three consumer electronics technology areas, VCR, HDTV and CD-i, showed from a different angle the great importance of strategic alliances to success or failure in product innovation.

From the VCR format battle, examined in Chapter 4, between JVC's VHS, Sony's Betamax and Philips' V2000, I drew the conclusion that industrial collaboration played a significant role in deciding the winner and losers. The success of VHS might not have been possible if JVC had not ensured a wide range of strategic partnerships with a range of Japanese as well as European firms to manufacture VHS machines on the one hand, and launched a free licence campaign to get support from the software industry on the other. In particular, JVC's joint venture agreements with Thomson and other European firms had made it possible for the former to successfully penetrate the European market. In contrast, Philips and Grundig failed to ensure similar industrial support from either European firms (e.g. Thomson) or Japanese and American firms for their V2000 system. Therefore, despite its alleged technological superiority to the VHS system, V2000 was eventually edged out of the competition by the mid-1980s.

Chapter 8 examined a completely different competitive scenario. Having experienced the V2000 failure, the Philips management changed their corporate strategy and began an ambitious strategy with an emphasis on strategic alliances with competitors, particularly the Japanese, to launch its CD-related optical disc technologies since the beginning of the 1980s. Firstly, Philips collaborated with Sony and successfully made the Compact Disc system a world standard in the early 1980s. Secondly, these two firms, together with a group of computer companies, extended their system and launched the CD-ROM format for data storage. Thirdly, Philips made further technological progress and developed CD-i as a home multimedia system. In contrast to the launch of the CD-Audio system, CD-i began to face fierce competition from a number of rival systems. In addition to its collaboration with Sony, mainly for getting technological support, Philips has made substantial efforts to establish strategic alliances and partnerships with other firms including hardware manufacturers and software designers all over the world. Finally, despite the uncertainties facing the CD-i system, Philips and Sony have entered another alliance to launch the MMCD format, which is now facing fierce competition from the SD Alliance led by Toshiba, Time Warner and Matsushita. Each side of the two competing consortia are hoping to make their own technology the only world standard for the emergent DVD market.

The 'hostile brotherhood'—competition and collaboration—between the leading firms in the consumer electronics industry since the early 1980s discussed in this book is well illustrated in Table 9.1 and Fig. 9.1. During the five major format battles (VCR, digital audio, HDTV, interactive multimedia and DVD), the four major firms—two from Europe (Philips and Thomson) and two from Japan (Sony and Matsushita)—have experienced both friendly (collaborative) and hostile (competitive) relationships from product to product. It is shown, for instance, in Table 9.1 and Fig. 9.1 that Philips and Sony have participated in launching new products in competition against rival technologies seven and six times, respectively, but there were only three occasions when the two companies collaborated for the same products. In the meantime, as shown in Table 9.1 and Fig. 9.1, firms' origin of country or region did not necessarily become a barrier for firms to collaborate for new technologies: as two leading European firms, Philips and Thomson have collaborated in two technology areas (i.e. HD-MAC in Europe and digital TV in the U.S.); otherwise they were backing rival systems (e.g. V2000 *versus* VHS during the VCR format battle and MMCD *versus* SD Disc in the currently ongoing format war for DVD). A similar relationship is also seen between the two leading Japanese firms, i.e. Sony and Matsushita. This suggests a significant difference between corporate strategies and government technology policy: the latter is intended mainly to promote domestic firms and their new technologies, whilst the former is designed solely to ensure a firm's commercial interests related to new technologies—

Table 9.1. Alliances for Selected Format Battles between Major Consumer Electronics Firms.

	VCR			Digital Audio			HDTV			Multimedia		DVD	
	VHS	Betamax	V2000	DCC	CD–DA	Mini Disc	HD-MAC	MUSE	Digital TV	CD-i	3DO	MMCD	SD Disc
Philips			X	X	X		X		X	X		X	
Thompson	X						X		X				X
Sony		X			X	X		X		X		X	
Matsushita	X			X		X		X			X		X

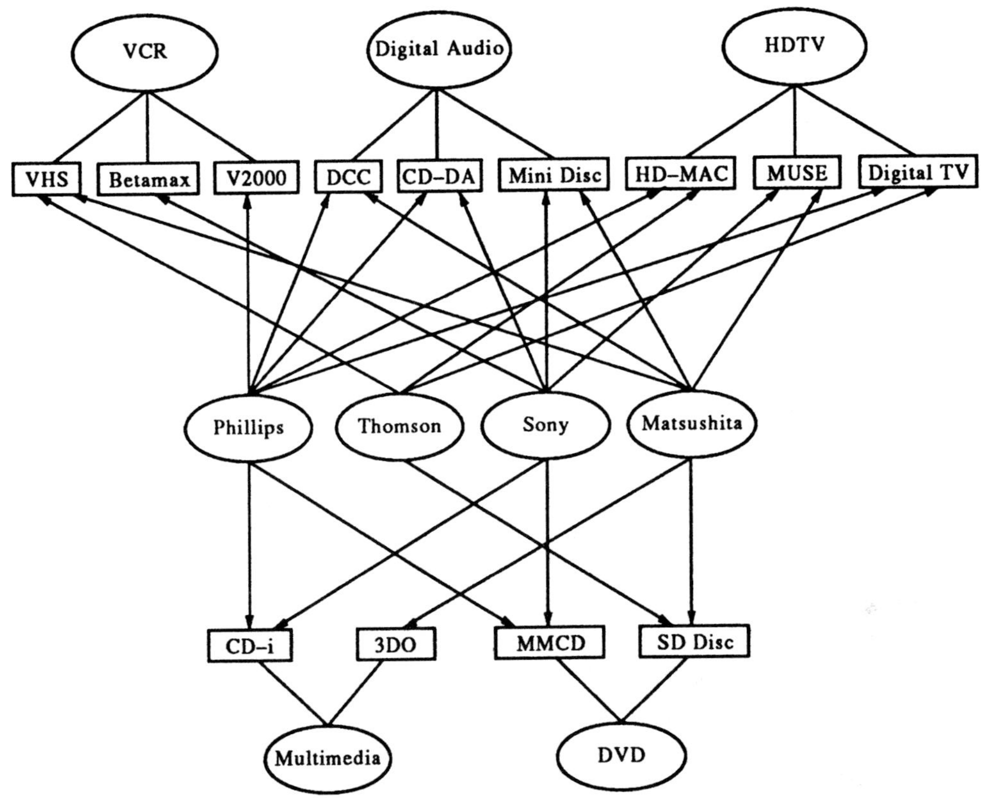

Fig. 9.1. Alliances for Selected Format Battles between Major Consumer Electronics Firms.

whenever and wherever it is necessary and convenient, firms, as in the case of Philips and Thomson, would have no hesitation in seeking collaboration or alliances with foreign firms in new technology areas regardless of government policy on national competitiveness. It seems that, as far as private firms are concerned, corporate competitiveness may be more important to pursue than national competitiveness. From time to time, policy-makers at either national or supranational government level tend to forget about this point.

Philips' industrial collaboration has also been extended to participation in large-scale European R&D consortia in the 1980s.[5]

Among many others, the Eureka 95 HDTV project, as examined in Chapter 6, has highlighted the company's new strategies for international competition in the consumer electronics industry. With strong political backing from the EU and some member states (e.g. France and The Netherlands), Philips, Thomson, Bosch and Thorn EMI joined forces to develop HD-MAC, intended as the European HDTV system, by setting up the EU95 project in 1986.

Membership of this consortium has been extended to many other European firms. Sharing a common objective to block the Japanese HDTV competition, Philips and Thomson, the two leading European manufacturers in consumer electronics, ended their long-standing hostility and entered a collaborative agreement for the first time. In addition to leading this European project, Philips and Thomson have been actively promoting digital HDTV R&D in collaboration with several American firms in another consortium from the late 1980s. Motivated primarily by the fear of losing the costly high-tech competition, the Philips/Thomson consortium and two other research groups signed a 'grand alliance' agreement to define a single digital HDTV system for the U.S. at the behest of the FCC in 1993.

The common element in the three important technological areas, i.e. VCR, CD-based interactive multimedia and HDTV, was that, firstly, industrial collaboration proved to be fundamentally important to the competition between rival technical standards in the consumer electronics industry. In other words, technological or product standardisation was in many cases achieved through interfirm collaboration. The victory of VHS and the failure of V2000 in the home video format battle was in part decided by the lack of industrial collaboration for the latter. The currently ongoing competition between Philips' CD-i and other rival systems in the newly emergent consumer multimedia industry has been largely characterised by interfirm collaboration and strategic partnerships. For instance, the demise of Commodore's CDTV system can be explained in part by the company's inability to get industrial support from the hardware manufacturers; this is in sharp contrast with Philips' wide range of strategic partnerships for its CD-i system with hardware manufacturers and software developers from many parts of the world. In the case of R&D for HDTV, global competition has been joined almost exclusively by allied industrial consortia within and from each of the Triad economies. In other words, no single company, small or big, was able to develop an HDTV system on its own; rather, large-scale industrial collaboration appeared to be necessary to share the costs and risks.

Secondly, the standardisation process for each hi-tech area examined in this book was closely associated with the availability of software. Philips' V2000 system, as well as the European launch of its LaserVision optical disc video technology, suffered from the lack of prerecorded software in the early 1980s. The competition between different consumer multimedia systems such as CD-i, CDTV(now CD-32) and 3DO has been largely focused on the development of software titles for each format. To some extent, the question of whether or not multimedia can be accepted on the consumer market will be determined by the number and quality of software titles available for purchase. The same can be said about the newly emergent DVD technologies. Apart from the technological competition between the MMCD camp led by Philips and Sony and the SD Alliance led by Toshiba and Time Warner, the support from the film studios and computer software houses will be a significant factor in deciding which technology is going to win the digital video format battle. The global competition for HDTV, as shown in Chapter 6, has been apparently focused on some technological points (e.g. digital *versus* analogue, terrestrial *versus* satellite transmission, compatible or incompatible with current TV systems). However, the ultimate success of any HDTV or digital TV system with any technical specifications will be largely decided by whether the broadcasters could provide the consumer with a choice of a wide-range of high-quality programmes at acceptable prices. In short, it was argued in this book that, to a great extent, the consumer electronics industry is a software-led sector; the principle of synergy between software and hardware, as introduced in Chapter 2, is applicable to each of the technological cases analysed in this book.

Thirdly, despite the existence of trade protectionism in each of the Triad economies, leading innovative firms in the consumer electronics industry have adopted a cross-border alliance strategy in the hope of making their new technologies or new products world standards over the last two decades. JVC had successfully brought its VHS technology by the early 1980s into Europe by collaborating with European firms. Philips had developed CD-i and decided to make it a world standard; but the most effective strategy adopted by the Philips management to realise this ambition was strategic alliances. The collapse of the HD-MAC system in Europe and the collaboration between the European (including Philips and Thomson) and American firms over recent years has made digital TV one step further towards a possible world standard.[6]

Finally, interfirm collaboration was, in one way or another, affected by government policy. Chapter 6 has argued that the EU95 HDTV consortium had become a conduit to carry out the EU's new technology policy, and the failure of HD-MAC caused a reassessment of EU policy. In response to the collapse of HD-MAC, the EU has been engaged in a process of making new policies and strategies leaning towards digital technologies, as suggested in Chapter 7, since 1993.

PROTECTIONISM *VERSUS* MARKET FORCES

Discussions in Chapter 2 have shown that there are many forms of trade protectionism and intervention over domestic industry exercised by national or supranational governments in the Triad economies. This has been reflected in the abundant literature contributed by many authors from different academic backgrounds. In particular, conflicting theories of industrial policy have been centred on whether the government, rather than market forces, should act to pick winners. By studying the experience of Philips and EU technology policy for HDTV, this book has analysed the attempts of the EU as well as some of its member states to improve the competitiveness of the European consumer electronics industry.

The shift of policy-making power from the national government level to the Community enabled the latter to intervene over trade, technology and industry from the mid-1980s. The failure of the European VCR technology, i.e. Philips' V2000 format, has heightened the competitive pressure upon the European consumer electronics industry from Japanese firms on the one hand, and, on the other, this in part prompted EU authorities to use more policy measures to protect domestic firms, despite the wide political difference and conflicts across different policy domains.

The VCR case reported in Chapter 4 has offered some evidence to assess the effectiveness of EU policies towards the industry. Firstly, the VER agreement negotiated between the EU and the Japanese government did lay down a quantitative ceiling for imported Japanese products; however, this measure failed to prevent further decline of the V2000 system and its eventual death in the marketplace. Secondly, after Philips had decided to cease production of its proprietary system, EU authorities opened in 1987 a series of antidumping investigations into the VCR sector, and provisional as well as definitive duties were imposed on some foreign companies. In addition to these EU procedures, the notorious 'Poitiers affair' occurred in France.[7] This was widely believed to be an *ad hoc* policy devised to curb VCR imports from abroad, despite the fact that there was no single French firm manufacturing VCRs at that time.

EU antidumping policy was not solely confined to the VCR sector. Table 5.3 indicated that the same policy had been applied to several other product areas of the consumer electronics industry including small-screen CTVs and CD players.

The role policy-makers play in promoting new technologies developed by domestic firms was most notable in the process of HDTV development in Europe from the mid-1980s. Research in Chapter 6 has shown that, initially promoted by the French government and later accepted by the European Commission (particularly DG-XIII), HDTV was seen as a strategic technology in Europe. In response to such political backing and the financial subsidies involved within the Eureka programme, Philips and Thomson have been leading the R&D team for HD-MAC, the proposed European system, from 1986 when the Japanese HDTV proposal was effectively blocked by the European governments at the CCIR conference held in Dubrovnik.

Contrary to the original expectation of the European Commission and the French government, the results of the EU95 HDTV project turned out to be technologically and politically disastrous.

First, HD-MAC was mainly an analogue, or otherwise known as 'not completely digital' or hybrid, system. This technological choice and the subsequent EU support may, arguably, be justified by the rapid technological progress achieved by the EU95 group within a period of about four years (1986–90), compared to about twenty years the Japanese had spent to develop their Hi-Vision system. However, the problem lies in that European HDTV policies continued to support HD-MAC in terms of proposed financial subsidies and coordination by the European Commission after General Instrument of the U.S. had achieved significant technological breakthrough in digital compression in 1990 and the HD-Divine group had demonstrated their fully digital HDTV system in Europe in 1992. The long-standing political manoeuvring over the European HDTV strategy between different power domains[8] had also added more confusion to the course of HDTV development in Europe.[9]

Secondly, the two leading European firms, Philips and Thomson, did not commit themselves exclusively to developing the HD-MAC system. On the contrary, they were actively engaged in digital HDTV R&D through an industrial consortium in the U.S. from the late 1980s. It was argued in this book that commitment to the MAC strategy had enabled Philips and Thomson to use Dutch and French government funds to help their HD-MAC research in Europe and, in the meantime, to invest their private resources heavily into digital research in the U.S. The collapse of HD-MAC in early 1993 did not appear to be a major setback to Philips or Thomson; both companies were at the forefront of digital TV including digital HDTV, as argued in Chapters 6 and 7, despite what happened in Europe. In stark contrast, the European Commission did run into political difficulties largely due to its exclusive support for the HD-MAC programme. In the end, the EU did agree to support widescreen programme making on any technological format with its newly revised HDTV strategy—a strategy without a focus. As discussed in Chapter 7, current EU policy is set to promote widescreen (16×9 aspect ratio) programme making on any new TV standard. This seems to be contradictory with another strand of the EU's new information society strategy, which is centred on the issues of technological convergence within the information and communications technology (ICT) sector with digitalisation and interoperability or interconnectivity as the key elements. However, the EU seems to have got it right this time. No matter what kind of TV standard is adopted, the importance of software, i.e. TV programmes, will continue to play a strategic role in determining the eventual success or failure of new technologies in the future. The same argument is equally applicable to the new ICT sector as a whole. The 'information superhighway' makes no sense at all without the information running on it.

The failure of the EU95 HDTV project has raised further doubts about whether government should pick winners between competing technologies, and whether bureaucrats are capable of making the right technological choice. As Chapter 6 has suggested, the EU was not alone in promoting R&D for HDTV. The Japanese government, *via* MITI and MPT, has been coordinating and financing various R&D consortia related to the Hi-Vision system since the mid-1980s. The FCC in the U.S. has also been coordinating research activities towards a single HDTV system, now a fully digital TV standard. The difference between the strategies adopted by the EU, Japan and the U.S. was that, firstly, the EU and the Japanese government have backed only one system from the very beginning of technological development, while the FCC has sponsored a competition to choose an HDTV standard. Secondly, both the EU and the Japanese government have committed substantial financial subsidies to their domestic firms, but the U.S. government has not offered any funding to any firm except for only one small-scale research project funded by the Pentagon's DARPA to develop high definition screen technology. Thirdly, foreign firms have been excluded from participating in the European and Japanese HDTV research activities sponsored by governments, whilst the FCC process allowed for European and Japanese firms, with both capital and technologies in their hands, to participate in HDTV competition in the U.S.

To be sure, HDTV has been widely identified as a strategic technology. However, public authorities in the EU, Japan and the U.S. have taken fundamentally different policies to influence the course of technological development. Consequently, the effects of these policies have turned out to be significantly different: the EU95 project was a *débâcle*; MUSE appears to be technologically out of date because it is essentially an analogue system;[10] digital TV, including digital HDTV, has been accepted as the future technology in the U.S., and it has an overwhelming effect on corporate strategies and public policies for HDTV elsewhere in the world, especially Europe and Japan. It is emphasised in this book that it is highly dangerous and costly if governments attempt to pick technological and/or corporate winners. The consequence of doing so inevitably pre-empts healthy competition and, sometimes, stifles further technological advances. Both the relatively pluralistic and flexible policy adopted by the FCC to coordinate the technological competition for digital HDTV without using subsidy in the U.S. and the rigid and protectionist technology policy adopted by the EU towards HD-MAC in Europe have demonstrated that it is misguided for governments and firms to fix upon specific technologies during periods of rapid technological transition, as they may be quickly superseded by new developments.

INDUSTRIAL LOBBYING: FIRMS AND PUBLIC POLICY

Government–industry relations, as this research has found, were not a one-way but an interactive process. Protectionist policies towards the consumer electronics industry in Europe were not exclusively initiated by policy-makers in the first place; rather, they were in many cases implemented following a series of industrial lobbying by domestic firms as well as their professional associations. As far as the consumer electronics industry is concerned, it seems that industrial lobbying has become a commonly used method by firms to influence government policy-making for the good of their vested business interests from the 1980s. Weidenbaum (1986) has suggested that 'offensive lobbying' is designed to get the company's views across to politicians and government bureaucrats in order to influence government legislation. It has been shown in this book that Philips, sometimes joined by others, lobbied EU officials in many ways since the early 1980s.

Antidumping measures, which are common in the European consumer electronics industry throughout the 1980s to the present date, begin with an investigation of alleged dumping cases submitted by European firms against leading Japanese, and increasingly other East Asian, suppliers. In late 1982, Philips and Grundig started their lobbying in Brussels and complained about Japanese firms' price cutting actions, which were believed by the two firms to be detrimental to the survival of the V2000 system. Partly in response to this investigation, and to the Voluntary Export Restraint agreement which followed, major Japanese manufacturers changed their strategies from relying on exports to direct investment in Europe. As argued in Chapter 4, industrial lobbying and the protectionist policies adopted by the Community did not prevent further decline of the European VCR industry. On the contrary, the increased volume of foreign direct investment has resulted in deepened 'Japanisation' of the European consumer electronics industry.

It is believed that the Philips office in Brussels functions as the company's 'embassy', whose major task is to maintain good contacts with EU authorities. Empirical research, as reported in this book, suggested that Philips keeps formal communications with the European Commission. On the one hand, Philips' position papers on major issues concerning technology and the industry were read by Commission officials within related sections; Community audio-visual policies and the Commission's communication papers were often drafted in direct consultation with European firms, including Philips, and their professional associations on the other hand. In addition, it was found that the EACEM (European Association of Consumer Electronics Manufacturers), in which Philips has been a leading member, had officially lobbied the European Commission to take further action to support HD-MAC.

In short, Philips has, on the one hand, adopted a strategy characterised by global strategic alliances with competitors for new technologies and new products from the beginning of the last decade; on the other hand, they were actively seeking political solutions at both national government and the Community levels in an effort to improve their competitiveness against foreign firms. In the meantime, the EU launched a series of protectionist policies intended to counter the decline of the European consumer electronics industry. Instead of encouraging competition on the domestic market, foreign firms, particularly the Japanese ones, have been excluded from most European R&D programmes. It seems there was a wide gap between the EU's inward-looking technology policy and the leading European firms' global strategies adopted since the mid-1980s, as far as the consumer electronics industry is concerned.

AGENDA FOR FURTHER RESEARCH

By studying the experience of Philips and the European consumer electronics industry, this book has argued that corporate strategies, interfirm collaboration and public policy had a significant impact on the success or failure of new technologies in different ways. While the empirical contributions derived from the case studies on Philips' historical expansion, its organisational restructuring, and the three technological areas, i.e. VCR, HDTV and CD-i, remain important to the understanding of such a complicated issue, the current study has also opened up avenues for further research.

One of the most interesting themes touched on in this book was the importance of interfirm collaboration to the course of technical change. It was argued in the book, in particular Chapters 6 and 7, that the development of HDTV technology and the emergent CD-based consumer multimedia industry from the last decade has brought many firms into alliances promoting different systems. The rapid technological convergence has also made the traditional sectoral demarcations increasingly irrelevant. Accordingly, the mode of competition between firms has been radically changed. It seems that industrial collaboration for technical change is not confined to the consumer electronics industry; on the contrary, it involves every branch of the ICT sector since the 1980s.

Recent development in IT-related new businesses, such as interactive TV, video on demand (VoD), video telephony and messaging, teleshopping, virtual reality (VR), and many other innovations built on digital compression technology and optical fibre cable networks, has made industrial collaboration a prerequisite for various firms to survive the fierce competition.

The Clinton administration has launched a nationwide campaign calling firms from the computer, telecommunications, and consumer electronics industries to join forces in order to build the so-called 'National Information Infrastructure (NII)' or the 'Information Superhighway' or 'electronic superhighway' in the U.S. The realisation, if it happens, of this new 'American dream' might bring about an integrated IT sector including information exchange and services, digital communication and digital entertainment. In response to the new wave of challenges from the U.S., the EU has launched the European Information Society programme, as represented by the Bangemann Report. In Europe, the Information Society programme has already been identified by most EU member states and many European regional authorities as a new opportunity to improve their technological and economic competitiveness. Among others, the Danish government has recently launched its 'Info 2000' programme in order to bring Denmark to the forefront of the march towards the information society. One step of the proposed 'Info 2000' programme is to provide optical fibre cable links between all of the country's municipalities. In the U.K., the Labour Party has organised a policy forum, headed by the Shadow Heritage Secretary Chris Smith, on the Information Superhighway, which it believes to be one of the key policy areas to win voters during the next general election. In parallel with these national schemes, the city of Amsterdam in The Netherlands and the city of Hull in the U.K. have launched the 'Digitale De Stad (DDS)' (the Digital City) and 'Virtual City' projects in 1994 and 1995, respectively. In addition to funding from the local Council, the DDS project has also secured financial support from the Dutch Ministries of Economic Affairs and Home Affairs. It is anticipated that, before long, the concept of 'Digital City' or 'Virtual City' or whatever it is called will prevail in Europe.

A firm may have competitive advantages in one or a few aspects but no single firm would be able to handle such an integrated and grand ICT sector. If industrial collaboration is the solution for firms, particularly industrial leaders, what are the policy implications? Should antitrust regulations and competition policies be further relaxed in response to this new development of technical change in the ICT sector? These important questions remain to be answered.

The trends observed in the 1980s and detailed in this book would seem to be accelerating in the 1990s and beyond, in particular the imperatives of interfirm collaboration which is likely to bring into the development of consumer electronics products a broader range of firms from related ICT sectors. The emergent corporate strategies of these new alliances, and the public policy responses, will continue to provide a fertile ground for future research.

NOTES

1. According to a report from the *Financial Times* (14 July 1995), on 13 July 1995, a date coincided with the fifth anniversary of Jan Timmer's tenure of Philips' President, Philips' shares have reached another peak at Fl 81.70, compared to the last historical record at Fl 73.60 achieved in 1969. The same source suggests Philips' 1995 first quarter net profit before extraordinary items more than doubled to Fl 544m ($353m).
2. *Financial Times,* 3 May 1995. The same source also suggests that the number of Philips joint ventures in China alone has increased to 14 in 1995.
3. Cawson *et al.* (1990) interpret strategic alliances between firms as hostile brotherhood or competitive collaboration.
4. Note that, nowadays, 'interfirm collaboration' is not only treated by theorists as a fashionable type of business practice but also favoured by politicians in charge of industrial policy. Whilst 'cartels', the extreme form of 'interfirm collaboration' (including division of markets and fixing price, etc.) are still regarded as anticompetitive behaviour and, therefore, illegal. However, in practice, the distinction between 'interfirm collaboration' and 'cartelisation' is not an easy one to make, in particular when law enforcement is involved.
5. See Table 5.13 for a list of European research programmes in which Philips has participated.
6. As indicated in this book, the MPEC-2 standard developed under the auspices of the ISO is posed to become the global standard for digital TV and digital video.
7. The French government decided in 1982 that all foreign-made VCRs coming to the country should pass through a tiny customs office in Poitiers.

8. For instance, the conflicts between the pro-HD-MAC faction, represented by DG-XIII of the European Commission and the French government, and the anti-HD-MAC faction represented by the U.K. government.
9. Political controversies have also been caused by the EU's MAC Directive of 1986 and the renewed version of 1991. A revised version was drafted in 1993 for approval by the European Parliament and Council of Ministers.
10. Therefore, the MUSE transmission standard is now facing a great uncertainty—it is highly possible that the system will be either completely dropped as what has happened to HD-MAC or subject to further technological upgrading, i.e. digitalisation of the entire system.

REFERENCES

Alic, J. (1990), 'Cooperation in R&D', *Technovation*, Vol. 10, No. 5, pp. 319–332.

Ansoff, H. I. (1965), *Corporate Strategy: An Analytical Approach to Business Policy for Growth and Expansion,* McGraw-Hill, New York.

Bangemann, M. (1992), *Meeting the Challenge: Establishing a Successful European Industrial Policy*, Kogan Page Ltd, London.

Bartlett, C. A. and Ghoshal, S. (1990), 'Matrix Management: Not a Structure, a Frame of Mind', *Harvard Business Review*, July–August, pp. 138–145.

Best, M. H. (1990), *The New Competition: Institutions of Industrial Restructuring,* Polity Press, Cambridge.

Bird, J. (1991), 'The Market for Advanced Television', *HDTV 92 and Future Television*, Proceedings of the Third Annual Conference, London, December.

BIS Mackintosh (1985), *The European Consumer Electronics Industry,* published by the Commission of the European Communities.

BIS Mackintosh (1988), *Digital Consumer Electronics.*

BIS Mackintosh (1990), *The Competitiveness of the European Consumer Electronics Industry*, prepared for the Commission of the European Communities.

Booz Allen and Hamilton, Inc. (1985), *EEC Consumer Electronics: Industrial Policy,* Final Report, Brussels, 24 June.

Borrus, M., Tyson, L. D. A. and Zysman, J. (1986), 'Creating Advantages: How Government Policies Shape International Trade in the Semiconductor Industry', in Krugman, P. R. ed., *Strategic Trade Policy and the New International Economics*, MIT Press, Cambridge, Mass., pp. 91–114.

Bouman, P. J. (1970), *Growth of an Enterprise: The Life of Anton Philips,* Macmillan, London.

Brander, J. A. (1986), 'Rationales for Strategic Trade and Industrial Policy', in Krugman, P. R. ed., *Strategic Trade Policy and the New International Economics*, MIT Press, Cambridge, Mass., pp. 23–46.

Bright, A. A. (1949), *The Electric-Lamp Industry: Technological Change and Economic Development from 1800 to 1947*, Macmillan, London.

Brown, A. *et al.* (1992), 'HDTV: High Definition, High Stakes, High Risk: A Report on High Definition Television', NERA (National Economic Research Associates), London.

Bruno, R. (1987), 'Making Compact Disks Interactive', *IEEE Spectrum*, November.

Carliner, G. (1986), 'Industrial Policy for Emerging Industries', in Krugman, P. R. ed., *Strategic Trade Policy and the New International Economics*, MIT Press, Cambridge, Mass., pp. 147–168.

Cawson, A. ed. (1985), *Organized Interests and the State: Studies in Meso-Corporatism,* Sage, London.

Cawson, A. (1986), *Corporatism and Political Theory*, Basil Blackwell, Oxford.

Cawson, A. (1989), 'European Consumer Electronics: Corporate Strategies and Public Policy', in Sharp, M. and Holmes, P. eds, *Strategies for New Technology*, Philip Allan, Hertfordshire.

Cawson, A. (1991), 'Running a High Tech Industry: Consumer Electronics', Unit 13 for Open University Course D212, *Running the Country*.

Cawson, A. (1992), 'Interests, Groups and Public Policy-Making: The Case of the European Consumer Electronics Industry', in Greenwood, J. *et al.* eds, *Organised Interests and the European Community*, Sage, London, pp. 99–118.

Cawson, A. (1992a), 'Sectoral Governance and Innovation: Private Interest Government and the EUREKA HDTV Project', paper presented at the *ESRC Conference on Government–Industry Relations*, Exeter, 20–22 May.

Cawson, A. (1995), 'High Definition Television in Europe', *The Political Quarterly,* Vol. 66, No. 2, April–June, pp. 157–173.

Cawson, A. *et al.* (1990), *Hostile Brothers: Competition and Closure in the European Electronics Industry*, Clarendon Press, Oxford.

Cawson, A., Haddon, L. and Miles, I. (1993), 'The Heart of Where the Home Is: The Innovation Process in Consumer IT Products', in Swann, P. ed., *New Technologies and the Firm: Innovation and Competition,* Routledge, London, pp. 242–264.

Cawson, A. and Holmes, P. (1991), 'The New Consumer Electronics', in Freeman, C., Sharp, M. and Walker, W. eds, *Technology and the Future of Europe: Global Competition and the Environment in the 1990s*, Pinter, London, pp. 167–182.

Chandler, A. D. (1962), *Strategy and Structure: Chapters in the History of the Industrial Enterprise,* MIT Press, Cambridge, Mass.

Chandler, A. D. (1990), *Scale and Scope: The Dynamics of Industrial Capitalism*, Belknap Press, Cambridge, Mass.

Chandler, A. D. (1991), 'The Functions of the HQ Unit in the Multibusiness Firm', *Strategic Management Journal*, Vol. 12, Special Issue, Winter, pp. 31–50.

Choy, J. (1989), 'Developing Advanced Television: Industrial Policy Revisited', *Japan Economic Institute*, 13 January, pp. 1–9.

Ciborra, C. (1991), 'Alliances as Learning Experiments: Cooperation and Change in High-Tech Industries', in Mytelka, L. K. ed., *Strategic Partnership: States, Firms and Multinational Competition*, Printer, London, pp. 51–77.

Cline, W. R. (1986), 'US Trade and Industrial Policy: The Experience of Textiles, Steel, and Automobiles', in Krugman, P. R. ed., *Strategic Trade Policy and the New International Economics,* The MIT Press, Cambridge, Mass., pp. 211–240.

Commission of the European Communities (1985), *The European Consumer Electronics Industry: Prospect and Policy (Draft)*, Brussels, 21 April.

Commission of the European Communities (1990), *Communication from the Commission to the Council and Parliament on Audiovisual Policy*, Brussels, 21 February.

Commission of the European Communities (1990a), *The European Community Policy in the Audio-visual Field: Legal and Political Texts*, Brussels–Luxembourg.

Commission of the European Communities (1990b), *Industrial Policies in An Open and Competitive Environment: Guidelines for a Community Approach*, Communication of the Commission to the Council and to the European Parliament, Brussels, 16 November.

Commission of the European Communities (1991), *MEDIA: Guide for the Audio-visual Industry (Edition 6)*, Brussels–Luxembourg.

Commission of the European Communities (1991a), *Improving the Functioning of Consumer Electronics Markets*, Brussels.

Commission of the European Communities (1992), *Proposal for a Council Decision on An Action Plan for the Introduction of Advanced Television Services in Europe*, COM(92) 154 final, Brussels, 5 May.

Commission of the European Communities (1993), *Digital Video Broadcasting: A Framework for Community Policy: Communications from the Commission and Draft Council Resolution,* Brussels, EC/6 COM(93)557.

Commission of the European Communities (1994), *Strategy Options to Strengthen the European Programme Industry in the Context of the Audiovisual Policy of the European Union,* Green Paper, COM(94) 96 final, Brussels.

Cooke, P. and Morgan, K. (1993), 'The Network Paradigm: New Departure in Corporate and Regional Development', *Environment and Planning D: Society and Space,* Vol. 11, pp. 543–564.

Cusumano, M. A., Mylonadis, Y. and Rosenbloom, R. S. (1991), 'Strategic Maneuvering and Mass-Market Dynamics: The Triumph of VHS Over Beta', Sloan School of Management working paper, No. 1991–048, Massachusetts Institute of Technology.

Cutts, R. L. (1992), 'Capitalism in Japan: Cartels and Keiretsu', *Harvard Business Review*, July–August, pp. 48–55.

Dai, X. (1994), *Corporate Strategies and Public Policy in the European Consumer Electronics Industry: A Case Study of Philips,* Unpublished Ph.D. thesis, University of Sussex, Brighton.

Dai, X. (1995), 'Corporate Structure and Global Competitiveness: Lessons from Philips', Paper presented at the Association of Business Historians 1995 Conference on *The Foundations of International Competitiveness,* Hosted by the University of Warwick at Cable and Wireless College, Coventry.

Dai, X. (forthcoming), 'Technical Convergence and Policy Divergence in the European Information Society: The Case of New TV Systems', in Wintle, M. J. ed., *Culture and Identity in Europe*, Avebury, Aldershot.

Dai, X. and Bottomley, A. (forthcoming), *Racing on the Information Superhighway: A European Perspective,* Avebury, Aldershot.

Dai, X., Cawson, A. and Holmes, P. (1994), 'Competition, Collaboration and Public Policy: A Case Study of the European HDTV Strategy', Sussex European

Institute Working Papers in Contemporary European Studies, No. 3, University of Sussex, Brighton.

Dai, X., Cawson, A. and Holmes, P. (1996), 'The Rise and Fall of HDTV: The Impact of European Technology Policy', *Journal of Common Market Studies,* June.

Dai, X. and Gao, S. (1995), 'Stategic Technology and Industrial Policy: Global Competition for HDTV and China's Strategic Choice', in *Proceedings of the Second Academic Conference of Young Scientists*. Sponsored by China Association for Science and Technology, The Chinese Science & Technology Press, Beijing, pp. 417–421.

Dai, X., Gao, S. and Gu L. (1994), 'Gao Qingxidu Dianshi de Quanqiu Jingzheng (The Global Competition for HDTV)', *Economic Reference Paper,* Beijing, 18 September.

Davis, J. and Daum, A. (1987), *Optical Disc Technologies in the European Community*, Knowledge Research Limited, 6 September.

Delapierre, M. and Zimmermann, J. B. (1991), 'Towards a New Europeanism: French Firms in Strategic Partnership', in Mytelka, L. K. ed., *Strategic Partnership: States, Firms and Multinational Competition*, Printer, London, pp. 102–119.

Dodgson, M. ed. (1989), *Technology Strategy and the Firm: Management and Public Policy*, Longman, Harlow.

Dodgson, M. (1993), *Technological Collaboration in Industry: Strategy, Policy and Internationalization in Innovation,* Routledge, London.

Dosi, G. (1984), *Technical Change and Industrial Transformation: The Theory and An Application to the Semiconductor Industry*, Macmillan, London.

Duddley, J. W. (1989), *1992: Strategies for the Single Market,* Guild Publishing, London.

Dupagne, M. (1990), 'High-Definition Television: A Policy Framework to Revive U.S. Leadership in Consumer Electronics', *The Information Society*, Vol. 7, pp. 53–76.

EIU (The Economist Intelligence Unit) (1983), 'Multinational Report: Philips and the Politics of the European VCR Market', *Multinational Business*, No. 3, pp. 26–30.

EIU (The Economist Intelligence Unit) (1983), 'Multinational Report: Whatever Happened to the Video Disc?', *Multinational Business*, No. 2, pp. 29–32.

EIU (The Economist Intelligence Unit) (1991), *High Definition Television Progress and Prospects: A Maturing Technology in Search of A Market*, Special Report No. 2189, Business International Ltd.

EU95 Directorate (1991), *Progressing towards HDTV*, published by Eureka 95 HDTV Directorate, June.

European Community Council Directive, *On the Adoption of Common Technical Specifications of the MAC/packet Family of Standards for Direct Satellite Television Broadcasting,* 3 November 1986, 86/529/EEC.

European Community Council Directive, 'On the Adoption of Standards for Satellite Broadcasting of Television Signals', 11 May 1992, *Official Journal of the European Communities,* No. L 137, 20 May 1992.

Evans, B. (1992), *Digital HDTV: The Way Forward,* IBC Technical Services Ltd., London.

Farrell, J. and Shapiro, C. (1992), 'Standard Setting in High-Definition Television', *Brookings Papers on Economic Activity: Microeconomics,* The Brookings Institution, Washington, D.C., pp. 1–93.

Feldman, T. (1991), *Multimedia in the 1990s*, BNB Research Fund Report, No. 54, the British Library.

Franko, L. G. (1976), *The European Multinationals: A Renewed Challenge to American and British Big Business*, Greylock Publishers, Stamford, Connecticut.
Freeman, C. (1982), *The Economics of Industrial Innovation*, Pinter, London.
Freeman, C. (1992), *Economics of Hope: Essays on Technical Change, Economic Growth and the Environment*, Pinter, London.
Freeman, C., Sharp, M. and Walker, W. eds. (1991), *Technology and the Future of Europe: Global Competition and the Environment in the 1990s*, Pinter, London.
Frost and Sullivan, Inc. (1990), *The European Market for Home Audio, Video and Television Equipment,* London, Fall.
Gao, S. and Dai, X. (1995), 'Zhanlue Jishu yu Zhanlue Xuanze: Gao Qingxidu Dianshi Guoji Jingzheng de Qishi (Strategic Technology and Strategic Choice: Implications of the Global Competition for HDTV)', Submitted to *Journal of Strategy and Management Studies,* Beijing, September–October.
Geddes, K. and Bussey, G. (1991), *The Setmakers: A History of the Radio and Television Industry,* BREMA, London.
Gerlach, M. L. (1992), *Alliance Capitalism: The Social Organization of Japanese Business,* University of California Press, Berkeley.
Ghoshal, S. and Bartlett, C. A. (1993), 'The Multinational Corporation as An International Network', in Ghoshal, S. and Westney, D. E. eds, *Organisation Theory and the Multinational Corporation,* Macmillan, pp. 77–104.
Grant, C. (1983), *The Home Video Revolution in Western Europe*, EIU (The Economist Intelligence Unit) Special Report No. 144.
Grant, W. (1982), *The Political Economy of Industrial Policy*, Butterworth, London.
Greenwood, J. *et al.* eds (1992), *Organised Interests and the European Community,* Sage Publications, London.
Gregory, G. (1986), *Japanese Electronics Technology: Enterprise and Innovation*, John Wiley & Sons. Originally published by The Japan Times Ltd.
Grossman, G. M. (1986), Strategic Export Promotion: A Critique', in Krugman, P. R. ed., *Strategic Trade Policy and the New International Economics*, MIT Press, Cambridge, Mass., pp. 47–68.
Hagedoorn, J. (1993), 'Understanding the Rationale of Strategic Technology Partnering: Interorganisational Modes of Cooperation and Sectoral Differences', *Strategic Management Journal,* Vol. 14, 1993, pp. 371–385.
Hamel, G. and Prahalad, C. K. (1993), 'Strategy as Stretch and Leverage', *Harvard Business Review*, March–April, pp. 75–84.
Handy, C. (1992), 'Balancing Corporate Power: A New Federalist Paper', *Harvard Business Review*, November–December, pp. 59–72.
Heerding, A. (1985), *The History of N. V. Philips' Gloeilampenfabrieken (Vol. 1): The Origin of the Dutch Incandescent Lamp Industry*, Cambridge University Press, Cambridge.
Heerding, A. (1988), *The History of N. V. Philips' Gloeilampenfabrieken (Vol. 2): A Company with Many Parts*, Cambridge University Press, Cambridge.
Helgerson, L. W. (1987), *Introduction to Optical Technology*, Image Technology Consumer Handbook, Association for Information and Image Management, Printed in the U. S.
Heskett, J. (1989), *Philips: A Study of the Corporate Management of Design,* Trefoil Publication, London.

Hill, R. (1982), 'Philips Falls Back upon Old-Fashioned Management Virtues', *International Management*, November, pp. 19–25.

Hobday, M. (1991), 'The European Semiconductor Industry: Resurgence and Rationalisation', in Freeman, C. *et al.* eds, *op. cit.*, pp. 80–94.

Hobday, M. (1992), 'External Operations in the European Semiconductor Industry: Corporate Strategies, Government Policies and Competitiveness', SPRU Memo, University of Sussex, Brighton.

Independent Television Association (1995), *Digital Terrestrial Television*, London, April.

Inglis, A. F. (1990), *Behind the Tube: A History of Broadcasting Technology and Business*, Focal Press, London.

Iuppa, N. V. and Anderson, C. (1987), *Interactive Videodiscs: New Tools and Applications*, Knowledge Industry, White Plains, New York.

Jenkins, B. (1991), 'Strategic Alliances in Telecommunications: The Role of States in Determining Competitive Advantage', in Mytelka, L. K. ed. *Strategic Partnership: States, Firms and Multinational Competition*, Pinter, London, pp. 167–181.

Johson, G. and Scholes, K. (1988), *Exploring Corporate Strategy,* 2nd ed., Prentice Hall, Englewood Cliffs, NJ.

Jorde, T. M. and Teece, D. J. eds (1992), *Antitrust, Innovation, and Competitiveness,* Oxford University Press, Oxford.

Jorde, T. M. and Teece, D. J. (1992a), 'Innovation, Cooperation, and Antitrust', in Jorde, T. M. and Teece, D. J. eds, *op. cit.*, pp. 47–81.

Julius, DeA. (1990), *Global Companies and Public Policy: The Growing Challenge of Foreign Direct Investment*, Pinter, London.

Kay, J. (1993), *Foundations of Corporate Success: How Business Strategies Add Value*, Oxford University Press, Oxford.

Kay, J. and Willman, P. (1993), 'Managing Technological Innovation: Architecture, Trust and Organisational Relationships in the Firm', in Swann, P. ed. (1993), *New Technologies and the Firm: Innovation and Competition,* Routledge, London, pp. 19–35.

Keen, B., 'Playing it again, Sony: The Double Life of Home Video Technology', *Science As Culture*, No. 1, 1987, pp. 7–42.

Kerkvliet, G. and van Well Groeneveld, W. (1987), *Industrial Design in Practice,* published by Akademie Industriële Vormgeving, Eindhoven.

Komiya, M. (1991), 'Japan's global HDTV strategy', *Advanced Television Markets*, Issue 1, November, pp. 10–11.

Krugman, P. R. (1986), 'Introduction: New Thinking about Trade Policy', in Krugman, P. R. ed., *Strategic Trade Policy and the New International Economics*, MIT Press, Cambridge, Mass., pp. 1–22.

Krugman, P. R. ed. (1986), *Strategic Trade Policy and the New International Economics*, MIT Press, Cambridge, Mass.

Kupfer, A. (1991), 'The US wins one in high-tech TV', *Fortune*, 8 April, pp. 50–54.

Lambert, S. and Sallis, J. eds (1987), *CD-I and Interactive Videodisc Technology*, Howard W. Sams & Co, U.S.

Lang, Y. H. (1990), *Dianshi zai Geming: Mingri de Dianshi Shijie* [TV Revolution: Tomorrow's World of Television], Cheng Chung Book Co., Ltd, Taiwan.

Lardner, J. (1987), *Fast Forward: Hollywood, the Japanese, and the Onslaught of the VCR*, W. W. Norton, New York.

Levy, J. and Samuels (1991), 'Institutions and Innovation: Research Collaboration as Technology Strategy in Japan', in Mytelka, L. K. ed., *Strategic Partnership: States, Firms and Multinational Competition*, Printer, London, pp. 120–148.

Line, R. (1990), *The International Electronics Industry,* The Economist Intelligence Unit, Special Report, No. 2050, December.

Link, A. N. and Tassey, G. (1987), *Strategies for Technology-based Competition: Meeting the New Global Challenge*, Lexington Books, Mass.

Metze, M. (1993), *Short Circuit: The Decline of A Once Great Company,* Minerva Press, London.

Michalet, C. A. (1991), 'Strategic Partnerships and the Changing Internationalisation Process', in Mytelka, L. K. ed., *Strategic Partnership: States, Firms and Multinational Competition*, Pinter, London, pp. 35–50.

Milgrom, P. and Roberts, J. (1992), *Economics, Organisation and Management,* Prentice-Hall, Englewood Cliffs, NJ.

Mintzberg, H. and Quinn, J. B. (1991), *The Strategy Process: Concepts, Contexts, Cases* (2nd edition), Prentice Hall, Englewood Cliffs, NJ.

Mintzberg, H. and Westley, F. (1992), 'Cycles of Organisational Change', *Strategic Management Journal*, Vol. 13, Special Issue, Winter, pp. 39–59.

Morgan, K. (1989), 'Telecom Strategies in Britain and France: The Scope and Limits of Neo-Liberalism and Dirigisme', in Sharp, M. and Holmes, P. eds, *Strategies for New Technology: Case Studies from Britain and France,* Philip Allan, Hertfordshire, pp. 19–55.

Morgan, K. (1991), 'Competition and Collaboration in Electronics: What are the Prospects for Britain?', *Environment and Planning A*, 23, pp. 1459–1482.

Morgan, K. and Sayer, A. (1988), *Microcircuits of Capital: 'Sunrise' Industry and Uneven Development*, Polity Press, Cambridge.

Morita, A. (1987), *Made in Japan*, Collins, London.

Morita, A. (1992), 'Partnering for Competitiveness: The Role of Japanese Business', *Harvard Business Review*, May–June 1992, pp. 76–83.

Mowery, D. C. (1988), 'Collaborative Ventures between the U.S. and Foreign Manufacturing Firms: An Overview', in Mowery, D. C. ed., *International Collaborative Ventures in U.S. Manufacturing,* Ballinger Publishing Company, Cambridge, Mass.

Mowery, D. C. (1991), 'International Collaboration in the Commercial Aircraft Industry', in Mytelka, L. K. ed., *Strategic Partnership: States, Firms and Multinational Competition*, Pinter, London, pp. 78–101.

Mytelka, L. K. ed (1991), *Strategic Partnership: States, Firms and Multinational Competition*, Pinter, London.

Mytelka, L. K. (1991a), 'Crisis, Technological Change and the Strategic Alliance', in Mytelka, L. K. ed, *op. cit.*, pp. 7–34.

Mytelka, L. K. (1991b), 'States, Strategic Alliances and International Oligopolies: The European ESPRIT Programme', in Mytelka, L. K. ed, *op. cit.*, pp. 182–210.

Mytelka, L. K. and Delapierre, M. (1987), 'The Alliance Strategies of European Firms in the Information Technology Industry and the Role of ESPRIT, *Journal of Common Market Studies*, 26, 1987, pp. 231–253.

National Consumer Council (1990), *Consumer Electronics and the EC's Anti-*

dumping Policy, International Trade and the Consumer Working Paper 1. March.

Nayak, P. R. and Ketteringham, J. M. (1986), *Breakthroughs!* Rawson Associates, New York.

Nelson, R. R. and Wright, G. (1992), 'The Rise and Fall of American Technological Leadership: The Postwar Era in Historical Perspective', *Journal of Economic Literature*, December, pp. 1931–1964.

Noble, G. (1988), 'The Social Significance of VCR Technology: TV or Not TV?', *The Information Society*, Vol. 5, pp. 133–146.

NTL (1995), *The Future of Television Transmission,* Winchester, March.

Ohmae, K. (1990), *The Borderless World: Power and Strategy in the Interlinked Economy*, Collins, London.

Oppenheim, P. (1992), *Trade Wars: Japan versus the West*, Weidenfeld and Nicolson, London.

Overkleeft, D. and Groosman, L. E. (1988), *The Dekker Perspective*, Graham and Trotman.

Parker, J. (1978), *The Economics and Innovation: The National and International Enterprise in Technological Change*, 2nd ed., Longman.

Pearson, G. J. (1990), *Strategic Thinking*, Prentice-Hall, Englewood Cliffs, N.J.

Peters, T. J. and Waterman, R. H. (1982), *In Search of Excellence: Lessons from America's Best-Run Companies,* Harper & Row, New York.

Philips Annual Report, 1956–1991.

Philips News, Eindhoven, 1972–1993.

Philips Post, London, 1953–1991.

Philips (1989), *Free Trade, Electronics and Europe*, a position paper, published by Corporate External Relations, Eindhoven, Philips International B.V., Eindhoven, September.

Philips (1991), *Conditions for Co-existence and Cooperation in the Electronics Industry—Towards A Sustainable Relationship between the European Community and Japan: A View from Philips Electronics*, published by Corporate External Relations, Philips International B.V., Eindhoven, September.

Philips (1991a), *Philips: A Century of Enterprise (1891–1991)*, a Philips-designed CD-I disc, Eindhoven.

Philips, *Philips in Europe*, published by Corporate External Relations, Philips International B.V., Eindhoven (No date).

Philips, *Europe's Industrial Vitality and the Role of A Vigorous Electronics Industry,* position paper, published by Philips Electronics N.V.

Philips, F. (1978), *45 Years with Philips*, Blandford Press, Poole, Dorset.

Porter, M. E. (1980), *Competitive Strategy: Techniques for Analysing Industries and Competitors*, Free Press, New York.

Porter, M. E. (1985), *Competitive Advantage: Creating and Sustaining Superior Performance*, Free Press, New York.

Porter, M. E. (1990) *The Competitive Advantage of Nations*, Macmillan, London.

Porter, M. E. (1991), 'Towards A Dynamic Theory of Strategy', *Strategic Management Journal*, Vol. 12, Special Issue, Winter, pp. 95–117.

Prahalad, C. K. and Hamel, G. (1990), 'The Core Competence of the Corporation', Harvard Business Review, May–June, pp. 79–91.

Preston, J. M. ed. (1988), *Compact Disc-Interactive: A Designer's Overview*, Kluwer.

Rosenbloom, R. S. and Abernathy, W. J. (1982), 'The Climate for Innovation in Industry: The Role of Management Attitudes and Practices in Consumer Electronics', *Research Policy*, 11, Horth-Holland Publishing Company, pp. 209–225.

Rosenbloom, R. S. and Cusumano, M. A. (1987), 'Technological Pioneering and Competitive Advantage: The Birth of the VCR Industry', *California Management Review*, Vol. XXIX, No. 4, Summer, pp. 51–76.

Rosenbloom, R. S. and Freeze, K. J. (1985), 'Ampex Corporation and Video Innovation', in Rosenbloom, R. S. ed. *Research on Technological Innovation, Management and Policy*, Vol. 2, JAI Press Inc., Connecticut, pp. 113–185.

Rothwell, R. and Zegveld, W. (1981), *Industrial Innovation and Public Policy: Preparing for the 1980s and the 1990s*, Pinter, London.

Rugman, A. and Verbeke, A. (1990), *Global Corporate Strategy and Trade Policy,* Routledge, London.

Schuknecht, L. (1992), *Trade Protection in the European Community,* Harwood Academic Publications, Chur.

Sharp, M. (1989), 'Corporate Strategies and Collaboration: The Case of ESPRIT and European Electronics', in Dodgson, M. ed., *op. cit.*, pp. 202–218.

Sharp, M. (1991), 'Europe: A Renaissance?', *Science and Public Policy*, Vol. 18, No. 4, pp. 393–400.

Silverstone, R. and Hirsch, E. eds (1992), *Consuming Technologies: Media and Information in Domestic Spaces,* Routledge, London.

Spencer, B. J. (1986), 'What Should Trade Policy Target?', in Krugman, P. R. ed., *Strategic Trade Policy and the New International Economics*, MIT Press, Cambridge, Mass., pp. 69–90.

Stanford-Smith, B. *et al.* eds (1987), *Report on The Intensive Workshop on CD-ROM*, A Joint Project by the CEC (DG XIIIB) and EURIPA, July.

Stocking, G. W. and Watkins, M. W. (1946), *Cartels in Action: Case Studies in International Business Diplomacy*, American Book–Stratford Press, New York.

Stocking, G. W. and Watkins, M. W. (1948), *Cartels or Competition? The Economics of International Controls by Business and Government*, American Book–Stratford Press, New York.

Sturmey, S. G. (1958), *The Economic Development of Radio*, Gerald Duckworth & Co. Ltd, London.

Sun, Z. (1984), *The Art of War,* Appendixed in Tao, H., *Sun Zi Bingfa Gailun* [Sun Zi's Art of War], The PLA Press, Beijing. Original 500 B.C.

Teece, D. J. (1986), 'Profiting from Technological Innovation: Implications for integration, Collaboration, Licensing and Public Policy', *Research Policy*, 15, p. 285–305.

Tyson, L. D'A. (1992), *Who's Bashing Whom? Trade Conflict in High-Technology Industries*, Institute for International Economics, Washington, D.C.

van de Klugt, C. J. (ex-President of Philips) (1989), *The Strategic Importance of Electronics for Europe*, speech to the European Parliament, Brussels, 7 March.

van Houten, S. (Executive Vice-President of Philips) (1991), *Why Industrial Research?* Lecture for the Royal Society of London, London, 2nd October.

Velia, M. and Dai, X. (1993), *Study on Consumer Electronics Prices,* Commissioned by DGIV of the Commission of the European Union, Coordinated by Holmes,

P. and Cawson, A., International Economics Research Centre, University of Sussex, Brighton, May.

Watson-Brown, A. (1987), 'The campaign for High Definition Television: a case study in Triad power', *Euro–Asia Business Review*, Vol. 6, No. 2, April, pp. 3–11.

Weidenbaum, M. L. (1986), *Business, Government, and the Public* (Third Edition), Prentice-Hall, Englewood Cliffs, NJ.

West, A. (1992), *Innovation Strategy*, Prentice-Hall, Englewood Cliffs, NJ.

Whiskin, R. (1989), 'Prospects for the Direct-to-Home Satellite TV Receiver Market in Europe', *International Journal of Technology Management*, Vol. 4, No. 1, pp. 108–113.

Ziemer, A. (1990), 'PALplus: the Downwards Compatible Enhancement of PAL', *HDTV 91 and Future Television*, Proceedings of the Second Annual Conference, London, December.

Zimmerman, M. (1985), *Dealing with the Japanese*, George Allen & Unwin, London

Zygmont, J. (1988), 'Compact-Disc: Companies Test New Frontier', *High Technology Business*, February.

INDEX